"*Conflict and Development* has been a key textbook for my teaching over the past years, with its balanced blend of critical perspectives, good examples, and accessible presentation. This new edition enhances its usefulness by being more contextual in analysis, more practice-oriented in its discussion, and more nuanced in its approaches to the post-Cold War peacebuilding. It is an excellent reference not only for students but for field practitioners and policymakers."

–**SungYong Lee**, *Professor, Soka University*

"This book is an excellent critical introduction to the subject area. The wide-ranging use of practical examples and case studies makes this invaluable for those who have a general interest in conflict and development as well as students. This 3rd edition adds Alasdair Gordon-Gibson as an author, who has had more than 20 years of experience delivering aid and development assistance with the Red Cross and Red Crescent Movement, the United Nations and others. This brings a more subtle analysis of aid and the humanitarian question to what is already an excellent book."

–**Paul Jackson**, *Professor of African Politics and Director of Research for the College of Social Sciences, University of Birmingham*

"With the growth of international efforts around the humanitarian, development and peace triple nexus, this classic volume on conflict and development is more relevant than ever before. The 3rd edition of *Conflict and Development* is an accessible volume, appropriate for learning in depth about how international development and humanitarian actors respond to conflicts. It includes helpful chapter summaries and discussion questions to stimulate debate around the many of the important questions being asked in international cooperation today."

–**Pamina Firchow**, *Associate Professor of Coexistence and Conflict, Brandeis University, and author of* Reclaiming Everyday Peace: Local Voices in Measurement and Evaluation after War

Conflict and Development

Fully revised and updated in its third edition, this timely book brings together the study of conflict and war and the problems surrounding the economic development of developing societies that are most prone to experiencing problems in moving on after war.

The book does so by reflecting on the issues surrounding war as it unfolds and after it has (in principle) 'ended', within the context of the history, present-day problems and future prospects. The book aims to highlight the possibilities, successes and failures of past and present policies that bring 'development' to countries and peoples that want to be more involved in deciding their own futures after conflict and war, and often find themselves subject to what can be seen as arbitrary and even alien ways of thinking and acting by institutions in which they theoretically have membership and agency but often do not in practice. The case studies have been fully updated to reflect changes and developments since the second edition of this text, and there are questions at the end of each chapter to promote reflection. This new edition presents a deeper dive into the history of conflict and the emergence of new theories and policy guidance about present and future options in the fields of conflict and development.

Accessible and engaging, this textbook is a pivotal resource for a nexus of subjects related to the often separated fields of conflict and development studies, as well as practitioners in this area.

Andrew J. Williams is Professor of International Relations (Emeritus) at the University of St Andrews. He specializes in the study of conflict and international history. His main current research interests include international conflict analysis and the international history of France. He has experience in the practice of conflict resolution and has worked as a consultant for the United Nations organizations, including United Nations Development Programme (UNDP) and UNITAR. He has published widely in key journals. His book *Liberalism and War* was published by Routledge in 2006 (second edition forthcoming 2024).

Alasdair Gordon-Gibson is an Honorary Lecturer with the University of St Andrews and has worked for over 20 years with the International Red Cross and Red Crescent Movement. He holds a PhD from the University of St Andrews with research focusing on examination of how the discourse around humanitarian principles and practice can be re-cast in the contemporary political environment. Publications include the book *Humanitarians on the Frontier: Identity and Access along the Borders of Power* (2021), which looks at the changing discourse around the aid and development sectors in contemporary conflicts.

Routledge Perspectives on Development

Series Editor: Professor Tony Binns, *University of Otago*

Since it was established in 2000, the same year as the Millennium Development Goals were set by the United Nations, the Routledge Perspectives on Development series has become the pre-eminent international textbook series on key development issues. Written by leading authors in their fields, the books have been popular with academics and students working in disciplines such as anthropology, economics, geography, international relations, politics and sociology. The series has also proved to be of particular interest to those working in interdisciplinary fields, such as area studies (African, Asian and Latin American studies), development studies, environmental studies, peace and conflict studies, rural and urban studies, travel and tourism.

If you would like to submit a book proposal for the series, please contact the Series Editor, Tony Binns, on: jab@geography.otago.ac.nz

South-South Development
Peter Kragelund

Gender and Development, 3rd edition
Janet Momsen

Aid and Development
John Overton and Warwick E. Murray

Theories and Practices of Development, 3rd edition
Katie Willis

Non-Governmental Organizations and Development, 2nd edition
David Lewis, Nazneen Kanji and Nuno S. Themudo

Development and Decolonization in Latin America, 2nd edition
Julie Cupples

Human Rights and Development
Peris S. Jones

Conflict and Development, 3rd edition
Andrew J. Williams and Alasdair Gordon-Gibson

For more information about this series, please visit: www.routledge.com/series/SE0684

Conflict and Development

Third Edition

Andrew J. Williams and
Alasdair Gordon-Gibson

Routledge
Taylor & Francis Group
LONDON AND NEW YORK

Cover: Beirut 1989, charcoal drawing by Pauline Ramsay

Third edition published 2025
by Routledge
4 Park Square, Milton Park, Abingdon, Oxon, OX14 4RN

and by Routledge
605 Third Avenue, New York, NY 10158

Routledge is an imprint of the Taylor & Francis Group, an informa business

First edition published by Routledge 2009
Second edition published by Routledge 2016

British Library Cataloguing-in-Publication Data
A catalogue record for this book is available from the British Library

ISBN: 978-1-032-40525-4 (hbk)
ISBN: 978-1-032-40523-0 (pbk)
ISBN: 978-1-003-35348-5 (ebk)

DOI: 10.4324/9781003353485

Typeset in Sabon
by SPi Technologies India Pvt Ltd (Straive)

This book is dedicated to Charlie Williams, Private, RAMC, 1896–1916, a working man who lost his life in a war over power and resources, like so many others before and since.

Contents

Boxes

Plates

Abbreviations

ADB	Asian Development Bank
AfDB	African Development Bank
AIIB	Asian Infrastructure Investment Bank
ANC	African National Congress
ATSG	Aid, Security, Trade, Governance
AU	African Union
BiH	Bosnia-Herzegovina
BRICS	Brazil, Russia, India, China, South Africa
CNN	Cable Network News
COIN	counter-insurgency
CPA	Coalition Provisional Authority
DAC	Development Assistance Countries
DDR	disarmament, demobilization and reintegration
DFID	Department for International Development
DRC	Democratic Republic of Congo
EAR	European Agency for Reconstruction
EBRD	European Bank for Reconstruction and Development
ECHO	European Commission Humanitarian Aid Office
ECOWAS	Economic Community of West African States
EU	European Union
FCDO	Foreign, Commonwealth and Development Office
FTSEI	Financial Times Stock Exchange Index
GDP	Gross Domestic Product
GNI	Gross National Income
GNP	Gross National Product
ICC	International Criminal Court
ICRC	International Committee of the Red Cross
IFI	international financial institution
IFRC	International Federation of Red Cross and Red Crescent Societies
IGO	intergovernmental organization
IMF	International Monetary Fund
INGO	international non-governmental organization
INSTRAW	(UN) International Research and Training Institute for the Advancement of Women

IS	Islamic State
ISIL	Islamic State of Iraq and the Levant
LDCs	least developed countries
LTTE	Liberation Tigers of Tamil Eelam
MDGs	Millennium Development Goals
MSF	Médecins Sans Frontières
NATO	North Atlantic Treaty Organization
NGO	non-governmental organization
NICs	newly industrialized countries
OCHA	Office for the Coordination of Humanitarian Affairs
ODA	Official Development Assistance
OECD	Organisation for Economic Co-operation and Development
OSCE	Organization for Security and Cooperation in Europe
PA	Palestinian Authority
PIOOM	Interdisciplinary Research Programme on Root Causes of Human Rights Violations
PLO	Palestine Liberation Organization
PMC	private military contractor
PRT	provincial reconstruction team
R2P	responsibility to protect
RUF	Revolutionary United Front (Sierra Leone)
SAP	structural adjustment programme
SDGs	Sustainable Development Goals
SSR	security sector reform
TRC	truth and reconciliation commission
UN	United Nations
UNAVEM	United Nations Angola Verification Missions
UNCTAD	United Nations Conference on Trade and Development
UNDP	United Nations Development Programme
UNICEF	United Nations Children's Fund
UNIDIR	United Nations Institute for Disarmament Research
UNMIK	United Nations Mission in Kosovo
UNRRA	United Nations Relief and Rehabilitation Agency
UNTAC	United Nations Transitional Authority in Cambodia
UNTAET	United Nations Transitional Administration in East Timor
UNWFP	United Nations World Food Programme
USAID	United States Agency for International Development
WCT	war crimes tribunal
WHS	World Humanitarian Summit

Introduction

In the eight years since the last edition of this book, published in 2016, much has changed, and much of what we said then is relevant now. Insofar as substantial new sections have been added to the text, we have taken the advice of many people, readers for Routledge, former students and professionals of various kinds. A lot of their suggestions have been adopted. As the saying goes, all omissions and views are our own. The following paragraphs give some idea of what we have tried to expand upon from previous editions.

First, since 2016, the political landscape of the planet, the West and non-Western countries has been transformed. In the West, there have been major changes in Britain, the United States and the European Union in particular. Populist politicians have swept to power and inflicted great damage on liberal democratic principles and policies. Among the other powerful states and blocs, there have been huge geopolitical changes, especially in China and Russia. In the developing world, new wars have disfigured the Horn of Africa, while other wars, especially in the Middle East, have continued or even got worse, as in Gaza. We have tried to reflect these changes while at the same time being aware that Hegel's famous 'owl of Minerva' has not yet flown on many of these conflicts. The book has tried to bring up to date as many conflict and war situations as we feel able to discuss with some personal authority, most notably in the Middle East, the Horn of Africa, Eastern Europe and Myanmar, as well as where we have a pretty good grasp of the available academic and biographical material. We cannot be experts on everything, of course, and accept that many will not agree with our choice of analysis. Conflicts and wars are (and this is natural) freighted with huge emotional, historical and physical significance for many people. We are more than aware of the irritation we might cause and have tried not to be overly judgemental or prescriptive in our analysis. Any student of conflict and development, analysed separately or together, treads through metaphorical (and sometimes actual) minefields.

Second, there have been many new economic challenges. The progress of globalization has been slowed considerably, partly by the agency of nature, a global pandemic the like of which we have not seen

DOI: 10.4324/9781003353485-1

since the First World War. The 'pale rider' of the 'Spanish Flu' (neither 'Spanish' nor a 'flu', at least not a normal one) of 1918–20 killed a probable 100,000,000 people, a much greater proportion of a much smaller population (Spinney 2017) before the existence of modern antibiotics, anti-viral drugs and the appropriate mass inoculation policies. The 2020 version of a new 'coronavirus' showed that the huge advances of globalization could spread a disease more easily but also lead to much faster and generally much more successful transnational cooperation. The aforementioned populist politicians in the West, some of them at least, managed to blame (or rather scapegoat) the Chinese for the spread of the virus as their predecessors had blamed the Spanish. The first real outbreak in Europe has been identified as originating in the Austrian resort of Ischgl, where not many Chinese citizens were seen dancing on the tables in this notoriously wild ski resort. Indeed, the 1918 outbreak was probably incubated in the trenches of the Western Front, not in then neutral Spain.

Also, on the macro-economic level, globalization has been slowed since the last edition by the reversal of a trend to free trade which had been evident until 2016. In that year, Britain voted to reduce its ties with its largest trading partner, the European Union. In 2020, President Donald Trump of the United States embarked on what was, in effect, a trade war with China over 'unfair' trading practices. Given that international organizations had backed freedom of trade as an effective brake on inter-state conflict since the Second World War and encouraged the lowering of trade barriers to help with development, this was a major issue in the global economy of the past few years. We will further discuss this driver of both conflict and war at several points in subsequent chapters. But the key emphasis of Chapters 1–3 is on those who make war, those who suffer from it and, on the other hand, those who try and manage it. This can be through the main agent of institutional globalization, the United Nations (UN) and its sister agencies, or through the vectors of capitalist activity, by which we mean entrepreneurial activities of many kinds. We do not most emphatically see ourselves as ideological warriors for or against these institutional or economic vectors. Neither are we going to shrink from the increasingly clear problems generated by liberal policies of the past, which many have now come to see as inappropriate for countries and populations that have been damaged by the over-enthusiastic neo-liberal economists of the intergovernmental organizations (IGOs) since the end of the Cold War. So we are unapologetically critical of both IGOs and politicians where we feel they have separately or together inflicted damage on innocent civilians. Students of security studies have now realized that war cannot be considered without reference to economic causes and consequences. To merely declare a war 'over', as did President George W. Bush in 2003, does not mean its 'over'. 'Reconstruction' (the subject of Chapter 5) must be done much more meticulously and fairly, or we will lose faith in its healing balm, so seemingly evident in Europe and Japan in the 1950s. After the reconstruction debacles in Afghanistan and Iraq, will we have more and more 'endless' or 'forever' wars, which are catastrophic in the developing world and sap the fibre of democracy in the developed part of the globe? Who cannot draw a straight line between the events of 9/11 in 2001 and the wars in the Middle East, the populist politicians of the West and the global malaise that now grips us all?

Third, we have tried to show how the transfer of knowledge and practice has been transmitted and often distorted both nationally and internationally by war and conflict processes. Hate as well as democracy have been facilitated by the enormous growth of the internet. The current (2022–present) war between Russia and Ukraine will undoubtedly be dubbed the 'First YouTube War'; never before has such detail been provided by the click of a laptop computer of the real-time destruction of thousands of young people in a war. This ultimately useless loss of life and suffering has been mirrored by still graphic footage of similar barbarism in Africa and elsewhere. One of our main aims as conflict analysts has always been to try and stop such activities; indeed, since the 1950s, this has been one major pillar of what has been attempted in the area of conflict resolution, the main emphasis in Chapter 4.

There are countervailing processes of a positive kind that we have particularly worked on in this volume. 'Humanitarianism' is one such focus, undoubtedly driven by one of the authors having worked for over 20 years in the International Red Cross and Red Crescent Movement. There is a growing understanding that humanitarianism is both more needed than ever and more criticized. Most of Chapter 6 looks at those twin imperatives, but the theme recurs in several places in the book.

Last, but not least, we have attempted to show how each of the main topics in the book has deep historical roots and use as much as we can of the extant appropriate historical literature. We have not been, indeed could not be, exhaustive. Where possible, we have given as much care to giving different historiographical leads as we can. This extension of the focus of the book has also been extended (again, as much as space and our competence allow) to more consideration of gender questions, some references to digital questions, a few 'new' conflicts and a downplaying of others. We also want to make a plea for a more 'imaginative' study of international relations as a wider field. We have tried to do this through more local voices, though the subject areas in this book have long been dominated by authors of white middle-class European backgrounds, like us. We freely admit we need more from those who are not.

If taken individually, development and conflict raise near-insoluble questions.

Development

- What constitutes development? Is it just about economic issues, or does it also extend to quality of life, culture and politics?
- How can we accommodate a rising world population (from around 8 billion people today to a projected 9.8 billion in 2050) with protection of scarce resources (UN DESA 2023) – a situation becoming increasingly acute with the pressures from a changing climate?
- What are the best ways to address fairness and inequality? Should we even bother?
- How can we measure development?

Conflict

- Can we pre-empt armed conflict by identifying the triggers and causes?
- Why do the costs of conflict fall more heavily on some groups?
- What can we do about persistent conflicts, like in Israel–Palestine? Are they simply destined to be with us for ever?
- What are the best ways for collectives of countries (like the UN or a coalition of regional powers) to deal with conflict?

When we bring conflict and development together, we also face tough questions. But by combining the two, we can see how it is difficult to address conflict without also addressing development. It also becomes clear that development issues can be the key to escalating and de-escalating conflict. Are peacebuilding and conflict resolution part of the new aid and development nexus? The links between conflict and development are seemingly endless. Countries that manufacture arms may want to maximize exports, but these exports may fuel conflict and oppression overseas. Southeast Asia is becoming increasingly militarized as states make claims on the South China Sea and the mineral wealth beneath it. A warming climate is making competition for access to resources in regions such as the Arctic and Antarctica more prominent. In recent times, the United States has become embroiled in costly and ultimately unwinnable wars in Afghanistan and Iraq. But it has only been able to do so by taking out enormous loans. So modern war depends on credit. Non-state armed groups like the Islamic State or the Taliban need money to pay fighters and buy arms. They need to exploit markets to raise these funds.

This book seeks to unpack the links between conflict and development. It is mindful, though, that the picture is very complicated and that we, as observers from the Global North, may not always have the best vantage point from which to see conflict and development in other parts of the world. We should be aware of the limitations of predominantly Western research methodologies and methods of explanation. We need to be aware that many of our discussions of conflict and development use peculiarly Western tools of analysis and occur at a level of abstraction far removed from the lived experience of those facing the challenges of conflict and development. Vibrant and diverse communities are rendered into figures in a spreadsheet or are described in the dry tones of economists and conflict analysts. Most development and peacebuilding policy (and the research that underpins it) is designed in the Global North but is directed at the Global South (Scholey 2006: 179–80; Cohen 2014). Moreover, Western analyses of the developing world and conflict contexts can be prone to stereotypes that overwrite a more complex reality. Thus we are all familiar with African women portrayed as '[a]lways poor, powerless and invariably pregnant' or African men as aggressive (Win 2007: 79).

Conflict and development – often studied separately

One of the oddities of social science research has been that theories of development and theories of conflict have largely evolved in isolation from one another. This was especially odd since development economics as an academic subdiscipline and area of policy expertise emerged in a post-Second World War context defined by violent conflict. Moreover, the lens of development studies was firmly focused on the developing states, many of which were prone to conflicts relating to decolonization, post-independence power struggles and proxy competition among Cold Warriors. The few development economists who considered the matter saw war as an interruption of development and surmised that development could not begin until war had ceased in a particular location (Thomas 2006: 186). As a result, most development specialists excluded countries experiencing violent conflict from their studies. In effect, conflict was written out of development.

We must not be too harsh on development theorists: many were economists more at ease with the study of a state's fiscal levers than its political machinations. Moreover, most of the development research was country specific and thus poorly placed to observe regional or international patterns, including those that contributed to violent conflict. Just as development theorists largely ignored conflict, the emerging subdiscipline of conflict studies had little to say – until relatively recently – about development issues. There were a few honourable exceptions among the conflict theorists (Gurr 1970; Azar 1990), but by and large, their lenses of inquiry overlooked the potential of development (and de-development, underdevelopment and uneven development) to contribute to both war and peace. The sole interest of many theorists was inter-state war and the interplay between military and political leaders in different states. Economic conditions within states were important only in terms of how they could sustain a state's ability to pursue war. The condition of people within states and how this might contribute to conflict and popular radicalization was often overlooked.

The not so splendid isolation of development and conflict studies from one another was no longer sustainable when the end of the Cold War witnessed an upsurge in civil wars. Not only were these post-1991 civil wars more visible than their Cold War predecessors, but the international system was also in flux, with national, regional and international mechanisms struggling to cope. In this context, there was a rush to explain the phenomenon of civil war, and the corpus of academic, policy-related and journalistic work exploring the linkages between conflict and development began to grow. Crucial in this process was a more sophisticated understanding of the nature of conflict and the long-term and often unsatisfactory nature of pacification strategies (Kaldor 2007). The orthodoxy that development was an integral part of peacebuilding spread very quickly and was notably promoted by UN Secretary-General Boutros Boutros-Ghali's 1992 *Agenda for Peace* document. The inclusion of peacebuilding into the aid and development nexus gained currency following the World Humanitarian Summit (WHS), convened by UN Secretary-General Ban Ki-moon, on 23–24 May 2016. There was also a growing realization of the links between the outbreak of violent conflict and underdevelopment and uneven development (Collier 2000; International Development Committee 2006: 8). As a result, many academics, policymakers and political leaders came to see development as a key to conflict prevention. The concept of human security, which gained prominence from the mid-1990s onwards, was highly influential in broadening conceptualizations of security to encompass issues traditionally regarded as germane to development or social improvement (United Nations Development Programme [UNDP] 1994; Shaw et al. 2006: 3–18).

The WHS in 2016 presented paradigms that it hoped would direct towards 'A New Way of Working' that included greater participation and resourcing of national structures and local actors in a humanitarian crisis (the 'Grand Bargain') (IASC 2022). Since then, the UN Secretary-General has pushed forward with a comprehensive set of reforms to the UN system, looking to rationalize the work of the Secretariat and its supporting structures in emergency response, humanitarian aid and development (One UN; UNDS Reform; The Quadrennial Comprehensive Policy Review (QCPR), etc., see Chapter 2). Together with other organizations, such as the International Committee of the Red Cross (ICRC) and the International Federation of Red Cross and Red Crescent Societies (IFRC), UN agencies are developing tools, policies and platforms of response to the changing dynamics in International Humanitarian Law (IHL) and emergency preparedness that are emerging from new technologies such as cyber warfare and artificial intelligence and the effects of climate change (see "Further reading"). The Red Cross and Red Crescent Movement is playing a key role in defining platforms of response and preparedness that engage its worldwide membership of 191 National Societies and an estimated 16 million volunteers.

About this book

Controversies abound in the study of conflict and development: are internal or external factors primarily responsible for a state's underdevelopment? Should aid agencies cooperate with warlords to distribute humanitarian goods? How can the security demands made by Western states be reconciled with human rights and political liberalization in developing world states? What are the implications for humanitarianism and the principles of neutrality and impartiality of growing civil-military cooperation? How far can the aid and development sectors remain distanced from the political influences in the contemporary environment of conflict and complex humanitarian emergencies? Can increasingly intrusive Western means of development programming and good governance be reconciled with indigenous and traditional norms? How can increasingly technocratic development and peace-support interventions address the affective or emotional dimensions of development and peacebuilding? How can powerful corporations be held to account? Can technology solve some long-term development problems? Do we, as researchers, have the correct tools with which to understand conflict and development issues? These questions, and many more, form the basis for this book.

The essential purpose of the book is to chart our understanding of the complex relationship between conflict and development. The book is written from a largely critical perspective: critical in the sense that it questions orthodox assumptions and conventional wisdom, but hopefully, the book does not forget that academic sophistry is of little help to those facing the very real hardships that attend to the problems of conflict and development. At heart, development and conflict revolve around people, yet social scientists (along with government planners, non-governmental organization (NGO) log-frames and financial models) have been particularly successful in writing people out of their analyses. This book recognizes that humanity, red and raw, plays an essential part in stories of conflict and development. Hate, rage, revenge, hopelessness, bitterness, ignorance, love, joy and mercy are as relevant to analyses of processes of conflict and development as academic conceptualizations.

Assumptions

The book is based on five assumptions about development that help with our understanding of the connections between conflict and development:

- Development is not necessarily a good thing; it can have negative and unintended consequences.
- Development can trigger and sustain violent conflict.
- Development is an uneven process.
- Development is not just about economic growth.
- Development can be targeted in ways that aid post-war reconstruction and reconciliation.

Cutting across all these assumptions is the issue of power: who holds it, how did they get it, and can it be taken off them (paraphrased from Benn 2001)? Power comes in many forms – political, economic, official, unofficial, hard (military), soft (diplomatic and cultural), earned, elected, seized, and coveted. Power and legitimacy do not always overlap: governments, big businesses and landowners may have official or coercive power, but they may be popularly despised. Local leaders (for example, clerics or tribal leaders) may have little economic or formal political power, but they may be respected within their community and have legitimacy. External actors, such as international organizations, donor governments or multinational corporations, may have immense power and capability, but again, this does not always translate into local acceptance.

When discussing the five assumptions that underpin this book, it is worth considering how they are modified by power. For example, in relation to the third assumption (development is an uneven process), it is worth thinking about what power factors make and maintain the unevenness of development. Do these factors derive from the internal dynamics of the country, or do they relate to international structures and the dominance of a narrative driven by the Global North?

The first assumption (that development is not necessarily a good thing) may initially seem Luddite or somehow antithetical to human advancement, especially since development is regarded by many as the means through which public goods (education and health care) and personal liberty (freedom of expression and action) can be attained. 'In everyday usage "development" is virtually synonymous with "progress"'

(Thomas 2006: 187). Despite the tremendous political, economic and moral power behind the orthodox position that development is always a 'good thing', critical observers must be prepared to judge development according to its actual impact and ambition. Such judgements will depend on the moral–ethical–political framework held by the individual, community or institution that makes the judgement and the vantage point from which they make their judgement. Put simply, where one sits will determine how one judges development.

If development is judged against the simple criteria of jobs created or the amount of electricity created (in the case of a hydroelectric dam), then it might be relatively easy to judge a scheme a success or failure. Yet a broader assessment framework is required – one that takes into account unanticipated outcomes and difficult-to-measure consequences, such as the loss of culture and scenery. If development is demonstrated to inflame conflict, degrade environmental conditions and have profoundly negative social and cultural consequences, then it is entirely reasonable that observers reflect this in their judgements. It is unreasonable to brand those who are against 'bad' or unjust development as being against all development.

This leads to the second assumption: development can trigger and sustain violent conflict. This is by no means always the case. Indeed, as will be demonstrated in later chapters, development can help prevent conflict and aid post-conflict reconciliation. Yet it is important to recognize the conflict-promoting potential of development processes whereby intergroup resource competition, population displacement, environmental degradation and the erosion of social structures that may have once restrained conflict may all contribute to violent conflict. These issues will be explored in detail in Chapter 1.

The third assumption, that development, by its very nature, is an uneven process, contributes much to explanations of the initiation, maintenance and ending of violent conflicts. A complex array of structural and proximate economic, political and geographical factors accounts for the uneven nature of development. Many of these factors will be discussed in later chapters. The unevenness of development (and resources and approaches to development) suggests that conflict will be 'inevitable', but as we will see, it is often the management of resources and development that matters. As important as unevenness is the *perception* of unevenness. In societies with identity-based divisions, groups may be anxious to get their 'fair' share of resources. In the period following the start of the war in Syria, many citizens in Lebanon, for example, saw charities ministering to Syrian refugees and believed that Syrians got preferential treatment when it came to health care (Harb and Saab 2014: 8).

The fourth assumption, that development is not just about economic growth, stems from the tendency of many observers (not just development economists) to overlook the social, political and cultural dimensions of development. Crucial in this regard is the type of development strategy pursued and the relative importance attached to redistribution and market freedom, as well as political and cultural development. A narrow economic lens could examine China's economic growth rates (an annual growth of just under 8 per cent in 2021 https://data.worldbank.org/country/china) and declare it a development success. But a more holistic approach might take account of the state's poor human rights record. Amnesty International's *State of the World's Human Rights Report 2021/22* observed,

> The year was marked by harsh crackdowns on human rights defenders and people perceived to be dissidents, as well as the systematic repression of ethnic minorities. The beginning of the year saw the start of the COVID-19 outbreak in Wuhan, which killed more than 4,600 people in China. People demanded freedom of expression and transparency after authorities reprimanded health professionals for warning about the virus. At the UN, China was strongly criticized and urged to allow immediate, meaningful and unfettered access to Xinjiang. Stringent restrictions on freedom of expression continued unabated.
>
> (Amnesty International 2022)

The final assumption is to recognize the potential of development to contribute to post-war reconstruction and reconciliation. Development can, in the correct circumstances, ease or cement a route out of conflict and division. In an ideal situation, a mutually reinforcing relationship can be established between development and reconstruction. Formerly divided peoples can come together for the joint pursuit of economic growth and social progress. This is not always the case, and poorly managed post-war reconstruction often has profound consequences for the nature of the post-war society, some of them negative.

Alongside these assumptions on development, it is worth noting that the main focus of this book is on civil war rather than inter-state war. This is primarily because war between states is a rare phenomenon, while civil war (albeit often internationalized) is more common. Most civil wars spill over national boundaries, with arms, refugees, fighters and resources crossing borders to give conflicts a messy transnational dimension. In 2021, for example, although 54 state-based conflicts were recorded (UCDP 2021) only one of them (between Russia and Ukraine) could be classed as inter-state. Importantly, our interest in conflict extends beyond the direct violence of overt war. Indirect or structural violence plays a key role in contexts of conflict and development. Such violence is often embedded in the socio-economic and politico-cultural behaviour of a society. It can be insidious, barely visible and taken for granted. It takes the form of discrimination in the provision of public goods and opportunities, the militarization of society and the prevalence of societal attitudes (sometimes encouraged by the state or other institutions) that certain groups are inferior to others. So this book proceeds by adopting holistic views of both conflict and development. Neither concept constitutes a neatly compartmentalized category. Instead, they are messy and ill-defined, and there is little agreement about the best way to pursue development and conflict transformation. Above all, both concepts relate to the most contrary and awkward species of all: humans. People don't always behave as we expect them to; they can act in ways that strike us as 'irrational', and they're not always as grateful to us as we feel they should be. People are often written out of analyses of conflict and development.

An array of factors helps write people out of many studies of peace, conflict and development: the media's need to compress thousands of individual experiences into a single narrative; the technocratic bias of policymaking in which units and spreadsheets are more manageable than people; and social sciences' move towards large-scale studies and their inability to deal with the affective dimension of human behaviour. Indicative of this 'writing out' of people is development and conflict-sensitive 'programming' as practised by many donor governments and agencies. The term 'programming' suggests a machine-like process whereby carefully regulated inputs are expected to have particular outputs. Just as war is often criticized for dehumanizing individuals and objectifying them into a lumpen enemy, there is a real danger that responses to conflict and underdevelopment have a similar effect. Most studies, apart from biography and some forms of anthropology, cannot hope to convey individual experiences. It is important that we recognize that people in conflict zones do not constitute an undifferentiated mass and that individuals often experience events and processes in very different ways. The shifting paradigms in humanitarian response and the inter-connection between aid, development and peace called for at the WHS mentioned earlier (the 'triple nexus', see Chapter 6) are framing new discussions about the social and the political nature of work in the sectors. There is growing scholarship on the need for a new way of working and a new way of thinking about the humanitarian identity and the relationships between stakeholders in a crisis that urge a shift away from the institutionalization and exceptionalism that some commentators have seen as the cause for dysfunction in contemporary responses (Donini 2010; DuBois 2018; Hilhorst 2018; Kapila 2022).

This chapter proceeds with brief overviews of the evolution of theories of development and conflict so as to provide a context for subsequent chapters. In relation to development studies, it is particularly important to understand the contemporary dominance of neo-liberalism and market-led solutions and the consequences of this for internationally sponsored development and peace-support interventions. The evolution of conflict studies shows how a more complex understanding of the causes and maintenance of conflict has emerged and how this has shaped contemporary conflict transformation interventions.

The evolution of development theory

Summarizing a multidisciplinary endeavour such as development studies is a difficult task, especially when the ultimate purpose of development and the optimum means of achieving it are hotly contested. Kothari (2005: 1) is rightfully critical of surveys of the discipline that 'articulate a singular theoretical genealogy'. Development studies has been peculiarly faddish, seizing upon theories and techniques at particular moments, only to discard them in favour of a new saviour theory or technique. Kothari (2005: 2) also warns against interpretations of development studies that regard 1945 as Year Zero, as though no development or thinking about development occurred before that year. Scrutiny of the means to achieve economic development has a long intellectual pedigree (Meier and Rauch 2000). J. M. Keynes was particularly prescient of the need for peace negotiations to take seriously the issue of long-term economic development (Davenport-Hines 2015). Dispirited after his ringside seat at the Treaty of Versailles negotiations following the First World War, he observed, 'Peace has been declared at Paris. But winter approaches' (Keynes 1920: 235):

> The Treaty includes no provisions for the economic rehabilitation of Europe, – nothing to make the defeated Central Empires into good neighbours ... nothing to reclaim Russia; nor does it promote in any way a compact of economic solidarity amongst the Allies themselves; no arrangement was reached at Paris for restoring the disordered finances of France and Italy, or to adjust the systems of the Old World and the New.
>
> (Keynes 1920: 235)

As will become clear in later chapters, lessons seem to have been learned by some, in that it is well recognized that the ending of wars provides unique opportunities for international political and economic intervention. Less clear, however, are the optimal types and extent of any intervention.

Although 1945 was not Year Zero, it was the year in which the contours of the modern international financial architecture were established. The World Bank and the Bretton Woods exchange rate mechanism (which facilitates international trade) date from this era. The Second World War also saw the United States re-establish itself as the predominant state in the world economy (Endres 2012). The decade and a half after the Second World War was the highpoint of economic planning during which development economists were convinced that '"good" scientific analysis would generate the "right answers"' (Harriss 2005a: 19). This was the period of positivist orthodoxy, in which the primary aim of development was accelerated economic growth, the primary agent was the state (mediated by the Bretton Woods institutions) and the primary means of achieving growth was careful analysis followed by a precise plan. The plan often involved raising rural productivity and transferring underutilized labour from the agricultural to industrial sectors (Leys 1996: 8). It was an era of optimism but was also tinged with colonial attitudes: if only those underdeveloped people would adopt scientifically proven methods of agriculture and economic organization then they will become developed – just like 'us'.

By the late 1950s and early 1960s, it was becoming clear that economic growth was difficult to achieve in many developing world contexts and that the fruits of any growth were rarely shared fairly. There was no shortage of economic theories, but as J. K. Galbraith (1964: 38) observed, '[I]t would be a mistake to identify complexity with completeness and sophistication with wisdom'. In other words, neat theories did not always work in real-world circumstances. The failure of the initial post-war development planning led to a new emphasis on modernization and technology transfer. The key here was the adoption of 'modern' forms of administrative and political organization in imitation of Western states and businesses, the transfer of Western knowledge and skills through education programmes and a 'green revolution' of increased agricultural productivity through the use of Western farming methods (Rostow 1960). This, in turn, sparked a radical critique in the late 1960s and early 1970s as it became clear that, in many cases, modernization strategies had made few appreciable differences to citizens in the developing world (Harriss 2005a: 19–25). According to left-wing critics, the '"modernising elites" were really … lumpen bourgeoisies, serving their own and foreign interests, not those of the people; world trade perpetuated structures of underdevelopment' (Leys 1996: 12). According to this view (usually called 'dependency theory'), modernization was a recipe for further immiseration and a structural dependence on Western economic powers (Frank 1967).

Global political trends, and specifically the Cold War, had a profound impact on development strategies. Several states, including Cuba, Ethiopia, Tanzania and Vietnam, adopted socialist development programmes, usually under the tutelage and protection of the Soviet Union. To differing degrees, 'scientific socialism' was mobilized in the service of development. This often involved the nationalization of industry, restrictions on foreign capital, land reform, and an enhanced role for the state in directing economic exchanges and initiatives (Clapham 1987). In most cases, the results were not good: the socialist 'reforms' tended to be disruptive and often reliant on state coercion and the global economy offered a poor fit for those not willing to reform. It is worth noting that the socialist experiment states such as Angola, Mozambique and Ethiopia were often the scenes of violent civil war, wars that were often stoked by the United States and its allies. The true fragility of these states' economic models did not become apparent until the collapse of the Soviet Union in 1991 and the withdrawal of subventions. Both dependency theory and the socialist model of development paled into the background with the coming tide of neo-liberalism.

Neo-liberalism is the belief that the market, freed from regulation, offers the solution to development problems. The neo-liberal revolution of the 1980s and beyond was in sympathy with wider political and economic changes. The political right was on the ascendant in the United Kingdom and the United States, and globalizing market forces meant that national and international controls over capital had been severely eroded. Traditional responses to state-directed development policy were no longer effective in an economic climate characterized by economic shocks, unstable commodity prices, international capital flight and increasingly powerful and mobile multinational corporations. The end of the Cold War meant not only the collapse of the Soviet Union as a material supporter of alternative models of economic development but also the collapse of the notion of 'an alternative'. There was no Plan B anymore. Former Soviet satellite states proved to be in no position to resist aggressive economic reform interventions (often called the 'shock doctrine') by international financial institutions. The rise of neo-liberalism was also assisted by abundant evidence of state incompetence and corruption throughout the developing world. Deepak Lal (1998: 65) noted how the 'old development economics ... implicitly assumed that the state was benevolent, omniscient, and omnipotent'. The 'Chicago School' (who claim intellectual authorship of neo-liberalism) was pushing at an open door at the headquarters of the World Bank, the International Monetary Fund (IMF) and right-wing governments. The essential neo-liberal argument was that the dead hand of the state and an invariably bloated public sector acted as a brake on economic development, while an unfettered market could act as an engine of development. In this view, the benefits of market-driven growth would trickle down and benefit all.

According to James Dorn (1998: 13), 'the real plight of underdeveloped countries is not market failure but government failure – that is, the failure of government to protect property rights, enforce contracts, and leave the market alone'. Champions of neo-liberalism were forthright in what needed to be done: the state must be pared back, the market freed from regulation, state assets should be privatized and exchange controls and industrial licences should be lifted. The salvation for underperforming economies lay in more exposure to the market, not protection from it. Peter Bauer (1998: 36) had little time for 'unfounded notions about Western responsibility for Third World backwardness'. For him it was

> abundantly evident throughout the Third World [that] the poorest and most backward societies and areas are those which have fewest commercial contacts with the West, and the most advanced are those with the most extensive and diversified contacts, including contacts with those bogeymen, the Western multinationals. Throughout the Third World the level of economic attainment declines as one moves away from regions with most Western contacts to the aborigines and pygmies at the other end of the spectrum.
>
> (Bauer 1998: 28)

Neo-liberal prescriptions, which were enthusiastically endorsed by global capital and the leading international financial institutions, were rolled out in former Soviet bloc states, often with catastrophic social consequences (Klein 2007: 180–4). Western governments, and by extension their development aid institutions, increasingly adopted market-led 'solutions' as the best way to assist societies in development and reconstruction.

Cloaked in a populist mantle of the empowerment of entrepreneurs and the cutting of public sector waste, neo-liberal truisms became the new orthodoxy. According to David Harvey,

> Neo-liberalism has, in short, become hegemonic as a mode of discourse. It has pervasive effects on ways of thought to the point where it has become incorporated into the common-sense way many of us interpret, live in, and understand the world.
>
> (Harvey 2005: 3)

In many Western European states, we have seen the dismantling of the post-Second World War social contract. State provision, particularly on welfare, has been pared back and with it the expectation that the state should provide. The intellectual dominance of neo-liberalism is essential to our understanding of the responses of leading states, international organizations and international financial institutions to the problems of conflict, development and peacebuilding. Market-led 'solutions' have been hardwired into the organizational culture and policy responses of donor governments and NGOs. Whether this manifests itself in micro-credit schemes to further women's empowerment or in the privatization of state resources as part of a post-war reconstruction programme, it has had a profound impact on the ethos of humanitarianism, development, and conflict amelioration policies. Somewhat ironically, the financial crisis that began in 2007, caused by a lack of control on lending to governments and individuals, reinforced neo-liberal 'solutions'. The language of austerity has been used to further cut back the state and normalize the notion that the market is the best, and most sustainable, route to economic salvation.

Under the guise of 'good governance', massive programmes of social, economic and political engineering have taken place in societies emerging from conflict and economic crisis. In many cases, relationships between citizens and the state, the state and the market, and the state and other states, have changed radically. For example, neo-liberal wisdom may demand that a state cuts its bureaucracy at the conclusion of a civil war to keep down inflation and achieve international competitiveness. Yet, the political loyalty of particular sections of the population may have been dependent on patronage from the state in terms of employment or access to public goods. Neo-liberal interventions may thus radically alter political bonds and have far-reaching consequences for political participation and stability, and public perceptions of political processes and institutions.

Unsurprisingly, neo-liberalism has attracted immense criticism, particularly in relation to its inability to address poverty and social exclusion. The near deification of the market and corporate power has, according to its critics, reconfigured the balance of power in many states, with public interests being demoted. Harvey (2005: 19) regards neo-liberalism as a 'system of justification and legitimation' aimed at reinforcing 'the capitalist social order'. In this view, it is a potentially authoritarian political model: 'The neoliberal state is necessarily hostile to all forms of social solidarity that put restraints on capital accumulation' (Harvey 2005: 75). While Dorn (1998: 14) regarded 'Chile's free market revolution' of 1973 as 'an example for the rest of Latin America', Harvey (2005: 8–9) was excoriating of Pinochet's US-backed coup on 'little September 11th' and the widespread repression that followed in the name of libertarian ideas. It is true that many champions of neo-liberalism are agnostic about the social and political costs of an unfettered market. Deepak Lal (1998: 68, 70) observed that 'The characteristics of good government are more important than its particular form' and 'it is by no means self-evident … that Western democracy necessarily promotes a market-friendly culture'. Critics remain unconvinced of the redistributive potential of the market and believe that politics do matter, particularly in relation to political commitments to social inclusion (Harriss 2005b: 228).

Certainly, there is a widespread understanding of the potentially pernicious effects of market-led programming in development and peacebuilding contexts (Chua 2004). The problem for critics of neo-liberalism is that their opponents have created a self-reinforcing *system* based on widely accepted norms of efficiency, cost-effectiveness and enterprise. The neo-liberal system is promulgated by corporate interests and international financial institutions, chimes with populist causes, has reconfigured the ethos of public sector institutions (from hospitals to universities) in the developed and developing world and is in alignment with the strategic interests of leading states. Critics are perfectly correct in pointing to the spectre of 'predatory disaster capitalism' that profits from the misery of others (Klein 2007), but the structures of the contemporary international political economy have been captured by neo-liberal forces and the system, until now, seems able to sustain itself (or at least defer or pass on the costs). It will be interesting to review how patterns of engagement and the identity of key donors and political stakeholders might change over the coming decade.

The international responses to the global financial crisis that began in 2007 are instructive. Governments all over the world were in thrall to markets and their promise of continuous economic growth (Blinder 2013). But light touch and negligent regulation of markets had given rise to a cannibalistic form of capitalism whereby capital flitted from country to country without responsibility to governments, citizens or any principle other than shareholder value (O'Toole 2010). When the fragility of this house of sand was revealed, governments swung into action. But rather than punish bankers, much of the burden was passed to citizen taxpayers. Banks were bailed out, and their debts were nationalized and made the responsibility of the taxpayer (Helleiner 2014; Wolf 2014). The social consequences were shocking and helped push the issue of inequality up the political agenda in a number of countries. In 2014, youth unemployment in Spain was 54.3 per cent and in Greece 59 per cent (House of Lords 2014: 16). The economic system, even though it was dysfunctional, was saved. Some additional regulations were introduced but, in general, a neo-liberal system was entrenched.

The predominance of neo-liberalism does not mean that development assistance aimed at emancipating populations, increasing opportunities and promoting redistribution in the developing world has come to an end. Certainly, neo-liberal structures and principles guide much development activity, but the development sector is flourishing. The United States alone spent almost $33 billion on development aid in 2014, although that amounted to just 0.19 per cent of Gross National Income (GNI). Sweden donates over 1 per cent of GNI. The face of development aid is changing fast though. Regional organizations, oil-rich Gulf and Arab states, Brazil, Türkiye (Turkey) and China (through infrastructure development) are major development and humanitarian actors operating in general outside of the international governance mechanisms and donor support groups dominated by Western humanitarian donors, and often with a regional focus (Binder, Meier and Steets 2010; Campbell and Hofmann 2015: 191–203; Mawdsley 2015: 204–14). Increasing emphasis is placed on integrated development interventions so that the economic, environmental, political and social dimensions of development are interlinked and mutually supporting (Baker 2006). The focus of much development and post-war reconstruction has been on poverty and 'pro-poor growth'. Dealing with poverty is politically unproblematic if economies are growing: the rich and the poor both do well, and the rich maintain their comparative advantage. More politically problematic, however, is dealing with inequality. For societies to become more equal, those with power, money and advantage have to give some of that up. Rich states and international organizations have kept inequality off the agenda by talking about poverty. Yet as the divide between rich and poor becomes ever more apparent, the issue of inequality is harder to suppress (Piketty 2014; Stiglitz 2015).

Sustainability, local participation and ownership and pro-poor initiatives are a common vein through contemporary development thinking and practice. The UN Millennium Development Goals (MDGs), as agreed by 189 governments in 2000, crystallized development priorities:

- Eradicate extreme poverty and hunger
- Achieve universal primary education
- Promote gender equality and empower women
- Reduce infant mortality
- Improve maternal health
- Combat HIV/AIDS, malaria and other diseases
- Ensure environmental sustainability
- Develop a global partnership for development

These were extremely influential, as they shaped development policies by international organizations, donor states and international non-governmental organizations (INGOs). A target date of 2015 was set, and many international organizations and donor states explicitly mentioned the MDGs in their programme goals. The MDGs were also significant in what they left out: tackling poverty was mentioned but inequality was not, and agriculture was not mentioned. The list was also apolitical and seemed incurious about the political economy that lies behind many developmental and humanitarian problems. There have been some successes, such as cutting global poverty, although this is due, in large part, to China's rapid economic growth. Since 2015, the

MDGs have been replaced by a more ambitious list of Sustainable Development Goals (SDGs). They are as follows:

1. End poverty in all forms everywhere
2. End hunger, achieve food security and improved nutrition, and promote sustainable agriculture
3. Ensure healthy lives and promote wellbeing for all at all ages
4. Ensure inclusive and equitable quality education and promote lifelong learning opportunities for all
5. Achieve gender equality and empower all women and girls
6. Ensure availability and sustainable management of water and sanitation for all
7. Ensure access to affordable, reliable, sustainable and modern energy for all
8. Promote sustained, inclusive and sustainable economic growth, full and productive employment and decent work for all
9. Build resilient infrastructure, promote inclusive and sustainable industrialization, and foster innovation
10. Reduce inequality within and among countries
11. Make cities and human settlements inclusive, safe, resilient and sustainable
12. Ensure sustainable consumption and production patterns
13. Take urgent action to combat climate change and its impacts
14. Conserve and sustainably use the oceans, seas and marine resources for sustainable development
15. Protect, restore and promote sustainable use of terrestrial ecosystems, sustainably manage forests, combat desertification and halt and reverse land degradation, and halt biodiversity loss
16. Promote peaceful and inclusive societies for sustainable development, provide access to justice for all and build effective, accountable and inclusive institutions at all levels
17. Strengthen the means of implementation and revitalize the global partnership for sustainable development.

The list is augmented by 169 targets and is interesting because many of the issues it tackles have a political dimension.

Although development activities have been subject to immense faddism since the late 1940s, and although there have been broad shifts in emphasis (from planning to market-led initiatives), several constants are worth noting. The first is that development is still largely a North to South enterprise: intellectually, practically and financially. The second constant is that macro-economic structures are still biased towards the Global North. There has been a general trend of wealth and capacity moving eastwards as China, Indonesia, Vietnam and others experience rapid growth. But global economic structures award advantages to already rich states. The third constant factor is that many of the states that were in the most need of development in the 1940s and 1950s (during the peak of development planning optimism) are still grossly undeveloped.

There is no great mystery about the causes and cures of underdevelopment. As Jeffrey Sachs (2008: 6) observed, 'Reaching the MDGs won't take miracles – we know how to keep children alive in malarial regions, we know how to increase food production'. The key issues are linked to political economies that perpetuate inequality. Dealing with inequality (as opposed to simply reducing poverty) requires those who have wealth and capacity to lose some of that. That is not a message that any political leader would find easy to sell. However, the UN Human Development Report 2021–22 indicates that for the first time in the 32 years that UNDP has been calculating it, the Human Development Index, which measures a nation's health, education and standard of living, has declined globally for two years in a row. Human development has fallen back to its 2016 levels, reversing much of the progress towards the SDGs (UNDP 2022).

Although this section has discussed development theory and policy, it is worth noting that populations in the Global South have not been sitting back waiting to be developed. Despite a global trading system stacked against them and often unsympathetic governments, people in Cambodia, Georgia, Uganda and many other countries have simply got on with the tasks of making a living, setting up businesses, consuming, sending their kids to school, migrating and investing in new technology. Governments may have helped with infrastructure and other initiatives, but by and large, people have just got on with the business of life and, in the process, have fuelled economic development (Bøås 2014).

For Indian households, the bicycle became a motorbike and now is becoming a car. For Ugandans, the mobile phone network allowed for the electronic transfer of money so businesses could expand. In Somalia, traders operating through Dubai made contact with counterparts in China and opened up new supply routes. The key point is that aside from macro-economic structures and conditions, people all over the world have used ingenuity, exploited opportunities and have sought to prosper. It is not clear that academic theories of development have recognized this tendency of people to get on with things, to have their own economic survival mechanisms, and to regard formal development projects and programmes as often irrelevant to their circumstances.

The evolution of conflict theory

It is worth noting that for many writers, violent conflict was regarded as a given (a mere by-product of Great Power interplay or the consequence of natural inclinations of humankind), and its causes were not subject to serious scrutiny. Military history, which depicted war as 'a deplorable necessity' (Creasy 1876: xi), thrived but did little to advance our understanding of the precipitants of conflict or its wider social or cultural impacts. More considered deliberations on the nature and causes of warfare occupied scholars from Sun Tzu (544–496 BCE) and Thucydides (460–395 BCE) to Saint Augustine (354–430) and Hugo Grotius (1583–1645), but their work did not constitute a united or recognizable field of conflict studies (Jacoby 2008: 8–12). Real-world 'traumas' such as the First World War (over 35 million casualties, the industrialization of warfare and the militarization of societies) or the unleashing of nuclear weapons gave renewed impetus to those attempting to systematize knowledge of conflict. For example, Lewis Fry Richardson (1950), a Quaker conscientious objector who served with the Friends' Ambulance Unit during the First World War, sought to apply mathematical analysis to warfare, particularly the propensity of armament programmes to lead to conflict. Yet Richardson and other pioneers of the study of conflict as a generic phenomenon, such as Quincy Wright (1942) and Georg Simmel (1955), constituted a distinct minority. Indeed, many of those who pursued 'peace studies' were derided as cranks, cowards, unpatriotic or 'religious nuts' (Rooney 2000: 16).

It was not until relatively recently that attempts to find general theories of conflict reached the academic mainstream (Azar 1990: 5). The subdiscipline of international relations, for example, has traditionally been concerned with wars between states and showed little inclination to investigate sub-state conflicts. It was not until the 1950s and 1960s that a recognizably modern strain of conflict research emerged. Miall (2007: 27) notes how much of this research attempted 'to capture the generic characteristics of conflict', stripping conflicts of their context in order to better examine the relationships between actors and the dynamics of conflict processes. Much of this research was influenced by game theory, behaviouralism and the application of social psychology to conflict (Axelrod 1990; Boulding 1990: 37–8). Rational choice and bargaining theory approaches to the study of conflict are now particularly popular (especially among North American and Scandinavian academics) and are prominent in leading academic journals of peace and conflict. These approaches have helped bring evidence and data into arguments that were often based on assertion (Regan 2013). Comprehensive databases of conflicts such as the Conflict Data Program at Uppsala, the Peace Accords Matrix or the Armed Conflict Database have allowed scholars and policymakers to study issues such as the optimal time to intervene in a conflict or the chances of particular types of conflicts having negotiated or violent outcomes.

Over time, more holistic understandings of conflict have emerged among policymakers and academics. This has been complemented by the 'mainstreaming' of peace and conflict research into a number of academic disciplines, with the establishment of research institutes, specialist journals and a greater acceptance of the systematic study of conflict in policymaking circles. There has been something of an intellectual emancipation of peace and conflict. No longer is it the preserve of a few pacifists and the closed shop of military and foreign policy officials. Indeed, after decades of being on the margins, the study of peace and conflict has even become popular. Christopher Mitchell (1994: 128) recalls how, as a reaction to the rash of civil wars in the 1990s, 'a range of scholars ... discovered that they have "really" been doing conflict resolution "all along"'. 'Ex-strategic theorists, military security experts, Sovietologists and area specialists' suddenly turned their attention to the problems of civil war. As will be outlined in Chapter 1, the rush to 'explain civil war' has variously focused on the importance of identity, regional factors and econometric indicators. The modern evolution of conflict studies has reflected real-world conditions. The post-Cold War upsurge in conflicts, greater interventionism by international organizations and INGOs, and the impact of globalization on public awareness of conflicts on the other side of the planet demanded a better understanding of conflict. The failure (or limited success) of some immediate post-Cold War international interventions demanded a further refinement of our understanding of conflict.

Specialist subfields in the study of conflict developed in the 1990s and beyond, with the comparative study of negotiated peacemaking processes (Darby and Mac Ginty 2000), gender and security sector reform (Özerdem 2010) gaining particular attention. A number of conflict resolution 'gurus', or respected practitioners such as John Paul Lederach, Roger Fisher, William Ury or Ben Hoffman, also gained greater prominence (Fisher and Ury 1991; Lederach 1995; Hoffman 2007). There was a growing consensus on the multidimensional nature of conflict, its increasingly transnational nature in a globalized context, the need to see beyond conflict manifestations to examine conflict causes, and – crucially – the importance of development issues in explaining conflict (Azar 1990: 2). Research on conflict has also recognized the protracted nature of many conflicts, the tendency of violent conflicts to reignite following periods of calm, and the difficult, long-term and costly nature of post-peace-accord peacebuilding. Indeed, 'no war, no peace' situations have become common in which parties agree to a ceasefire but fail to push for a comprehensive peace process (for example, in Sri Lanka, Israel–Palestine and Colombia at various times in the 1990s and 2000s) or reach a peace agreement but fail to move towards a widespread reconciliation (for example, Northern Ireland or Lebanon) (Mac Ginty 2006).

The increased research on peace and conflict was often part of desperate attempts to understand conflict. Much research has been sponsored by governments, international organizations and development agencies. Theories of peace and conflict have largely bifurcated into problem-solving and critical approaches (Cox 1981; Pugh 2013). The problem-solving approach is interested in remedying conflict manifestations. It is often policy orientated and is attractive in that 'something is being done'. The critical approach, on the other hand, is less easily satisfied. It concentrates on the structural factors that undergird violent conflict and is suspicious of peace-support interventions that may staunch fighting but leave power and economic relations largely unchanged (Schmid 1968). The critics accuse the problem-solvers of being incurious about the factors that start and maintain conflict. The problem-solvers accuse the critics of constantly complaining but never getting around to doing anything. In this view, the system of intervention is far from perfect, but we must work with what we have (Paris 2010).

Importantly, there has been an elision of many international strategies to deal with conflict and underdevelopment, with the WHS in 2016 marking calls for an institutional shift towards engaging peacebuilding with the aid and development nexus. Major documents by international organizations show an awareness of how conflict and development often have a symbiotic relationship and should be tackled together. Many of the same strategies that are deployed in societies emerging from civil war can also be found in societies free from civil war but suffering from underdevelopment. Development strategies in societies emerging from civil war might be modified so as to be 'conflict sensitive', but they essentially amount to the same thing. Over the past few decades, there has been increased emphasis on the 'local ownership' of 'solutions' on the understanding that if local communities feel invested in a programme or project, then it is likely to be more effective and sustainable. Arguments rage about the extent to which this 'local turn' in development and peacebuilding planning is rhetoric or denotes a more fundamental shift: the promotion of the orthodoxy of neo-liberalism, 'good' governance reforms, and the use of aid conditionality or selectivity to encourage conformity. (Richmond 2012; Kluczewska 2019; Gordon-Gibson 2021). 'Frontline' humanitarian organizations, such as ICRC and Médecins Sans Frontières (MSF), voiced their concerns over a closer integration of actors, policy and practice called for at the WHS (MSF declined the invitation to attend the meeting in Istanbul).

Alongside the critical strain to the literature on peace, conflict and development, other scholars and policymakers have been vexed at how to 'win' the War on Terror and subjugate opponents in Afghanistan, Iraq and elsewhere. The United States and the United Kingdom saw development and post-war reconstruction as a key means of winning 'hearts and minds', as the United States had termed it during the Vietnam War. That concept has not born the test of time because of that war, though it is still used (Gawthorpe 2018). The United States engaged in a direct war of theories and practices in development with the USSR when it was still believed in Washington that the Soviet experience of development was one that provided a direct threat to Western views on the area. The United States was not entirely wrong, as is evidenced by the popularity until the late 1980s of Soviet expertise and aid in many 'Third World' countries, most notably India, Egypt, Ethiopia and many others. China, which had fallen out with the USSR in the 1960s, had started its development road in conjunction with the Soviet Union (Seitz 2018; see also Chapter 5).

Since the Vietnam War, at least, we have seen the securitization of development or the incorporation of development into security strategies. This has been most visible in the provincial reconstruction teams at work in Afghanistan and Iraq, whereby soldiers become 'armed humanitarians', with obvious consequences for ideas of neutrality and impartiality which some associated with humanitarianism. Newly coined phrases from the wars in Syria and Ukraine, such as 'Lethal Aid' and 'Humanitarian Corridors' that aim to 'humanize' the conflict serve to muddy the narrative further. Critics say that Western states are more interested in order and security than development and reconstruction. Egypt has received $51 billion in military aid since 1979, as of 2020 figures, and since 1985, none of that has to be repaid. One report says Egypt had at that time received over 1,300 Abrams main battle tanks, of which most have never been used, compared with the hundred or so set to the Ukraine where they are desperately needed. By way of comparison, Egypt received a not-inconsiderable sum of $30 billion of *non*-military aid from 1978 to 2021, in which year the sum was $125 million (POMED 2020 US Embassy, Cairo 2021). The sums were effectively (and successfully) intended to keep Egypt as an ally of the United States and keep it out of any more wars with Israel.

A conflict-specific nomenclature has developed, with a battery of prefixes and suffixes fine-tuning our understanding of the concepts of peace and conflict. Thus 'peacekeeping' denotes traditional UN 'blue helmet' troop deployments, usually to police an agreed separation between antagonists. 'Peacemaking' suggests a more active form of intervention or activity, possibly including force to compel parties to negotiate or make concessions. 'Peacebuilding' was originally regarded as an activity that occurred after a peace accord was reached in order to support that accord. It covered a range of political, social, economic and cultural activities and so was cognisant of the links between conflict, peace and development. The strict delineation of peacebuilding as an activity that occurs after a violent conflict ceases no longer holds.

'Peace implementation' usually refers to both the fulfilment of the provisions of a peace accord and attempts to provide an environment (for example, security, minority return, a tax base to fund social spending) to enable the implementation of the terms of the accord. 'Peace fixing', if we may be permitted to coin a new term, is the common activity of internal and external parties to a peace accord returning to the negotiating table to modify the accord to react to problems. In the post-9/11 period, there has been a slight draw back from the use of the word 'peace' by a number of leading states as they promote security agendas (often under the term 'stabilization'), but the recent call for peace to form a new 'triple nexus' with aid and development illustrates the continued linkage.

There has also been a transition in the suffixes used in relation to conflict. 'Conflict resolution' was overtaken by 'conflict management' when some critics suggested that conflicts could not be definitively resolved. Instead of finding solutions, societies had to recognize the integral nature of conflict to human societies and thus find non-violent ways of managing the conflict. But the term 'conflict management' was criticized for assuming that some actors would be managers (often powerful or well-resourced external actors) while other actors would be managed (often indigenous, less powerful actors). The term 'conflict transformation' is now current (doubtless to be supplanted by another term in the coming years). Conflict transformation again recognizes that conflict is part and parcel of human existence but aims at transforming relationships between individuals, groups and institutions from destructive to constructive bonds. We have already mentioned neologisms such as 'Lethal Aid' that are forming part of the new vocabulary.

Peacebuilding

Peacebuilding, as a catch-all term that is much used in policy documents, and indeed this book, deserves elaboration. The term is often more 'described than defined', comes with or without hyphens and still does not appear in the *Oxford English Dictionary* (Holt 2011: 32; see also Jeong 2005). Johan Galtung describes his version of 'peacebuilding' as 'building structural and cultural peace') possibly as vague as his core idea of 'structural violence'; John Paul Lederach sees it encompassing the 'full array of processes, approaches and stages needed to transform conflict toward more sustainable, peaceful relationships' (Galtung 1996; Lederach 1997, both quoted in Holt 2011: 32–3).

Consequently, we might say that it has changed in line with historical context, both in the 'target' areas where conflict is a problem and in the institutional reaction to these areas, for the UN itself is also subject to a certain amount of definitional imprecision when trying to explain what peacebuilding aims to do. But what is common to all peacebuilding approaches is a realization that violent conflict (not conflict *per se*) 'is inflicting immense damage on the societies and economies of the developing world'. Collier and Hoeffler have been quoted as estimating that a 'typical' developing country civil war inflicts at least US$64.2 billion of economic and social damage (Collier and Hoeffler 2004, quoted in Addison 2005: 406).

Originally, many regarded peacebuilding as a post-conflict activity – after violence had ceased. The chronic nature of conflicts (for example, in parts of the Horn of Africa) means that it is difficult to maintain an argument that actors should wait until conflict ends before trying any ameliorative effort. Thus peacebuilding has become a generic term for conflict response interventions before, during and after violent conflict. An interest in conflict prevention in the 1990s (Boutros-Ghali 1992; UN 2000; Carnegie 1997) has been superseded by international concern for resilient states and communities, able to withstand destabilizing pressures. There is difficulty identifying where peacebuilding, conceptually and practically, begins and ends. Certainly, there is consensus that conflict demands multidimensional responses and that local voices have to be incorporated into the design and implementation of the response (Addison 2005; Richmond 2006: 74).

The ultimate aim of peacebuilding can be regarded as 'sustainable' or even a 'stable' peace, classically defined by Karl Deutsch as being when there is the 'real assurance that the members of [a] community will not fight each other physically but will settle their disputes in some other way' (Deutsch quoted in Kacowicz et al. 2000: xi). This has been refined as where '[political] leaders craft political institutions that will sustain the peace and foster democracy in ethnically divided societies after conflicts such as civil wars' (Roeder and Rothchild 2005: 3). These ambitions beg a lot of questions, particularly regarding where power might lie, and the 'rights' of Western states to intervene in societies and states thousands of miles away (Chabal and Daloz 1999, 2006: vii).

Fundamentally, and this is a major theme throughout this book, peacebuilding raises questions about the balance between local, national, regional and international actors. In many circumstances, national, regional and international actors may have power and capacity, but they may lack legitimacy among local populations who have to endure the privations of conflict, de-development and underdevelopment. Obviously, there are immense practical issues involved here, but crucial to the issue is how we see the world. Do we accept terms like 'peacebuilding', 'stability' and 'international community' uncritically, or do we take them apart and see what they are made of? This book hopes to contribute to the process of taking apart ideas and practices. Patrick Chabal's challenge still stands. He said that Western development economists and political commentators have always seen 'Africa [as the] product of its own imagination rather than that of a serious interest in what actually happens on the continent' (Chabal in Werbner and Ranger, 1996: 36) The challenge is therefore how Western-dominated IGOs and states (our 'hardware') can change their 'software' to accommodate the different realities of developing countries with vastly different cultures and problems than our own.

Structure of the book

Chapter 1 advances our understanding of conflict and development by examining the links between poverty, profit and violent conflict. It outlines the main theories that explain the escalation and maintenance of conflict and discusses the political economy of violence whereby conflict becomes a profitable activity and therefore – in some cases – sustainable. The chapter also examines the connection between the presence and exploitation of natural resources and conflict. Chapter 2 provides an overview of the institutional architecture that shapes contemporary peace and development: states, the market, international organizations and international financial institutions. Contemporary conflict and development take place in a context of complex multilateralism and hyper-globalization in which multiple actors can be involved in the same conflict and/or development process, often using proxies and sometimes acting in contradictory ways. The chapter seeks to make sense of the institutional architecture of conflict and development using the debates on 'liberal peace', and it considers how changing trends in international relations might alter patterns of engagement.

In contrast to Chapter 2 and its focus on top-down institutions, Chapter 3 deals with people as citizens, the displaced, victims, antagonists, consumers of public goods, men or women, unified in civil society or fragmented into particularistic groups. The elixir for those who wish to promote conflict, peacebuilding or development is popular participation as a means of legitimizing their enterprise. This chapter will explore the impact of conflict and development on people, and the strategies that ethnic entrepreneurs, political leaders, aid agencies and others use to mobilize groups and legitimize their activities. It examines how the social dynamic of aid and development can interact with the political.

Chapters 4, 5 and 6 are essentially intended to show how policymakers and academic students of conflict alike have tried to conceptualize the problems that arise after conflicts have arisen and, in some cases, 'ended'. We have summed it up as 'transitions' from conflict and war to a kind of 'peace' (Organization for Economic Cooperation and Development [OECD] 2007: Foreword). Chapter 4 looks at thinking on conflict resolution (though we prefer the term 'transformation') and the techniques and methods that have emerged since the end of the Cold War to try to damp down conflicts and deal with their psychological, social and other 'deficits', as Miall et al. (1999) and Ramsbotham et al. (2005) have put it. This will encompass an examination of different approaches to ending conflict, including newer approaches such as truth and reconciliation commissions. Chapter 5 looks at the evolving notion of 'reconstruction' after wars, and in particular takes issue with the belief that we can have a 'one-size-fits-all' approach to such efforts. It will in particular flesh out what we called earlier the 'increasingly technocratic [nature of] development and peace-support interventions'. The alphabet soup of reconstruction now includes acronyms like DDR (disarmament, demobilization and reintegration) and SSR (security sector reform) that have to be understood by any neophyte (or advanced) student of conflict and development. We will attempt to sieve the soup in such a way as to make clear both the underlying rationale for such terms and practices but also their implementation. In this chapter, we will also look in more detail at some of the human and structural problems involved in 'reconstructing' a society. What are the problems to do with the health of the population that need to be considered after a war, for example? Finally, Chapter 6 comprises an overview of the difficulties of using 'aid' as a panacea during, but mainly after, war. The focus here will be on those delivering aid, and the reaction of those to whom it is delivered, which is not always positive. We consider that this dilemma is one of the keys to understanding what is both wrong, and right, with current development policies in the developing world. Outlining new donors and patterns of engagement, this chapter will explore issues surrounding the expansion in delivery of aid, its utility, and the criticism surrounding classical and contemporary expressions of 'humanitarianism'. Along the way, text boxes are used to illustrate points. Chapters will end with summaries, discussion questions and suggested further reading. Since many of the organizations and issues mentioned in the chapters have content-rich websites, e-resources are annotated at the end of each chapter.

Summary

- Theories of conflict and development have largely evolved in isolation from one another.
- More recently, there has been a greater cross-fertilization between ideas and policy approaches to conflict and underdevelopment.
- Particular worldviews have dominated thinking about conflict and development. At the moment, neo-liberal economic ideas influence development thinking and, in the aftermath of 9/11, security and order are playing a prominent role in peace promotion.

- More complex understandings of conflict and development have taken root in the policy and scholarly worlds.
- There is a growing realization that there are no 'quick fixes'.

Discussion questions

- How could 'success' in development be defined?
- How can we make sure that development and peace-support interventions do not become new forms of imperialism?
- Consider if there are realistic alternatives to neo-liberal development models.

Further reading

Excellent surveys of the evolution of development theory and practice can be found in C. Leys (1996) *The Rise and Fall of Development Theory*, Oxford: James Currey; and U. Kothari (ed.) (2005) *A Radical History of Development Studies: Individuals, institutions and ideologies*, London: Zed. A neo-liberal assault on Keynesian or state-led development strategies can be found in J. Dorn, S. Hanke and A. Walters (eds) (1998) *The Revolution in Development Economics*, Washington, DC: Cato Institute. O. Ramsbotham, T. Woodhouse and H. Miall (2011) *Contemporary Conflict Resolution: The prevention, management and transformation of deadly conflicts*, 3rd edition, Cambridge: Polity, provide an excellent survey of literature and approaches to conflict. Major edited volumes with contributions from leading authors on peacebuilding and humanitarianism are R. Mac Ginty (ed.) (2013) *Routledge Handbook of Peacebuilding*, London: Routledge; and R. Mac Ginty and J. Peterson (eds) (2015) *Routledge Companion to Humanitarian Action*, London: Routledge. Global initiatives on climate change and responses the humanitarian consequences can be found at IFRC climate centre https://www.climatecentre.org/. See also climate charter https://www.climate-charter.org/.

Useful websites

A goldmine of material on conflict and conflict transformation can be found at Beyond Intractability (www.beyondintractability.org/), Open Democracy (https://www.opendemocracy.net/), International Alert (http://www.international-alert.org/) and International Crisis Group (http://www.crisisgroup.org/). The Humanitarian Practice Network provides online reports on development and humanitarian issues: www.odihpn.org/. See also the Global Humanitarian Assistance website, particularly its annual reports: https://devinit.org/resources/global-humanitarian-assistance-report-2023/.

UN documents on the Sustainable Development Goals can be found here: https://sustainabledevelopment.un.org/topics/sustainabledevelopmentgoals. The *Guardian* newspaper has an excellent resource of articles and data on development: http://www.theguardian.com/global-development.

Uppsala Conflict Data Program https://ucdp.uu.se/

Bibliography

Addison, T. (2005) Conflict and peace building: Interactions between politics and economics. In Addison, T. (ed.) Special Issue of *Round Table: The Commonwealth Journal of International Affairs* 94(381): 405–11.

Amnesty International (2013) China, Annual report 2013: The state of the world's human rights. Available at: http://www.amnesty.org/en/region/china/report-2013 (accessed 1 February 2015).

Axelrod, R. (1990) *The Evolution of Cooperation*. London: Penguin.

Azar, E. (1990) *The Management of Protracted Social Conflict: Theory and Cases*. Aldershot: Dartmouth.

Baker, S. (2006) *Sustainable Development*. London: Routledge.

Bauer, P. (1998) The Disregard of Reality. In Dorn, J., Hanke, S., and Walters, A. (eds) *The Revolution in Development Economics*. Washington, DC: Cato Institute, pp. 25–39.

Benn, T. (2001) Hansard: Record of parliamentary debates, column 510, 22 March. Available at: http://www.publications.parliament.uk/pa/cm200001/cmhansrd/vo010322/debtext/10322-13.htm (accessed 1 February 2015).

Binder, A., Meier, C., and Steets, S. (2010) 'Humanitarian Assistance: Truly Universal? A mapping study of non-Western donors.' *GPPi Research Paper Series No. 12*. Berlin: Global Public Policy Institute.

Blinder, A.S. (2013) *After the Music Stopped: The Financial Crisis, the Response and the Work Ahead*. London: Penguin.

Bøås, M. (2014) *The Politics of Conflict Economies: Miners, Merchants and Warriors in the African Borderland*. London: Routledge.

Boulding, K. (1990) Future Directions in Conflict and Peace Research. In Burton, J. and Dukes, F. (eds) *Conflict: Readings in Management and Resolution*. London: Macmillan, pp. 35–47.

Boutros-Ghali, B. (1992) *An Agenda for Peace Preventive diplomacy, peace-making and peace-keeping*. Report of the Secretary-General pursuant to the statement adopted by the Summit Meeting of the Security Council on 31 January 1992.

Campbell, S. and Hofmann, S. (2015) Regional Humanitarian Organisations. In Mac Ginty, R. and Peterson, J. (eds) *Routledge Companion to Humanitarian Action*. London: Routledge, pp. 191–203.

Carnegie Commission on Preventing Deadly Conflict (1997) *Preventing Deadly Conflict*. New York: Carnegie Corporation of New York.

Chabal, P. and Daloz, J.-P. (1999) *Africa Works: Disorder as a Political Instrument*. Oxford: James Currey.

Chabal, P. and Daloz, J.-P. (2006) *Culture Troubles: Politics and the Interpretation of Meaning*. London: Hurst.

Chua, A. (2004) *World on Fire: How Exporting Free Market Democracy Breeds Ethnic Hatred and Global Instability*. New York: Anchor.

Clapham, C. (1987) Revolutionary socialist development in Ethiopia. *African Affairs* 86(343): 151–65.

Cohen, S.B. (2014) The challenging dynamics of global north–south peacebuilding partnerships: Practical stories from the field. *Journal of Peacebuilding and Development* 9(3): 65–81.
Collier, P. (2000) *Economic Causes of Civil Conflict and their Implications for Policy*. Washington, DC: World Bank.
Collier, P. and Hoeffler, A. (2004). Greed and grievance in civil war, *Oxford Economic Papers* 56(4): 563–95.
Cox, R.W. (1981) Social forces, states and world orders: Beyond international relations theory. *Millennium 10*: 126–56.
Creasy, E. (1876) *The Fifteen Decisive Battles of the World: From Marathon to Waterloo*. London: Bentley.
Darby, J. and Mac Ginty R. (eds) (2000) *The Management of Peace Processes*. Basingstoke: Macmillan.
Davenport-Hines, R. (2015) *Universal Man: The Seven Lives of John Maynard Keynes*. London: Collins.
Donini, A. (2010) The far side: The meta functions of humanitarianism in a globalised world. *Disasters*, *34*(S2): S220–S237.
Dorn, J. (1998) Competing Visions of Development Policy. In Dorn, J., Hanke, S., and Walters, A. (eds) *The Revolution in Development Economics*. Washington, DC: Cato Institute, pp. 1–21.
DuBois, M. (2018) 'The new humanitarian basics' *Humanitarian Policy Working Group, ODI*.
Endres, A. (2012) *Architects of the International Financial System*. London: Routledge.
Fisher, R. and Ury, W. (1991) *Getting to Yes: Negotiating Agreement Without Giving In*. London: Penguin.
Frank, A.G. (1967) *Capitalism and Underdevelopment in Latin America*. New York: Monthly Review Press.
Galbraith, J.K. (1964) *Economic Development*. Cambridge, MA: Harvard University Press.
Galtung, J. (1996) *Peace by Peaceful Means: Peace and Conflict, Development and Civilization*. London: Sage.
Gawthorpe, A. (2018) *To Build as Well as Destroy: American Nation Building in South Vietnam*. New York: Cornell University Press.
Gordon-Gibson, A. (2021) *Humanitarians on the Frontier: Access and Identity along the Borders of Power*. Lanham, MD: Rowman and Littlefield.
Gurr, T. (1970) *Why Men Rebel*. Princeton, NJ: Princeton University Press.
Harb, C. and Saab, R. (2014) *Social Cohesion and Intergroup Relations: Syrian refugees and Lebanese nationals in the Bekka and Akkar*. London: Save the Children.
Harriss, J. (2005a) Great Promise, Hubris and Recovery: A Participant's History of Development Studies. In Kothari, U. (ed.) *A Radical History of Development Studies: Individuals, Institutions and Ideologies*. London: Zed, pp. 17–46.
Harriss, J. (2005b) Do Political Regimes Matter? Poverty Reduction and Regime Differences across India. In Houtzager, P. and Moore, M. (eds) *Changing Paths: International Development and the New Politics of Inclusion*. Ann Arbor, MI: University of Michigan Press, pp. 204–32.
Harvey, D. (2005) *A Brief History of Neoliberalism*. Oxford: Oxford University Press.
Helleiner, E. (2014) *The Status Quo Crisis: Global Financial Governance after the 2008 Meltdown*. Oxford: Oxford University Press.
Hilhorst, D. (2018) Classical humanitarianism and resilience humanitarianism: Making sense of two brands of humanitarian action. *Journal of International Humanitarian Action* 3: 15.

Hoffman, B. (2007) *The Peace Guerilla Handbook*. York: Newmathforhumaity.
Holt, S. (2011) *Aid, Peacebuilding and the Resurgence of War: Buying Time in Sri Lanka*. London: Palgrave Macmillan.
House of Lords (2014) *Youth Unemployment in the EU: A Scarred Generation?* London: The Stationery Office.
IASC (2022) About the Grand Bargain. Available at: https://interagencystandingcommittee.org/about-the-grand-bargain#:~:text=The%20Grand%20Bargain%2C%20launched%20during,efficiency%20of%20the%20humanitarian%20action (accessed 31 August 2023).
International Development Committee (2006) *Conflict and Development: Peacebuilding and post-conflict reconstruction. Sixth Report of the Session 2005–06*, Vol. 1. London: The Stationery Office.
Jacoby, T. (2008) *Understanding Conflict and Violence: Theoretical and Interdisciplinary Approaches*. London: Routledge.
Jeong, H.-W. (2005) *Peacebuilding in Post-Conflict Societies: Strategy and Process*. Boulder, CO and London: Lynne Rienner.
Kacowicz, A.M., Bar-Siman-Tov, Y., Elgström, O., and Jerneck, M. (eds) (2000) *Stable Peace Among Nations*. Lanham, MD: Rowman and Littlefield.
Kaldor, M. (2007) *New and Old Wars: Organized Violence in a Globalized World*, 2nd edn. London: Polity.
Kapila, M. (2022) Why is modern humanitarianism failing, even as we give more and more aid? *The National*. https://www.thenationalnews.com/opinion/comment/2022/12/28/why-is-modern-humanitarianism-failing-even-as-we-give-more-and-more-aid/
Keynes, J.M. (1920) *The Economic Consequences of Peace*. London: Macmillan.
Klein, N. (2007) *The Shock Doctrine: The Rise of Disaster Capitalism*. London: Penguin.
Kluczewska, K. (2019) How to translate 'Good Governance' into Tajik? An American Good Governance Fund and norm localisation in Tajikistan, *Journal of Intervention and Statebuilding*, *13*(3): 357–376.
Kothari, U. (2005) A Radical History of Development Studies: Individuals, Institutions and Ideologies. In Kothari, U. (ed.) *A Radical History of Development Studies: Individuals, Institutions and Ideologies*. London: Zed, pp. 1–13.
Lal, D. (1998) The Transformation of Developing Economies: From Plan to Market. In Dorn, J., Hanke, S., and Walters, A. (eds) *The Revolution in Development Economics*. Washington, DC: Cato Institute, pp. 55–74.
Lederach, J.P. (1995) *Preparing for Peace: Conflict Transformation across Cultures*. Syracuse, NY: Syracuse University Press.
Lederach, J.P. (1997) *Building Peace: Sustainable Reconciliation in Divided Societies*. Washington, DC: United States Institute of Peace Press.
Leys, C. (1996) *The Rise and Fall of Development Theory*. London: James Currey.
Mac Ginty, R. (2006) *No War, No Peace: The Rejuvenation of Stalled Peace Processes and Peace Accords*. London: Palgrave.
Mawdsley, E. (2015) 'Non-DAC' Humanitarian Actors. In Mac Ginty, R. and Peterson, J. (eds) *Routledge Companion to Humanitarian Action*. London: Routledge, pp. 204–14.
Meier G.M. and Rauch, J.E. (2000) *Leading Issues in Economic Development*. New York: Oxford University Press.
Miall, H. (2007) *Emergent Conflict and Peaceful Change*. Basingstoke: Palgrave Macmillan.
Miall, H., Ramsbotham, O. and Woodhouse, T. (1999) *Contemporary Conflict Resolution*, 1st edn. London: Polity.

Mitchell, C. (1994) Conflict Management. In Groom, A.J.R. and Light, M. (eds) *Contemporary International Relations: A Guide to Theory*. London: Pinter, pp. 128–41.
O'Toole, F. (2010) *Ship of Fools: How Stupidity and Corruption Sank the Celtic Tiger*. New York: PublicAffairs.
OECD (2007) *Handbook on 'Security Sector Reform'*. Paris: OECD.
Özerdem, A. (2010) Insurgency, militias and DDR as part of security sector reconstruction in Iraq: How not to do it, *Disasters 34*(1): S40–S59.
Paris, R. (2010) Saving liberal peacebuilding. *Review of International Studies 36*(2): 337–65.
Piketty, T. (2014) *Capital in the Twenty-First Century*. Harvard, MA: Harvard University Press.
POMED (2020) Fact sheet – U.S. Military assistance to Egypt: Separating fact from fiction. https://pomed.org/publication/fact-sheet-u-s-military-assistance-to-egypt-separating-fact-from-fiction/ (accessed 12 November 2023).
Pugh, M. (2013) The Problem-solving and Critical Paradigms. In Mac Ginty, R. (ed.) *Routledge Handbook of Peacebuilding*. London: Routledge, pp. 11–24.
Ramsbotham, O., Woodhouse, T. and Miall, H. (2005) *Contemporary Conflict Resolution*, 2nd edn. London: Polity.
Regan, P.M. (2013) Quantitative Approaches. In Mac Ginty, R. (ed.) *Routledge Handbook of Peacebuilding*. London: Routledge, pp. 183–93.
Richardson, L.F. (1950) *The Statistics of Deadly Quarrels*. Pittsburgh, PA: Boxwood.
Richmond, O. (2006) The problem of peace: Understanding the 'liberal peace'. *Conflict, Security and Development 6*(3): 291–314.
Richmond, O. (2012) Beyond local ownership in the architecture of international peacebuilding. *Ethnopolitics 11*(4): 354–75.
Roeder, P.G. and Rothchild, D. (eds) (2005) *Sustainable Peace: Power and Democracy after Civil War*. Ithaca, NY and London: Cornell University Press.
Rooney, A. (2000) *My War*. New York: PublicAffairs.
Rostow, W.W. (1960) *The Stages of Economic Growth: A Non-communist Manifesto*. Cambridge: Cambridge University Press.
Sachs, J. (2008) Promises, promises. *Developments 40*: 6.
Schmid, H. (1968) Peace research and politics. *Journal of Peace Research 5*: 217–32.
Scholey, P. (2006) Peacebuilding Research and North–South Research Relationships: Perspectives, Opportunities and Challenges. In MacLean, S., Black, D., and Shaw, T. (eds) *A Decade of Human Security: Global Governance and New Multilateralisms*. Aldershot: Ashgate, pp. 179–92.
Seitz, T. (2018) *The Evolving Role of Nation-Building in US Foreign Policy: Lessons Learned, Lessons Lost*. Manchester: Manchester University Press.
Shaw, T., MacLean, S., and Black, D. (2006) Introduction: A Decade of Human Security: What Prospects for Global Governance and New Multilateralisms? In MacLean, S., Black, D. and Shaw, T. (eds) *A Decade of Human Security: Global Governance and New Multilateralisms*. Aldershot: Ashgate, pp. 3–18.
Simmel, G. (1955) *Conflict: The Web of Group Affiliations*. New York: Free Press.
Spinney, L. (2017) *Pale Rider: The Spanish Flu of 1919 and How It Changed the World*. London: Vintage.
Stiglitz, J. (2015) *The Great Divide*. New York: Allen Lane.
Thomas, A. (2006) Reflections on Development in a Context of War. In Yanacopulos, H. and Hanlon, J. (eds) *Civil Peace, Civil War*. Milton Keynes: Open University Press, pp. 185–205.

UN DESA (2023) World population projected to reach 9.8 billion in 2050, and 11.2billionin2100.Availableat:https://www.un.org/en/desa/world-population-projected-reach-98-billion-2050-and-112-billion-2100 (accessed 31 August 2023).
UNDP (1994) *Human Development Report.* Oxford: Oxford University Press.
United Nations (2000) *Report of the Panel on United Nations Peace Operations.* New York: United Nations, A/55/305.
UNDP (2006) *Human Development Report 2006: Beyond Scarcity – Power, Poverty and the Global Water Crisis.* New York: UNDP.
UNDP (2022) Human Development Report 2021/2022. *Uncertain times, unsettled lives. Shaping our future in a transforming world.* New York: United Nations Development Programme.
Uppsala Conflict Data Program (2021) https://ucdp.uu.se/year/2021
US Embassy, Cairo (2021) The United States provides $125 million in economic assistance to Egypt. https://eg.usembassy.gov/the-united-states-provides-125-million-in-economic-assistance-to-egypt/ (accessed 12 November 2023).
Werbner, R. and Ranger, T. (eds) (1996) *Postcolonial Identities in Africa.* London: Zed.
Win, E. (2007) Not Very Poor, Powerless or Pregnant: The African Woman Forgotten by Development. In Cornwall, A., Harrison, E. and Whitehead, A. (eds) *Feminisms in Development: Contradictions, Contestations and Challenges.* London: Zed, pp. 79–85.
Wolf, M. (2014) *The Shift and the Shocks: What We've Learned and What We Still Have to Learn from the Financial Crisis.* London: Penguin.
Wright, Q. (1942) *A Study of War.* Chicago, IL: Chicago University Press.
DuBois, M. (2020) 'The Triple Nexus – Threat or Opportunity for the Humanitarian Principles?' Discussion Paper, Centre for Humanitarian Action (2020).
Guardian (2007) Iran's secret plan to force the US out. *Guardian*, 22 May.
Harding, T. (2008) A year in Helmand: 4m bullets fired by British. *Daily Telegraph*, 13 January.
ICRC (2016) Humanitarian summit, given by Hugo Slim, Head of Policy at the International Committee of the Red Cross. https://www.icrc.org/en/document/icrcs-statement-world-humanitarian-summit (accessed 12 November 2023).
Pettersson, T. and Wallensteen, P. (2015) Armed conflicts, 1946–2014. *Journal of Peace Research* 52(4): 536–50.
Troy, G. (2007) *Morning in America: How Ronald Reagan Reinvented the 1980s.* Princeton, NJ: Princeton University Press.
World Bank (2013) GDP growth (annual %), *World Bank Indicators.* Available at: http://data.worldbank.org/indicator/NY.GDP.MKTP.KD.ZG (accessed 1 February 2015).

1 Poverty, profit and the political economy of violent conflict

Introduction

There are two basic views on the economics of violent conflict. In the first view, economic development offers a ladder out of conflict. This ladder operates at the international and individual levels. States that trade together have an interest in stability and continued prosperity and so have no interest in the disruptions that come with war. Similarly, this thinking goes, individuals have an interest in peace. Individuals are likely to behave rationally and so would personally avoid conflict and encourage their political leaders by voting for moderate parties to avoid conflict. Rational individuals would see economic development as a route out of conflict. In short, 'free markets made free men' and free men would not be foolish enough to become involved in war (Mandelbaum 2002).

The second view on the economics of conflict points towards the predatory nature of capitalism and how the deep inequalities it causes can fan the flames of conflict. Marxist economists in particular blame capitalism for the causes of both poverty and war. The classic statement of this was by V. I. Lenin: 'All wars are inseparable from the political systems that engender them', and the dominant system of the developed world was capitalism. Capitalism was, he further wrote, 'the highest form of imperialism', thus creating in his mind the link between war and imperialism (Lenin 1917, 1964). The West, in this reading, did not 'develop' the South but rather 'underdeveloped' it, breaking up existing structures of society and economic exchange and, in their place installing predatory economic and militarized colonies that exploited and depleted the places they colonized (Rodney, 2018, 2022). Such thinking has been at the origin of many 'revolutionary wars', which Lenin encouraged.

This view of economics as a cause of violent conflict also points towards the profiteering of warriors. Whether arms manufacturers, roadside bandits levying a war 'tax' or those running blood diamond or oil-smuggling operations in war zones, there are hefty profits to be made from war. Moreover, the 'normal' trade routes are used to transmit war, as they are supposed to enhance peace in classical liberal

DOI: 10.4324/9781003353485-2

theory. This posits that the more economic interactions there are between states, the less they are likely to engage each other in war (Angell, 1910; Barbieri, 2002). Globalization has helped this process considerably in the last 30 years. In such cases, peace is a threat to the livelihoods of combatants and traffickers on an often regional basis (Nordstrom, 2004, 2007; Pugh et al., 2004, 2008).

So, opinion is polarized on the role of economics (saviour or villain?) in the outbreak of civil and (increasingly) interstate war. What can be agreed upon is the symbiotic negative effect of war on poverty levels. As Hugo Slim put it, '[M]ost people experience war as poverty rather than as battle.... It is the civilian, not the wounded soldier, who stands at the centre of the moral frame we put around war' (Slim 2022: xxxiii, quoted in Plaut and Vaughan 2023: 14).

This chapter will examine the often-contradictory literature on conflict and development, using examples to illustrate the linkages between conflict, poverty and profit. The chapter is divided into five sections. The first section examines the various economic and development-related arguments on the causation and escalation of violent conflict. The second section delves more deeply into the complex relationship between economics and civil war. Section three examines the factors behind conflict maintenance and the peculiar economic dynamics that sustain war. An anthropological lens is particularly useful in illustrating the ways in which individuals, communities, armed groups, businesses and states variously prosper, starve or 'get by' during war. The fourth section focuses on corruption and considers how the term is often used selectively and is unhelpful as an analytical category. The fifth section concentrates on the resource environments often found in the sites of violent conflict and considers the extent to which the presence, absence and distribution of resources such as diamonds or water can fuel or calm violent conflict.

Conflict causation and escalation

There is no universal theory of conflict causation. The sheer variety of conflict actors, environments and dynamics makes impossible a general theory of conflict or, indeed, of conflict management. Each conflict has a peculiar 'conflict DNA'. This may be a partial match with other conflicts but will contain factors specific only to that conflict. Conflicts have multiple causes that interact in highly specific ways according to the context. They can have primary causes that take precedence over secondary causes, but the variegated nature of human polities, economies and societies means that a single factor cannot spark a conflict. Different factors will have different weights at different stages of a conflict trajectory. For example, a single atrocity or grievance may prove inflammatory at the outbreak of a civil war, but it becomes overtaken by other factors that sustain the conflict in the longer term. The task for the conflict analyst is first to identify conflict causes (plural) and then establish the connections between them. This is rarely an easy task since the public rhetoric used by protagonists may mask truer motives or because protagonists (such as the Lord's Resistance Army in northern Uganda or state-sponsored militias in Darfur) may be secretive and offer few public clues as to their motivations.

Complicating matters even further is that antagonists operating in the same conflict may have very different motivations and have assumed particular acuity in the light of the war in Eastern Europe between Russia and Ukraine. This applies within and between groups. Consider the recent wars in the former Soviet Union. Many commentators have said, including the ones writing this book, that geopolitics, the cause of many local and even global conflicts concerning entire nation states, was a thing of the past. After all, since the end of the Cold War in 1990, there had been few interstate conflicts. The most cited were those between Eritrea and Ethiopia in 1998–2000 and what was seen at the time as border skirmishes between Russia and Georgia (2008) and Ukraine (2014). But in the recent past, it has become clear to many commentators that Russia had embarked on a much more significant war against Ukraine in 2014, which may be described as 'imperial' and which certainly includes an attempt at the annexation of large areas of Ukraine (2014–ongoing). The motivations for this are also multiple, ranging from unwise Western incursions into what Russia sees as its sphere of influence to the personal ambitions of Russian president Vladimir Putin. Many 'frozen' conflicts, as they were called for decades, now seem to be 'hot' again. In the neighbourhood of the former Soviet Union (FSU), these include the aforementioned Georgia and Ukraine but also Moldova's 'internal/external' conflict with Transdniestria and several others in the Caucasus region (Galeotti, 2022).

Similarly, the Israeli–Palestinian conflict, for many years seen as an essentially internal conflict between two groups who live in close proximity with each other, can be seen as part of a much wider geopolitical struggle for the religious and secular heart and soul of the entire Middle East. Is it not possible to see the 'local' war between Israel and Palestine as just an important part of a global war? The attacks on New York, Nairobi and elsewhere in the early 2000s involved a coalition of Western states, as well as active interventions by Iran in Syria and Pakistan in Afghanistan. We can see the ultimate expression of this in the emergence of transnational Islamic movements like *Al Qaeda*, responsible for the 2001 attacks in New York and elsewhere, and ultimately the 'Islamic Caliphate', also known as 'ISIS', both of which are still a real threat to the region and beyond (Gattas 2020). Some conflict theorists have coined the expression 'intractable conflicts' to indicate both the difficulties in solving them but also the complexity of their make-up (Mitchell 2014).

The conflict which has (re-)erupted between Israel and Hamas (which has governed Gaza since 2006) with the latter's surprise invasion of Israel on 7 October 2023, has more than demonstrated that local and non-local issues are combining to make the situation increasingly poisonous for all concerned. The conflict has causes that, depending on the commentator, date back to the creation of the state of Israel in 1947 or the establishment of the British League of Nations Mandate in the 1920s. The actors are both local and international, the causes deeply embedded in different views of history and economics as well as ideological positions in the region and beyond it (Barr 2011; Fawcett 2023). For one (reasonably) neutral view on the present, October 2023, situation, see CFR (2023).

We might also cite the amalgam of regional alliances and potential causes of geopolitical war in the Horn of Africa, ones that really date back to the 1970s. The independence of Eritrea from Ethiopia (often described as a form of 'secession') in 1991 was after a war that started in the 1970s and killed many hundreds of thousands on both sides. This was followed by arguments over territorial boundaries, access to the ports of the Red Sea coast and a myriad of mutual insults, culminating in a bloody war between Ethiopia and Eritrea in 1998–2000.

Whatever trigger we might find for these various wars, they certainly help to maintain entire populations in poverty due to the inevitable effect they have on family life and health, among other indicators. One of the most distressing aspects has been the proliferation of some of these conflicts of gender-based violence. The Tigray war, which as of 2023 can be said to have been concluded as an armistice rather than a peace treaty, has seen appalling accounts of rape and the indiscriminate murder of civilians. Much of this has been kept from the world's press. Western governments long wished to see Ethiopia as a beacon of press freedom and human rights (Eritrea Focus/Oslo Analytica 2021). Its president was a recipient of the Nobel Peace Prize in 2019. Meanwhile, Eritrea has one of the worst human rights records on the planet (US State Department 2021). The use of rape as a weapon of war by Russian troops in Bucha, Ukraine, has also been corroborated by outside sources, including captured Russian soldiers (Coynash 2022; UN 2023).

Food as a weapon of war

Even food is a potential weapon of war. The international, even intercontinental, economic implications of the wars currently underway in Eastern Europe, Africa and elsewhere became very apparent with the blockade of the Black Sea by the Russian navy, which prevented Ukraine from exporting its grain, causing the risk of widespread famine across swathes of Africa, which relies to a great extent on Russian and Ukrainian wheat. A deal brokered by Türkiye in 2022 allowed grain to be exported, though it was a fragile truce and was suspended. Ukraine itself evoked the danger of a new *Holodomor*, a reference to the Soviet era famine of 1932–33, where several million died of hunger, usually seen as having been caused by Stalin to control what he saw then (and Putin sees now) as Ukrainian nationalism (Snyder 2011; Applebaum 2017). In Israel, the Israeli authorities often use their power to blockade Palestinian areas that are seen as the sources of violence on Israeli settlers and shows the use of economic pressure to achieve security gains. UN agencies are often obliged to provide food aid to areas thus affected.

Until the events of October 2023 in Gaza, around 500 lorries of food and humanitarian assistance a *day* were provided for Gaza by the UN and ICRC; after the conflict started, Israel largely closed the border to people and goods. Poverty thus can and is inflicted when Palestinians who cannot cross the border see their income severely curtailed. This form of 'collective punishment', such as the periodic closures of the border between Gaza and Israel, has been condemned by UN Human Rights Rapporteur Michael Lynk and the Israeli commentator Alex Lederman. Lynk attacks what he sees as 'the destruction of Gaza's economy, causing immeasurable suffering to its two million inhabitants' (OHCHR, 2020; Laderman, 2023; for 'Food as a Weapon of war, see Additional Readings). What is euphemistically called 'economic statecraft', the use of economic instruments of foreign policy, such as sanctions against states or populations deemed to be conflictual, has a real impact on livelihoods beyond those engaging in violence but caught up in it often by no fault of their own (Baldwin, 2020).

The net results for the civilian populations of all these areas are the same: whoever carries out the violence for whatever the cause. There are direct civilian casualties, refugee flows (for example, over five million in the case of Ukraine as of the end of 2022), as well as the previously discussed problems of hunger and rape. We might ask what leads individual human beings to do such things to each other – a 'love of cruelty' maybe? A prominent psychiatrist suggests that 'deep in human psychology, there are urges to humiliate, torment, wound and kill people' (Glover, 2001). States cite their national interest, geopolitics, 'justice' and the like. It is the civilian populations that do most of the suffering.

So the links between the 'local' causes of war and the 'international' ones are complex. At the group level, recruits may be motivated to join Palestinian militant organizations for any number of reasons: spiritual fulfilment, to avenge a grievance, peer pressure, family tradition, belief in their political and strategic aims, to boost personal esteem or to earn an income. All these motivations seem to arise in any conflict situation. Israelis may join the Israeli Defence Force because of a desire to defend their state and community, to gain new skills and further their career, to continue a family tradition and fulfil conscription obligations or to avoid being branded deviant or cowardly by not joining up. Russian Jews have long fled the repressive policies of Russia to find safety in Israel, and Americans have come to Israel to support their ideals of the Jewish Homeland. They bring with them presuppositions of many kinds that sometimes clash with the beliefs of both many Palestinians and many Israelis, causing additional sources of conflict.

There is a cacophony of motivations swirling around the conflict area. Some of these may be economic, but many are not. Trying to construct an intelligible narrative thread out of all of these motivations is a difficult task but one that is necessary for the conflict analyst. Although Palestinian militants and the Israeli Defence Force share the same 'battlefield' and pledge themselves to the destruction of the other, their conflicts have very different aims. For Israel, the conflict is largely justified in terms of defence and security (Box 1.1).

Box 1.1 Edward Said's explanations of the Israel-Palestine conflict

Edward Said (1935–2003), the former New York–based academic who was born in Palestine, struggled with his internal misgivings about Israeli, Palestinian and US actions in his place of birth. He always rejected the idea that the conflict was in effect a 'mirror' as he pointed to the gross power imbalance between the two main groupings. But he also blamed both the United States and the surrounding Arab regimes for not doing more to help resolve the conflict while being a major inspiration behind a change in literary studies (now usually called 'postcolonial studies'), especially with his book *Orientalism* (Said 1978) which has also continued to cause a furore in the West. He can thus be seen as both an analyst of the conflict and, to some extent, a participant. His ideas have also had a major impact not only on the Palestinian perspective on the conflict but also on cultural studies much more widely and on the question of global justice.

Sources: Said (1978); Morefield (2022)

But Said is only an extreme version of the danger (if danger it be) of a conflict analyst getting too close to his subject matter and taking sides. The conflict analyst is faced with a bewildering array of 'evidence' and, ultimately, must make a judgement call on which factors they believe to be most significant in the escalation and maintenance of a conflict.

Given that the focus of this book is on conflict and development, the bulk of this section has concentrated so far on development and economic-related explanations for conflict causation and maintenance. Yet there are entire subfields of literature on conflict causation that make little reference to development, economics, profit or poverty. Non-economic or non-development-related explanations for the outbreak of violent conflict include the often overlapping categories of biological disposition (Simmel 1955), psychology (Tajfel 1978), religion (Appleby 2000), identity (Sen 2006), ethnicity (Connor 1994; Young 2003), nationalism, ideology, history and ancient hatreds, bad neighbours, manipulative leaders (Brown 1997), the security dilemma (Posen 1993), cultural dysfunction (Kaplan 1994), the nature of the state (Tilly 1985) and incompatible worldviews (Huntington 1998). Many of these explanations regard economic factors as contingent or providing a context in which the primary factor operates. Thus, for example, an ethnic group may develop an elaborate self-narrative of grievance, how its rights are denied and how it is distinct from other identity groups in the context of declining economic conditions. These conditions may provide the backdrop or even a tipping point for a slide into violent conflict. A group may become convinced of its own 'relative deprivation', especially if inequality is visible along religious, ethnic or racial lines (Jacoby 2008: 103–13). But, in such explanations, economic and development-related factors are secondary and only come into play when stimulated by other factors or if a prior existing condition (entrenched ethnic or religious difference) is in place.

The important point to bear in mind is that conflicts are caused by a combination of factors. Those who promote economic or development-related explanations for the outbreak of violent conflicts need to take account of non development related explanations and how these interact with development-related factors. The philosopher and economist Amartya Sen makes the point that poverty on its own is not enough to cause conflict. He recalls his own childhood memories from Calcutta during the 1943 Bengal famine and 'the sight of starving people dying in front of sweetshop windows with various layers of luscious food displayed behind glass windows, without a single glass being broken, or law and order being disrupted' (Sen 2006: 143). Other factors, especially 'the illusion of singular identity ... in a world so obviously full of plural affiliations', were required to transform inequality and destitution into violent conflict (Sen 2006: 175). The philosopher Martha Nussbaum has developed Sen's approach to work in the sector of international development and formulated an interpretation of the 'Capability Approach' that prioritizes the entitlement to a dignified life as the goal for human welfare (Nussbaum 2006).

Conflicts do not just happen. Just because a society is ethnically, racially or religiously fissured does not mean that conflict will follow. Seattle (a diverse, multicultural city) is a more common model than Sarajevo (one with a history of ethnonational conflict but also a past of being a diverse, multicultural city when it was in Yugoslavia till 1991). Indeed, given the multiplicity of identity groups that claim to be distinct from others, there is remarkably little violent conflict on the planet, and there is some evidence that it has declined over time (Brubaker and Laitin 1998; Pinker 2012). This suggests two points. The first is that many human societies have developed everyday peace systems that manage or suppress differences, often in non-violent ways (Mac Ginty 2014). The second is that violent conflict requires active instigation agents, particularly if latent tensions are to be escalated into overt violence. These instigation agents may take the form of political or community leaders who purposively inflame and mobilize their supporters or circumstances – such as an assassination or shock election result – that agitate already tense group sensibilities (Zartman 2005: 268–73). Ukraine, Syria, Yemen, Nigeria, Lebanon and many other societies have experienced extended periods of calm (though not peace in a holistic sense) only for civil war to develop. This required active steps and reactions by political, military and community leaders.

As was alluded to in our 'Introduction', the roots of such problems are generally explainable by a reference to the notion of 'power'. Power has, of course, been a reality since the beginning of human relations. Hence it is also an idea that we can use in the context of this book. It was re-conceptualized by the American sociologist C. Wright Mills at the beginning of what we can now see as the 'development business' in the 1950s: 'The powers of ordinary men are circumscribed by the everyday worlds in which they live, yet even in these rounds of job, family and neighbourhood they often seem driven by forces they can neither understand nor govern' Mills 1956: 3). This was summed up by Stephen Lukes as 'power is at its most effective when least observable' (Lukes 2005:1). This notion of the link between power and the 'everyday' experience of power is a vital one for the scholar of both conflict and development and one to which we will return. It is as true for the worker in New York as it is for the farmer in Sri Lanka and of relations between the elites and the general population. Understanding the real or perceived abuse of such power is a powerful tool for understanding why conflict erupts and also how that understanding can help us see ways of transforming it into more positive channels. Consequently, it dominates in often subtle ways the discourse of the subject matters dealt with in this book.

Plate 1 Mehreb Bridge Eritrean-Ethiopian border. Prisoner and detainee repatriation. Copyright ICRC. Photographer: Marçal Izard. 29/11/2002.

Economics and civil war

Unsurprisingly, economists have been at the forefront of arguments that conflict causation can be explained by economics. Paul Collier and a number of collaborators have produced a corpus of studies that link the onset of civil war to economic factors (Collier and Hoeffler 2002; Collier et al. 2003). Many of these studies are based on econometric modelling and are attractive because they allow analysts to avoid considering nebulous and difficult-to-define factors, such as identity or historical grievances. Two main arguments have been advanced under what has been termed the 'greed thesis' or economic explanations of violent conflict:

- Economic factors can act as predictors of violent conflict (or help identify civil war-prone societies).
- Combatants are motivated by economic predation.

The first argument identified economic factors – usually at the national level – that make a society prone to civil war. In particular, the level of per capita income, its rate of growth and the structure of the economy (especially its dependence on commodity exports) were identified as the key risk factors for the onset of civil war (on a dataset of 52 civil wars in the 1960–99 period). Collier and colleagues found that a doubling of per capita income halved the risk of civil war and that when commodity exports account for 25 per cent of Gross Domestic Product (GDP), the risk of civil war is 33 per cent, as opposed to an 11 per cent risk of civil war if commodity exports are at 10 per cent (Bannon and Collier 2003: 2–3). Findings such as these have encouraged governments and policymakers to promote poverty reduction and economic diversification programmes – often based on free-market remedies – as part of conflict prevention strategies (Brauer and Dunne 2012).

The second argument identified the profit motive – or 'greed' – as the primary motor behind civil war. Collier noted that 'conflicts are far more likely to be caused by economic opportunities than by grievance', but that rebel organizations will engage in a public discourse of grievance 'since they are unlikely to be so naive so as to admit to greed as a motive' (Collier 2000a: 91–2). Thus, 'civil wars occur where rebel organizations are financially viable', with the ability of antagonists to generate revenue being the principal reason why civil wars break out in some locations and not in others (Collier 2000b: 2). Münkler (2005) reinforces the view that civil war is economic rationalism taken to the extreme: the availability of weapons and untrained young men makes contemporary civil war 'downright cheap' and 'highly lucrative'. 'In the short term the force used in them yields more than it costs and the long-term costs are borne by others' (Münkler 2005: 74, 77). In this view, civil war conforms to a straightforward business model that seeks to maximize resource extraction through banditry, 'taxation' or the trafficking of diamonds, timber or people. It also aims to reduce costs by overlooking social responsibilities to citizens and cutting the costs of running a regular army. By boosting income and cutting costs, profit will be maximized. 'The entrepreneurs of the new wars' often emerged from the criminal underworld and used nationalist or ethnic movements as convenient vehicles from which to pursue their business interests (Münkler 2005: 80). William Reno's analyses of the civil wars in Sierra Leone and Liberia paint a dystopian picture in which political leaders drop all pretence of maintaining a functioning state that offers basic public services and protection to citizens (Reno 1997a, 1997b). Instead, leaders formed alliances with business organizations (often from overseas) to extract mineral resources. Ultimately, the civil wars in both territories resembled privatized conflict, with control of mineral resources the key aim (Keen 1998).

The claim by Collier and others that economic motivations must be given precedence over non-economic issues (such as identity) amounted to an intellectual drone strike aimed at scholars who held that ethnicity or religion held the key to the onset of civil war. They responded in kind and were particularly annoyed that the economic rationalism arguments found favour with the world's main international financial institutions (IFIs), which were playing an increasing role in the management of conflict and post-war reconstruction. The econometric methodologies favoured by the greed theorists matched the bias of the IFIs towards rational, quantifiable explanations for social phenomena and the technocratic, free-market remedies they favoured. Critics of the greed thesis pointed out that societies experiencing civil war were a poor environment for the collection of statistics and so urged caution over the datasets employed to suggest that certain countries offered a permissive economic context for the onset of civil war (Cramer 2002; Bensted 2011; Newman 2014: 22–3). Collier was also criticized for using proxies, using high male unemployment as a proxy for greed rather than grievance (Kandeh 2005: 96). But those who rejected the greed thesis or the economic explanations for civil war had two more serious objections. The first was that the greed thesis located the causes of war inside states and conveniently absolved external (mainly Western) actors from any blame (Demmers 2012: 112). This was especially the case in relation to the iniquitous international trading regimes that condemned many developing world states to prolonged economic retardation. The single-state lens also tended to ignore regional dynamics, such as the flow of weapons and people across a border or interference from a neighbouring state, which often contributed to violent conflict. The second objection to the greed thesis was that its proponents mistook correlation for causation (Nathan 2005; Spear 2012: 232). Few denied that a permissive economic environment could encourage conflict or that a self-sustaining political economy of war could develop. What they did object to was the argument that economic factors were the *primary* engine of war. Instead, they argued that political and identity factors were the key initiation agents of war and that economic factors often subsequently came into play to change the nature and aim of the conflict.

Over time, the greed *or* grievance academic debate has given way to a consensus that greed *and* grievance are responsible for the outbreak of civil war (Ballentine and Nitzschke 2003; Murshed and Tadjoeddin 2009). The precise weight to be afforded to each is still contested, though since this weighting will change from conflict to conflict, it is sensible to avoid building a general theory of conflict. We can say that most civil wars take place in poor countries, though poverty and inequality *per se* are not sufficient factors in the outbreak of civil war. Moreover, certain economic characteristics (such as low growth and a dependency on commodity exports) predispose societies to civil war. But a permissive environment does not amount to a causation factor. Certainly, economic factors can enable civil war, but for combustion to occur, the economic factors need to spark with other factors (Homer-Dixon 1994).

It is also important to note that development, rather than offering a ladder out of conflict, can contribute to conflict. It is also usually the case that development may well raise the standards of living of a good proportion of society, often those destined to become the middle class, while leaving a poorer section of society, often a clear majority, even further behind. This process is often referred to as 'unequal development'. Liberal economists have a tendency to refer to the alleviation of poverty in terms of rising Gross National Product (GNP), where the activities of urban elites do indeed generate more wealth for a country, which is supposed to 'trickle down' to the less fortunate. The actual result is more often an even greater disparity between rich and poor. This has often then been exacerbated by international organizations (like the IMF, see Chapter 2) who have lent money to developing countries and expected them to maintain financial discipline so that they can be repaid, a process known as 'structural adjustment' to help the development of 'market-friendly' economic conditions, conditions which are almost impossible when a person or group find themselves in 'unspeakable poverty' (Payne 2005: 11–14). Further complicating matters, the poor often belong to one social or even racial group, the populations that are descendants of the Inca and other indigenous people, as in Latin America, for example. In Columbia, discussed below, many of the indigenous poor farmers make the most of the opportunities offered by the cultivation of coca, the base substance that is exported as cocaine. The government and the US Federal Drug Authority (FDA) burn their crops and demand they grow coffee instead. The reason is that they are far more valuable, in limited quantities, medically approved and even widely chewed to alleviate altitude sickness in the high Andes mountains. The UN, the Colombian authorities and the FDA have decreed that coca is an illegal substance; the result has been a local civil war against the largely white population of the urban elites who wish to be well treated by the United States and to benefit from the fruits of globalization. In effect, demand for 'coke' by Wall Street bankers has killed tens of thousands of Columbian *campesinos* and middle-class Columbians alike before we even consider the thousands of Americans who have died due to drug abuse and violence on the streets of US cities (Box 1.2).

Box 1.2 Conflict and natural resources

In Peru, the *Sendero Luminoso* Marxist group recruited heavily among indigenous farmers and was brutally repressed by the army of the urban elites. The ensuing war, which lasted from 1980 to 1992, killed over 70,000 and displaced hundreds of thousands. Fighting has now re-erupted after an indigenous leader, the leftist president, was deposed and his deputy put in his place. The Lima elite sent troops to the provincial town of Juliaca when a revolt started to protest in the countryside, and many were killed, though only one member of the police died. The roots of the uprisings of 1980 and now are much the same: mining operations that benefit the middle class and destroy the environment and livelihoods of poor peasants. Gold and copper are taken from the ground to no benefit for the local economy. Elections do not seem to help the problem, with seven presidents in the last six years. The peasants evoke their ancestors, like peasant leader Tupac Amaru; 'We must fight on…' while asking for either the EU or Russia to come to their aid (Phillips 2023).

Plate 2 The coca plant. Wiki Commons Coca (Erythroxylum coca) en Meta (Colombia) 2.

Many of the social processes associated with development create conditions in which conflict is less easily restrained or is more easily escalated. As Box 1.3 and the example of cattle-raiding in Kenya and Tanzania show, urbanization, environmental degradation, the breakdown of family structures and a lessening of respect for traditional sources of dispute resolution may all create conditions permissive for violent conflict. Yet, to some, these social processes may simply be the by-products of social progress. China's rapid economic development illustrates the potential of development to contribute to conflict. An aggressive state-led development programme has resulted in the displacement of millions of people (well over a million people were displaced as part of the Yangtze Dam project: BBC News 2012), severe environmental degradation, land confiscations and the perception among many rural peasants that they are the collateral damage in the country's economic liberalization. Significant levels of social

Box 1.3 Cattle-raiding in Kenya and Tanzania: development escalating conflict

Pastoral communities in Kenya and Tanzania have a long history of inter-tribal cattle-raiding (Fleischer 1998). Traditionally, the cattle-raiding was sustainable in that relatively small numbers of cattle were taken and casualty figures among the raiders and herders were low because traditional weapons were used. Often cattle raids were linked with rites-of-passage ceremonies, with adolescents using the raids as an opportunity to prove their valour (Kenya Human Rights Commission 2010). In recent years, however, development-related changes in society have transformed the character of cattle-raiding. Small arms transfers from neighbouring conflict regions have also added to the lethality and tended to lead to the more hot-headed youth rapidly replacing the elders, with a corresponding rise in gender-based violence. This is very marked in the Karamoja region of Kenya and Uganda (Odhiambo 2003; Karamoja Resilience Support Unit 2022). As a result, the fallout of cattle-raiding – in terms of casualties and displacement – has increased markedly. There has been an increasing monetisation of exchange, with the result that cattle-raiders are stealing cattle to sell to urban-based criminal gangs rather than for the traditional reasons of individual/group esteem and subsistence pastoral farming. Cattle-raiders killed at least 32 Kenyan police officers in one ambush in 2012 (Macharia 2012). An increasingly urban and aspirational population is fuelling the demand for a meat-based diet, a demand that entrepreneurs are keen to satisfy. In addition to these development-related drivers, the intensity and effect of cattle-raiding have escalated as traditional weapons are being replaced by firearms (readily available from conflicts in the region).

unrest have been reported in China, a trend that has continued as its stellar growth rates have slowed (Göbel and Ong 2012: 8). 'Education and vocational training camps' are reported to have forcibly detained large numbers of the Uyghur population in Xinjian Province since 2017 (BBC News 2020; OHCR 2022) The key point is that development often produces dislocation and uncertainty, and may materially disadvantage some groups, encouraging them to view their status in relation to other groups. In such circumstances, the restraints on conflict may lessen.

The political economy of conflict maintenance

In some cases, the economic rationale of long-running violent conflicts is so apparent that it is possible to think that armed groups are motivated only by profit. Economic rationales often become more visible once a conflict is established and once markets and entrepreneurs have determined ways in which to exploit the opportunities of war. It is perfectly possible to assert that many of these market 'routes' were established before colonial invaders arrived and have been put to use ever since, with only the merchandise and the arms being used changing (Cramer 2006). Colonialism also did much to encourage 'ethnic fragmentation', which damaged the local political economy and contributed to poor economic performance and uneven development. This leads to poor people exploiting any economic or other advantage they might have to make money (Easterly and Levine 1997, quoted in Cramer, n.d.: 2).

Here, we might think of the illegal opiate production in Syria, the illicit trafficking of oil out of Libya and rebel groups that run mines in the Democratic Republic of Congo. But we can also think of the international arms manufacturers who have an interest in states and armed groups using their weapons so that they need new supplies. We can also think of large corporations like G4S that run prisons in Israel that hold Palestinians (Neate 2014). When established, conflict economies can become self-perpetuating, and entrepreneur-combatants may see few incentives to explore an end to the conflict. Economists have developed sophisticated models to show how looting and other forms of economic predation provide a powerful motive for combatants. But such models reveal little of the human character of civil war economies and the trials faced by citizens in times of war. Anthropologists, on the other hand, have succeeded in illustrating the extraordinary lengths to which individuals, families and communities go in order to survive (Walker 2013; Bøås 2015). These studies show the adaptability of humans in their attempts to 'get by' and the extraordinary complexity of civil war economies (McIlwaine and Moser 2004; Nordstrom 2004, 2007).

A common misconception is that civil war economies are very much removed from the faraway mainstream economies of the developed world. The shiny shopping malls and online banking systems of the post-industrialized West seem a million miles away from civil war economies in which many economic transactions are illegal, unregulated or conducted under duress. But as anthropologist Carolyn Nordstrom (2008) shows, combatants and civilians in the midst of civil wars are often closely connected with the globalized international economy. This insight is important as it suggests that Western states and financial institutions – and indeed Western consumers – are complicit in the perpetuation of civil war economies in the developing world. Nordstrom (2007, 2008) argues that all civil wars rely on technologies (arms, communication, money transfers) and networks (trading partners and political supporters) that are, to some extent, international and transnational. As a result, the shadow economies of the civil war environment come into contact with the licit international economy: 'illicit profiteering must make use of legal production, transport, and monetary institutions' (Nordstrom 2008: 290). As David Keen (1998: 42) observes, '[E]ven bandits need to sell what they steal'. The globalized 'buy/sell now, ask questions later' free market makes it easier for the licit and illicit economies to interact, and the sheer complexity of international markets (with multiple brokers) means that the paper trail from manufacturer/grower to consumer is easily obscured. Nordstrom makes the case that rather than being marginal to the world economy, the apparently 'illicit' and 'shadowy' civil war economies are in fact central to it. Warlords and conflict entrepreneurs convert their illegally made profits into legal investments in the formal international economy or demand the same consumer goods that Western shoppers aspire to. Vast sums of money of dubious origin lubricate the international financial markets. Thus, drug barons' money from Afghanistan and Colombia, once suitably laundered, is invested alongside the pension funds of Anglican bishops. In 2017, it was estimated that transnational crime generated $1.6–$2.2 trillion annually.

Much of this was the illegal extraction of resources such as crude oil, mining, fishing and wildlife, as well as the more well-understood trade in illegal drugs. These activities represent a net loss to the treasuries of developing countries. The Nigerian government tried to counter this illegal traffic, as well as the activities of multinational companies that export, legally, vast amounts of oil and minerals as raw materials. The plan, elaborated by President Jonathan Goodhand in 2013, wished to see a huge programme of transformation of raw materials into semi-processed products in Nigeria, which it was estimated would add 3 per cent to the national GDP by 2015. This failed to make any measurable impact. The Nigerian government has just

(October 2023) tried to repeat the effort, but already there are doubts that such efforts will bring better-paid employment and some profits to Nigeria, a country wracked by poverty and massive violence in the North (Olorounbi and Orjinmo 2023).

It is also clear that all of these are dangerous activities generally carried out by the poorest in developing countries. Worse still is organ trafficking, from the bodies of the same poor people, which alone counted in 2017 for between $840 and $1.7 billion, and human trafficking, $150 billion of illegal revenue and untold suffering (Global Financial Integrity 2023). These figures are undoubtedly much higher today given the greatly increased levels of forced migration from Africa to Europe.

On top of this is the vast worldwide underground economy of dodging taxes and regulations. Just as consumers in war-torn societies demand goods and services from developed economies, consumers in the developed world demand goods from the sites of civil war. The high street shops and websites selling mobile phones (in 2023, 98 per cent of adults in the United Kingdom owned a mobile phone (Ofcom 2023)) are just one end of a network of economic exchange stretching from the coltan mines of the Democratic Republic of Congo (DRC) from where an essential component in phone circuitry is extracted (Bøås 2015). Similarly, despite extensive international regulatory mechanisms, Sierra Leone's 'blood diamonds' manage to reach apparently legitimate retail outlets. A report published by the High-Level Panel on International Financial Accountability (FACTI Panel), established by the president of the UN General Assembly in 2020, estimates $500 billion in losses to governments each year from profit-shifting enterprises; $7 trillion in private wealth hidden in haven countries, with 10 per cent of world GDP held offshore and money laundering of around $1.6 trillion per year, or 2.7 per cent of global GDP (FACTI 2021). The COVID-19 pandemic threatened to reverse hard-won progress in developmental gain. Forecasts for 2022 estimate that 75 million more people than expected prior to the pandemic will be living in extreme poverty. Rising food prices and the broader impacts of the war in Ukraine could push that number even higher (UN Sustainable Development Goals Report 2022).

Corruption

Western analysts, and particularly the Western news media, can be shrill in their condemnation of corruption in developing world and civil war contexts. Certainly, kleptocracy by ruling cliques and routine skimming by state functionaries can reach staggering proportions. Mohammed Soharto, Ferdinand Marcos and Mobutu Sese Seko are reputed to have embezzled a collective $50 billion during their respective reigns in Indonesia, the Philippines and Zaire (Denny 2004). The WikiLeaks US diplomatic cables contained the rumour that former Sudanese President Omar al-Bashir squirrelled away $9 billion in overseas bank accounts (Hirsch 2010). But there is a big difference between the elite level siphoning off large sums of money and the more common use of the underground economy by individuals and families who are simply trying to get by. The peculiar economic context of societies experiencing civil war means that we need to reassess our understanding of 'corruption' and 'illegal' market activities. Can corruption be said to exist if the formal economy has broken down and people need to rely on informal market mechanisms simply to survive?

The formal economy may be so dysfunctional (and often overpriced) that citizens have no choice but to operate in the informal sector. In cases of state collapse, there may be no legal economy at all. More commonly, the state is weak and only able to regulate a fraction of economic activity within its borders. Estimates of the size of the informal economy in Africa range from 50 to 80 per cent of GDP (Benjamin 2014). The UN's review of corruption in the country in 2012 was not hopeful (UNODC 2012). Following the overthrow of the Taliban in 2001, the new government in Afghanistan made massive strides in introducing a tax system. By 2013, it was able to raise $2.5 billion through tax revenues, customs duties and mining, but this was less than half of the $7 billion needed for government expenditure. One news report summed up the views of many Afghans on the notion of paying taxes:

> [U]nlike developed countries where personal income tax generates a sizeable chunk of revenue, most Afghans scoffed at the idea of giving the government some of their meager earnings. 'It's not a good government,' said moneychanger Abdurrahman Arif, 28, as he held a wad of soiled notes and scanned for customers. 'I don't pay tax. The rich people don't and the government should go to them before they come to me'.
>
> (Houreld 2013)

One of the unfortunate corollaries of this logic was that the Afghan state (until 2022 at least) had to rely on Western aid to keep up with social payments, such as education for women and girls, as the population resented paying taxes.

This proved to be but one of the reasons that the Western – backed Afghan state collapsed so easily in 2022. Jodi Vittori of the Carnegie Endowment assessed that corruption also bought about the end of the regime by a slow degradation of the armed forces from within. It is worth quoting her at length:

'[C]orruption in the Afghan National Defense and Security Forces gradually undermined the U.S.-backed forces' willingness and ability to fight before the dramatic final downfall'. The United States' government's own 'Office of the Special Inspector General for Afghanistan Reconstruction' concluded as early as 2017 that the occupation forces had taken the easy option of not training a proper new Afghan military and law enforcement structure but had rather in effect hired the existing warlords. Vittori's damning conclusion was that:

> 'Provided with copious amounts of arms and other resources plus impunity for eye-watering levels of criminality, warlord-led networks used violence, patronage, and bribery to help morph the Afghan security sector into an arena for their own armed muscle'. In effect: 'corruption and poor governance were helping to fuel insecurity and provide recruits and resources to a resurgent Taliban ... the security situation continued to deteriorate'.
>
> (Vittori 2021)

Just as the formal, monetized and regulated economy is a way of life for most people in many Western states, the informal economy is a socially embedded behavioural and entirely rational norm in many civil war and post-civil-war societies. It makes sense to use non-patented medicines when patented medicines are either unavailable or exorbitantly priced. The formalization of the medical industry through the protection of pharmaceutical patents would not be in the interests of the vast majority of citizens because it would entail rocketing prices. The careless branding of certain economic activities as 'corrupt' or 'illegal' says as much about the Western worldview as it does about the activities themselves (Brown et al. 2004; Cheng and Zaum 2011). This is not to deny that corruption takes place and that it poses a real hazard to development and donor activity. Instead, it is to caution against the unthinking extension of Western yardsticks to non-Western, war-torn contexts. Box 1.3 illustrates the absurdity of some Western norms in war-affected societies. Kolstad et al. (2008) stress the importance of distinguishing between different types and scales of corruption. There is a difference between administrative informality and petty corruption found in everyday exchanges in a remote town where state officials feel the need to augment their wages and large-scale frauds perpetrated by political leaders or senior bureaucrats. Simplistic moral and ethical judgements may not always be sensitive to the context in which corruption takes place. We should also be alert to the potential for international assistance to use and reinforce existing clientelistic and patronage networks. What begins as the rational use of 'local systems of disbursement' can reinvigorate networks that are less than transparent and may even reinforce warlord politics.

In some cases, international connections have served to prop up weak states and their corrupt patronage networks. Bayart (2000) notes how many African leaders have become adept at fobbing off international donors with the message they want to hear:

> [D]emocracy, or more precisely the discourse of democracy, is no more than yet another source of economic rents, comparable to earlier discourses such as the denunciation of communism or of imperialism in the time of the Cold War, but better adapted to the spirit of the age. It is, as it were, a form of pidgin language that various native princes use in their communication with Western sovereigns and financiers. Senegal, one of the main recipients of public development aid in sub-Saharan Africa, is a past master in this game of make-believe. It is no exaggeration to say that the export of its institutional image … has replaced the export of groundnuts.
>
> (Bayart 2000: 226)

Essentially, the argument runs that international support has allowed corrupt regimes to continue systems of neo-patrimonialism and defer redistributive political and economic reform.

Legitimate monies and resources often reach civil war societies in the form of external donor aid. Once there, these funds risk fuelling the conflict. In an ethnically divided society, the infusion of external assistance is likely to be jealously scrutinized by all sides to make sure that their group gets 'its share'. Lebanon provides a complex example. The country played host to over one million Syrian refugees fleeing the civil war. Many of the refugees received basic care, including health care, from a range of UN bodies, bilateral donors and INGOs. But many Lebanese looked on with a deepening sense of jealousy. The Lebanese health-care system is largely privatised, and so some of the care available to refugees is not available to those in the host country (Amnesty International 2014). This resentment felt by the host community was refracted through the sectarian lens of an already divided Lebanon, causing further tensions.

External humanitarian and development agencies are often placed in an invidious position in war-torn societies. Do they sit on the sidelines, refusing to give assistance for fear that aid may fall into the 'wrong hands', or do they muck in and hope that their efforts help the genuinely needy despite the risks? Médecins Sans Frontières (MSF) staff in Myanmar/Burma faced this problem during the years of the military dictatorship. The organization wanted to treat malaria and AIDS sufferers in extremely poor areas but needed military permission to do so. Senior staff were faced with the dilemma of 'playing golf with the generals' (a way of talking to them informally) or sitting on their hands. The danger of dealing with the generals was that it might confer legitimacy onto military thugs (Terry 2011: 109–11). Civil war contexts are unlikely to leave ethical principles unscathed, yet that means material and symbolic resources being at the mercy of combatants. Moreover, in some contexts, the sheer scale of donor assistance can distort the economy. As one observer noted, '[O]ther than the state itself, the aid business is today the single biggest employer in most African states' (van de Walle 2001: 58).

Just as there is a political economy of war, there is a political economy of humanitarianism and development assistance. Perhaps this is most visible in the micro-economies that spring up to service international humanitarian workers in war-affected societies. The cluster of Western- style bars and fast-food outlets are often identifiable because of the white 4 × 4s parked outside. But at the macro-economic level, as will be discussed in later chapters, international economic and development interventions are often deliberately aimed at reshaping the war-affected economy to reconnect it with the formal global economy. Many of the liberal economic 'reforms' actually lead to the further immiseration of citizens: state employees are sacked ('rightsizing bureaucracy'), debts run up by previous regimes must be paid ('respecting international financial obligations'), prices rise as exchanges are formalized and monetized ('regulation') and indigenous businesses cannot compete with cheap imports ('the global free market'). Kiely (2007: 434) notes how 'liberalisation undermines the capacity of developing countries to develop dynamic comparative advantages', yet liberalization seems to be the main tool in the international toolbox, at least for now.

The chief point of this section is that as war further distorts economies, people in war-affected societies (combatants and civilians alike) will take extraordinary measures to survive, and this may involve activities that Western observers may judge as 'corrupt' or 'illegal'. But outside observers may be hypocritical in such judgements: civil war economies are hardwired into the very fabric of the formal international economy. Consumer demand in Western states and the international economic structures erected by Western states influence the choices and constraints faced by people on the ground in civil war societies. While we can paint an abstract picture of the economic impact of civil war (according to Collier (2004) the 'typical' civil war costs $50 billion). it is important that we recognize the human experience of civil war and how many people are brutalized and humiliated by civil war economics, whether by being trafficked, being forced to sell family heirlooms or living in an environment in which theft is regarded as a normal survival mechanism (Mac Ginty 2004). As importantly, anthropologists such as Ivana Maček (2009) have presented accounts of how these communities can turn these shared experiences into more positive symbols of solidarity in a world turned upside-down.

Natural resources and conflict

As already noted, economists have claimed that an economic dependency on natural resource exports increases the likelihood of the outbreak of civil war. But the mere presence of natural resources does not lead to armed conflict. As Cramer (2006: 117) observes, '[S]carcity and violence are a product of social relations rather than inherent in the relative abundance of a particular good, object or resource'. In other words, it is the management of the resources that really matters. It is notable that Norway, with its abundant natural gas supplies, high taxes and social provision, tops most global living standards tables. Yet, despite being 'blessed' by nature's largesse, a significant number of resource-rich states are chronically poor. Prominent in the list of mineral-dependent countries (reliant on extractive minerals for 25 per cent of non-tangible exports) were the DRC, Somalia and Libya – all the sites of conflict (Hagland 2011). We mentioned Nigeria earlier, a country that has abundant natural resources and one of the worst insurgency situations in Africa. It is the labour practices (voluntary or coerced), distribution of licences (open competition or patronage) and destination of profits (public or private coffers) that will determine whether states face a resource curse or windfall. Patterns of landownership are particularly important in developing world contexts, as access to land (and the quality of that land) may afford subsistence and thus some measure of autonomy in economic matters (Miall 2007: 125–9).

Crucially, the perception of the management of resources is important. In a number of cases, minority ethnic, nationalist or religious groups have pursued grievances stemming from allegations that the state was plundering 'their' natural resources with few obvious benefits in return. Thus the Acehnese in Indonesia, Muslims in Mindanao (Philippines), Christians in southern Sudan and the Ogoni in the Niger Delta (see Box 1.4) have all engaged in secessionist conflict with the state and have campaigned for their 'fair share' of returns from resource exploitation. Indeed, a recurring narrative in debates for Scottish independence concerns who has the right to the receipts from North Sea gas: the Scottish Government or the whole of the United Kingdom? Although natural resources play a crucial role in these conflicts, it is incorrect to conceive of them as pure 'resource wars'. Instead, the conflict arises from a complex mix of the presence of resources, the pattern of resource exploitation, the perception of the benefits of that exploitation and identity affiliations. If the stakes are high, identity affiliations can mutate, with groups and individuals attaching increasing weight to the purity of their ethnic group and rediscovering (or inventing) their 'unique' history (Wilmer 2002: ix). In such ways, conflicts become 'ethnicized', and exclusion from the benefits of natural resources may provide a powerful impetus to escalate the conflict.

Box 1.4 Oil extraction in Nigeria's Niger Delta

Nigeria's oil-rich Niger Delta region has experienced significant levels of conflict, criminal violence and environmental damage for several decades. The central government has encouraged major oil corporations to exploit the oil reserves, but residents in the Delta region claim that any benefits bypass local communities and that pollution has seriously affected the quality of life. Armed criminal gangs regularly steal oil from pipelines and kidnap foreign oil workers for ransom, while other local groups have been vocal in their condemnation of the profiteering, corruption and environmental disregard of 'imperial Abuja' or have encouraged labour unrest. On top of this, there have been clashes between rival ethnic groups. Oil company attempts to co-opt tribal chiefs through payments have made chieftaincies extremely lucrative, sparking conflict and changing community perceptions of chiefs. In response to the oil-related tension and violence, the Nigerian government has variously declared a state of emergency and sent in the army, attempted to buy off local leaders, reassured foreign oil companies, promised to invest a greater share of oil in the region and occasionally mounted prosecutions against corrupt officials (including one against two rear-admirals accused of stealing an oil tanker: Clayton 2005). The violence has seriously disrupted oil production, but the potential rewards for the government, local and national political leaders, and overseas oil companies are simply too great for anyone to contemplate withdrawing from the region (Obi and Rustad 2011).

Sources: Clayton (2005); Obi and Rustad (2011)

Oil, power and climate change

Winston Churchill's observation that 'God put the West's oil under Middle Eastern feet' is a reminder of competitive geo-strategic interests in natural resources that make resource-rich developing world states prone to intervention by powerful states (again, see Barr 2011). Quite simply, advanced and developing economies are both dependent on oil, and it is unlikely that non-fossil fuel replacements on the scale necessary will be available any time soon. And both suffer from the resultant harm done by the global warming so engendered. Arguably, migration flows, desertification, hotter summers, wildfires, the spread of disease and so on affect both the 'North' and the 'South', but it is more likely that the North can find ways to mitigate some of these problems through access to technology and capital. As this book was being finalized, a major conference on the subject was about to be held in Dubai, hopefully entitled 'Bridging the Gap to the Future' (WGES 2023). Churchill was one of the main proponents of the move from coal to oil due to his desire to update the propulsion systems of Royal

Navy battleships and defend the interests of the (then) British Empire. It is unlikely he even vaguely considered the developmental or environmental results of such actions. Once again, war and development have proved strange bedfellows.

Until the replacement of fossil fuels with sustainable forms of energy is feasible (a topic for another volume), the ways of life, politics and economics of the rich and poor of the planet would be utterly unsustainable if ready access to oil were not secured. The production of oil and gas has been a major driver of globalization since at least the First World War (Yergin 1991; Barr 2011, 2018) and also a major driver of political conflict. In the main, the market has been successful in procuring oil, and major Western states are not as dependent on oil from conflict-affected areas as they were in the 1970s. The United States was the world's largest oil producer in 2022 (Rapier 2022). But although most oil-producing Gulf States are compliant with Western economic and geopolitical strategy, Saudi Arabia and Russia are not. This has become more obvious since the Biden Administration took office in early 2020 when Biden made clear his displeasure about Saudi human rights abuses. The Ukraine war of 2022 led to massive disruption to world oil and gas markets due to Russia cutting off supplies to Europe and US-imposed sanctions on Russian oil companies.

And as demand for oil in the developing world (especially India and China) surges, pressure has grown on apparently finite oil resources increases. Both the market and states can be expected to act in self-interested ways to ensure continued access to energy. Western oil companies have made vast profits on the back of this turmoil, which has led to calls for 'windfall' taxes. It has also accelerated the installation of more and more 'green' technology, such as solar and wind farms. These can help the economies of certain developing countries that have access to wind and sun resources, which might lead to increased GDP, but, as with oil in Nigeria and Iraq, not necessarily to increase income for the section of the population that cannot benefit.

Angola is a good example of how oil has enriched a narrow elite and further impoverished the wider population in a developing nation. The latest GDP per capita figures are very low for an oil-producing nation, at $1935.5 (barely above Haiti's, which has no oil at all.) Although GDP accelerated sharply to 15 per cent *per annum* for more than a decade after 2000, from 2016 and 2020, a lengthy recession left a third of the population below the poverty line (less than $2.15 a day), with food inflation at 24 per cent a year in July 2022 (and even 31.6 per cent in 2021), an indicator that particularly hits the poorest. The World Bank has urged the government of Angola to do more to 'transform the state-led and oil-funded economic model into a sustainable, inclusive, private-sector-led growth model' (World Bank 2022). It is incumbent upon us to ask if oil cannot be as much a curse as a benefit to developing countries, as the example of Nigeria also shows (see Box 1.4).

After all, the example of the Netherlands, which is currently attempting to close down the biggest gas field in Europe near Groningen on environmental grounds, has often been cited as having 'the Dutch disease' of 'bad' development. As the Netherlands got richer from gas revenues, the Dutch industry suffered and, in some cases, left for more congenial places with lower exchange and salary levels. Likewise, Britain claimed the discovery of North Sea oil in the 1970s as a huge boon, but it also accelerated the de-industrialization of large areas of Northern England and Scotland, which in turn caused social and political turmoil. The Scottish demand for independence from the United Kingdom really dates from this period. Oil and gas make countries uncompetitive due to the resultant over-valuation of their currency. Some countries in the Global North have managed to mitigate the effects of Dutch disease by investing a large proportion of oil and gas revenues in a 'sovereign wealth fund' of which the most famous is Norway. In the Middle East, Kuwait has the largest *per capita* fund, which has been invested in domestic production facilities and lucrative investments abroad. Many developing countries have found that it is much too easy for corrupt government elites to 'put their hand in the till'; in other words, the fund has fuelled corruption (Karl 1997). Notwithstanding that, a new type of sovereign wealth funds is now emerging in countries that are not necessarily 'rich' in the aforementioned sense but are trying to emulate their success – Türkiye, Romania, India and Bangladesh have all attempted to set them up in recent times (Milhench/NRGI 2017). Newly discovered oil fields are prompting other states to do the same thing, notably Kenya and Mozambique (see, for example, NRGI 2021).

As the cases of post-Saddam Iraq, Colombia and Nigeria show, consumer demand is so great that complexes of private companies and state bodies will come together to create oases of petrochemical calm in the midst of wider conflict just to ensure that oil continues to flow.

Water wars?

Such attitudes might also lead to 'water wars' as neighbours feel they are being robbed of a vital resource. Water is certainly a regional problem that could well be a driver of future conflict in the Horn of Africa and more widely. The absence of water in many parts of North Africa, the Middle East and the Horn meant that for much of human history, the population was kept low by an inability of the population to go beyond a certain area of cultivation. The area south of Aswan in Egypt was a desert until the building of the Aswan Dam in 1970 opened up a huge lake south of Aswan (Lake Nasser) and increased the cultivatable

Plate 3 The Aswan Dam.

Source: Alain Petit/Getty Images.

area of the Nile in Egypt. The discovery of oil across much of the area created the possibility of desalinization plants, which has led to a many-fold increase in the population of much of the Persian Gulf, and so on. This in turn has increased emigration, as much of the benefit of these developments has not improved the economic fate of many of the poorer inhabitants. This has in turn led to conflict between the southern and northern shores of the Mediterranean, as well as within the local populations.

In a new source of potential conflict, Ethiopia has just (2022) completed The Grand Ethiopian Renaissance Dam (GERD), which many of its neighbours assert will reduce water flows the length of the Blue and White Nile Rivers (so stretching from Lake Victoria which borders Uganda, Tanzania and Kenya – the source of the Blue Nile to the Mediterranean coast in Egypt). Any conflict over these water sources and rivers could rapidly turn into a pan-African conflict. Ethiopia has proved itself to be very keen to defend its rights as a regional hegemon in recent times (as in the brutal war against its province of Tigray). The whole area is volatile, with shifting allegiances, such as Ethiopia's re-engagement with Eritrea after a long period of hostility. Eritrea helped Ethiopia by invading Tigray in the recent hostilities, which killed over 600,000 people and displaced many more. Alex de Waal called it 'the world's most lethal conflict, characterised by extreme brutality and the use of mass starvation as a weapon'. Meanwhile, Eritrea has been engaging in repression of its own people (many of whom fled to Ethiopia,

Plate 4 Adi-Harush Refugee Camp, Ethiopia: many of the refugees are Eritrean.

Source: Peter Turnley/Getty Images.

Sudan and far beyond) and warfare with all its other neighbours on a periodic basis since 1993 (Plaut and Vaughan 2023). The GERD has also been described as an existential issue for Egypt and Ethiopia: 'this conflict does not revolve only around physical resources but also extends to the very identity of the two states' (Ayferam 2023; Al Jazeera 2023).

Conclusions

There is a growing consensus among policymakers, academics and others that the escalation, maintenance and transformation of conflict are linked to development. The precise nature of these linkages is still debated. In truth, there is no exact science linking conflict to development; the constellation of variables is simply too great. Despite the apparent sophistication of econometric modelling or the 'seen it all' world-weary cynicism of development practitioners, conflict interventions and development programming contain a good deal of guesswork and finger-crossing. Western donor governments, such as the Department for International Development (DFID) in the United Kingdom, have historically shown a commitment to conflict-sensitive programming or development interventions calibrated to have a minimal or positive impact on a conflict situation. Unfortunately, this commitment has been reduced in recent years. DFID was merged with the Foreign and Commonwealth Office in September 2020, giving it the new moniker of the Foreign Commonwealth and Development Office. In the United

Kingdom, since 2020, aid for development has been cut to 0.5 per cent of Gross National Income (GNI) from 0.7 per cent, leading to a 21 per cent reduction in spending (House of Commons 2022).

There has developed a belief in certain political circles in the rich North that both poverty and conflict in developing countries are intractable, and this has led to what might be called 'development' or even 'compassion' fatigue. But all commentators agree that the results of conflict and poverty provoke patterns of emigration that must be 'dealt with' somehow. Whether it be Eritreans fleeing compulsory conscription in the army and ingrained poverty or inhabitants of the Sahel region of Africa fleeing Islamic militias and the loss of their traditional grazing grounds for cattle, the European Continent has been forced to face up to facts and devise (largely ineffective) policies to deal with the problem (European Parliament 2022). The root causes are undoubtedly domestic to the countries from which they flee, and it appears that development programmes delivered *in situ* often fail to reduce such flows of people. But the disarray about what should be done is palpable. The British Government decided in 2021 that sending the 'boat people' who made it to the English Channel to Rwanda after they landed would be a good idea; hardly anyone who studies or deals with the problem in the field agrees.

Not all failures to address such problems can be attributed to desperate politicians or, indeed, their parliamentary constituents. The volatility and lack of regulation in many war-torn or post-conflict societies mean that the precise effect of development inputs is unknown. The bottom line is that development or reconstruction assistance is a resource, and political and militant actors will act rationally in attempting to maximize their access to, or benefit from, those resources. Despite good intentions, development inputs may actually fuel conflict. Moreover, and as will be discussed in Chapter 2, many of the factors that influence development in a war-torn society will be external to that society. Just as many citizens may feel powerless in the midst of conflict, they may also feel that development is a process that is *done to* them.

The relationship between security and development is fraught with thorny questions, especially in the post-9/11 world in which the 'security imperative' is easier to justify among many audiences than a 'development imperative'. When the Islamic State (IS) occupied large parts of Iraq from 2013 onwards, the Western response was one of bombing from the air, not of examining development alternatives that could defeat IS through livelihoods and economic advancement among the populace. Development and reconstruction require a certain level of order. At its most benign, this order can take the form of institutionalization and regulation, in effect, a process of 'normalization' whereby the uncertainty of a violent context is replaced by the certainty of stability. But in a less benign scenario, there can be unacceptable costs associated with 'stabilization', such as restraining civil liberties or empowering private security contractors. In effect, a process of development and post-war reconstruction can be as disempowering as the war once was. If a holistic view of development is taken, in which development extends far beyond the narrow confines of economic growth, then the process of development might also involve de-development.

Summary

- Explanations of the causes of violent conflict have focused on issues of 'greed' (economic causes) and 'grievance' (such as identity).
- Most scholars argue that a mixture of greed and grievance factors contributes to violent conflict.
- The factors that cause a violent conflict may differ from the factors that sustain a conflict.
- Resources alone, such as oil, diamonds or water, do not cause conflict. What is important is the nature of the extraction and management of those resources.
- The effects of Climate Change are provoking emerging patterns of displacement and migration.
- Just as violent conflict can distort an economy, so too can aid and peace support, with issues of 'corruption' gaining increasing attention in recent years.

Discussion questions

- Do you find that the criticisms of the greed thesis of conflict causation are justified?
- What are the economic factors that can sustain violent conflict once it has started?
- Can there be corruption if the formal economy has broken down?
- Why and how is food used as a weapon of war?
- Is the discovery of oil in a developing country likely to be a blessing or a curse?

Further reading

There is an enormous literature on conflict causation. A good starting point for the econometric perspective is P. Collier et al. (2003) *Breaking the Conflict Trap: Civil war and development policy*, Washington, DC: World Bank and Oxford University Press, and P. Collier (2007) *The Bottom Billion: Why the poorest countries are failing and what can be done about it*, Oxford: Oxford University Press. Wider studies of conflict that make connections between development and conflict include works by Mark Duffield, Paul Jackson, Eleanor O'Gorman, Mary Kaldor, David Keen and Christopher Cramer. A particularly useful way of following developments in conflict zones is to search for blogs by people living there or by aid workers. A few years after the conflict, autobiographies might emerge as well that give insights into living amidst conflict.

Useful websites

On Food as a Weapon of War and related issues, see: *Report of the UN Special Rapporteur on the right to food, Section V. 'Right to food in armed conflict and protracted crises' (2020 UN Doc A/HRC/46/33 pages 17–19)* https://documents-dds-ny.un.org/doc/UNDOC/GEN/G20/355/23/PDF/G2035523.pdf?OpenElement*; Israel Policy Forum* https://israelpolicyforum.org/2023/02/01/the-folly-of-collective-punishment-as-a-terror-deterrent/; https://www.un.org/en/genocideprevention/documents/atrocity-crimes/Doc.33GC-IV-EN.pdf; Useful websites. The World Bank has many documents on Fragility, Conflict and Violence: http://www.worldbank.org/en/topic/fragilityconflictviolence. The Uppsala Conflict Data Program contains probably the most comprehensive source of conflict statistics (http://www.pcr.uu.se/research/UCDP/), while the Global Peace Index (http://www.visionofhumanity.org/#/page/our-vision) collects country-wide information on development and peace. Useful information can also be found via the International Crisis Group (www.crisisgroup.org) and its conflict-tracking website CrisisWatch (https://www.crisisgroup.org/crisiswatch), Peace Direct (peacedirect.org) and International Alert (http://www.international-alert.org), Uppsala Conflict Data Program https://ucdp.uu.se/. On resource exploitation, details of attempts to reduce the trade in blood diamonds can be found at www.kimberleyprocess.com/. Details of the anti-corruption organization Tiri can be found at www.tiri.com. For useful discussions and links to community resilience to climate change in Southeast Asia, see Cook, A. (2023). Climate change, environment and humanitarian action in Southeast Asia. ICRC Humanitarian Law and Policy https://blogs.icrc.org/law-and-policy/2023/03/28/climate-change-environment-humanitarian-southeast-asia/

For general reading on global climate change initiatives, see https://journals.plos.org/climate/ and https://www.climatecentre.org/

Bibliography

Al Jazeera (2023) 10 September: 'Filling of Grand Renaissance dam on the Nile complete', Ethiopia says: https://www.aljazeera.com/news/2023/9/10/filling-of-grand-renaissance-dam-on-the-nile-complete-ethiopia-says (accessed 25 October 2023).

Amnesty International (2014) Syrian refugees in Lebanon desperate for health care amid international apathy. Amnesty International, 21 May. Available at: https://www.amnesty.org/en/articles/news/2014/05/syrian-refugees-lebanon-desperate-health-care-amid-international-apathy/ (accessed 26 October 2023).

Amnesty International (2015) *Amnesty International Report 2014/2015: The State of the World's Human Rights*. London: Amnesty International. https://www.amnesty.org/en/documents/pol10/0001/2015/en/ (accessed 26 October 2023).

Angell, N. (1910) *The Great Illusion: AA Study of the Relation of Military Power to National Advantage*. London: William Heinemann (and many other imprints since)

Applebaum, A. (2017) *Red Famine: Stalin's War on Ukraine, 1921–33*. London: Allen Lane.

Appleby, R.S. (2000) *The Ambivalence of the Sacred: Religion, violence and reconciliation*. Lanham, MD: Rowman and Littlefield.

Ayferam, G. Carnegie Endowment (2023) 19 January 2023: 'The Nile dispute: Beyond water security'. https://carnegieendowment.org/sada/88842 (accessed 25 October 2023).

Baldwin, D. (2020) *Economic Statecraft*. Princeton, NJ: Princeton University Press.

Ballentine, K. and Nitzschke, H. (2003) *Beyond Greed and Grievance: Policy Lessons from Studies in the Political Economy of Armed Conflict*. New York: IPA Policy Report.

Bannon, I. and Collier, P. (2003) Natural Resources and Conflict: What Can We Do. In Bannon, I. and Collier, P. (eds) *Natural Resources and Violent Conflict*. Washington, DC: World Bank, pp. 1–16.

Barbieri, K. (2002) *The Liberal Illusion: Does Trade Promote Peace?* Chicago: University of Michigan Press.

Barr, J. (2011) *A Line in the Sand: Britain, France and the Struggle that Shaped the Middle East*. London: Simon and Schuster.

Barr, J. (2018) *Lords of the Desert: Britain's Struggle with America to Dominate the Middle East*. London: Simon and Schuster.

Bayart, J.F. (2000) Africa in the World: a history of extraversion. *African Affairs* 99: 217–67.

BBC News (2012) China's Three Gorges Dam may displace another 100,000. 18 April. http://www.bbc.co.uk/news/world-asia-china-17754256 (accessed 27 October 2015).

BBC News (2020) Xinjiang: China defends 'education' camps. 17 September. https://www.bbc.co.uk/news/world-asia-china-54195325 (accessed 31 August 2023).

Benjamin, N. (2014) Informal economy and the World Bank. World Bank Research Working Paper no. 6888, May.

Bensted, R. (2011) A critique of Paul Collier's 'greed versus grievance' thesis of civil war. *African Security Review* 20(3): 84–90.

Bøås, M. (2015) *The Politics of Conflict Economies: Miners, Merchants and Warriors in the African Borderland*. London: Routledge.

Brauer, J. and Dunne, J.P. (2012) *Peace Economics: A Macroeconomic Primer for Violence Affected States*. Washington, DC: USIP Press.

Brown, E., Cloke, J. and Sohail, M. (2004) Key myths about corruption. Briefing paper for a workshop on corruption and development presented at the Development Studies Association Annual Conference, Westminster, London, 6 November.

Brown, M.E. (1997) The Causes of Ethnic Conflict: An Overview. In Brown, M.E., Coté, O.R., Lynne-Jones, S.M., and Miller, S.E. (eds) *Nationalism and Ethnic Conflict*. Cambridge, MA: MIT Press, pp. 3–25.

Brubaker, R. and Laitin, D.D. (1998) Ethnic and nationalist violence. *Annual Review of Sociology 24*: 423–54.

CFR – Council on Foreign Relations CFR (2023) What is Hamas? https://www.cfr.org/backgrounder/what-hamas#chapter-title-0-6

Cheng, C. and Zaum, D. (eds). (2011) *Post-Conflict Peacebuilding: Selling the Peace?* London: Routledge.

Clayton, J. (2005) Nigerian admirals pay the price for stealing captured oil tanker. *Times Online*, 8 January. http://www.timesonline.co.uk/tol/news/world/article409606.ece (accessed 13 June 2008).

Collier, P. (2000a) Doing Well out of War: An Economic Perspective. In Berdal, M. and Malone, D. (eds) *Greed and Grievance: Economic Agendas in Civil Wars*. Boulder, CO: Lynne Rienner, pp. 91–111.

Collier, P. (2000b) *Economic Causes of Civil Conflict and their Implications for Policy*. Washington, DC: World Bank.

Collier, P. (2004) Development and Security. 12th Bradford Development Lecture, University of Bradford, 11 November.

Collier, P., Elliot, V.L., Hegre, H., Hoeffler, A., Reynal-Querol, M., and Sambanis, N. (2003) *Breaking the Conflict Trap: Civil War and Development Policy*. Washington, DC: World Bank and Oxford University Press.

Collier, P. and Hoeffler, A. (2002) *Greed and Grievance in Civil Wars*. Working Paper Series. Oxford: Centre for the Study of African Economies.

Connor, W. (1994) *Ethnonationalism: The Quest for Understanding*. Princeton, NJ: Princeton University Press.

Coynash, H. (2022) 'Raped, shot and thrown in a mass grave by the Russian invaders who turned Bucha into hell' Kharkiv Human Rights Protection Group, Ukraine. https://khpg.org/en/1608810427 (accessed 16 February 2023).

Cramer, C. (2002) Homo Economicus goes to war: Methodological individualism, rational choice and the political economy of war. *World Development 30*(11): 1845–64.

Cramer, C. (2006) *Civil War is Not a Stupid Thing: Accounting for Violence in Developing Countries*. London: Hurst.

Cramer, C. (n.d.) SOAS Economics Working Paper 73: 'Civil War is Not a Stupid Thing'; Exploring growth, distribution and conflict linkages'. https://www.soas.ac.uk/sites/default/files/2022-10/economics-wp073.pdf (accessed 7 February 2023).

Demmers, J. (2012) *Theories of Violent Conflict: An Introduction*. London: Routledge.

Denny, C. (2004) Suharto, Marcos and Mobutu head corruption table with $50bn scams. *The Guardian*, 26 March.

Easterly, W. and R. Levine (1997) Africa's growth tragedy: Policies and ethnic divisions. World Bank Discussion Paper, mimeo.

European Parliament (2022) Georgiana Sandu, 'Immigration Policy'. https://www.europarl.europa.eu/factsheets/en/sheet/152/immigration-policy

Fawcett, L. ed. (2023) *International Relations of the Middle East*, 6th edn. Oxford: Oxford University Press.

Fleischer, M. (1998) Cattle raiding and its correlates: the cultural–ecological consequences of market-oriented cattle raiding among the Kuria of Tanzania. *Human Ecology 26*(2): 547–72.

Galeotti, M. (2022) *Putin's Wars: From Chechnya to Ukraine*. London: Bloomsbury.

Gasper, D. (2004) *The Ethics of Development*. Edinburgh: Edinburgh University Press.

Gattas, K. (2020) *Black Wave: Saudi Arabia, Iran and the Rivalry that Unravelled the Middle East*. London: Hachette/Headline.
Global Financial Integrity (2023) Transnational crime and the developing world. https://gfintegrity.org/issue/transnational-crime/ (accessed 16 August 2023).
Glover, J. (2001) *Humanity: A Moral History of the Twentieth Century*. London: Pimlico.
Göbel, C. and Ong, L.H. (2012) *Social Unrest in China*. London: Chatham House.
Guáqueta, A. (2007) The way back in: reintegrating illegal armed groups in Colombia then and now. *Conflict, Security and Development* 7(3): 417–56.
Hagland, D. (2011) *Blessing or Curse? The Rise of Mineral Dependence among Low and Middle Income Countries*. Oxford Policy Management Group Report, December. Oxford: OPMG.
Hirsch, A. (2010) WikiLeaks cables: Sudanese President 'stashed $9bn in UK banks'. *The Guardian*, 17 December.
Homer-Dixon, T. (1994) Environmental scarcities and violent conflict: evidence from cases. *International Security*, *19*(1): 5–40.
Houreld, K. (2013) Afghans warned: The taxman is coming after you. Reuters, 31 March. http://www.reuters.com/article/2013/03/31/us-afghanistan-tax-idUSBRE92U00Q20130331 (accessed 8 September 2015).
Huntington, S. (1998) *The Clash of Civilizations and the Remaking of World Order*. London: Touchstone.
International Crisis Group (n.d.) *CrisisWatch*. https://www.crisisgroup.org/crisiswatch
Jacoby, T. (2008) *Understanding Conflict and Violence: Theoretical and Interdisciplinary Approaches*. London: Routledge.
Kandeh, J. (2005) The Criminalization of the RUF Insurgency in Sierra Leone. In Arnson, C. and Zartman, I.W. (eds) *Rethinking the Economics of War: The Intersections of Need, Greed and Creed*. Baltimore, MD: Johns Hopkins University Press, pp. 84–106.
Kaplan, R. (1994) The coming anarchy. *Atlantic Monthly*, February.
Karamoja Resilience Support Unit (2022) *The Return of Conflict in Karamoja, Uganda: Community Perspectives*. USAID/Irish Aid: https://karamojaresilience.org/wp-content/uploads/2022/11/TUFTS_2250_KRSU_return_of_conflict_in_Karamoja_V3_online.pdf (accessed 7 February 2023).
Karl, T. L. (1997) *The Paradox of Plenty: Oil Booms and Petro-states*. Berkeley: University of California Press.
Keen, D. (1998) *The Economic Functions of Violence in Civil War*. Adelphi Paper 320. London: International Institute of Strategic Studies.
Kenya Human Rights Commission (2010) *Morans No More: The Changing Face of Cattle Rustling in Kenya*. Nairobi: Kenya Human Rights Commission.
Kiely, R. (2007) Poverty reduction through liberalisation? Neo-liberalism and the myth of global convergence. *Review of International Studies 33*: 415–34.
Kolstad, I., Fritz, V., and O'Neil, T. (2008) *Corruption, anti-corruption efforts and aid: Do donors have the right approach?* Dublin: Advisory Board for Irish Aid. https://odi.org/en/publications/ (accessed 18 June 2008).
Laderman, A. (2023) The folly of collective punishment as a terror deterrent. Israel Policy Forum. https://israelpolicyforum.org/2023/02/01/the-folly-of-collective-punishment-as-a-terror-deterrent/ (accessed 8 August 2023).
Lenin, V.I. (1917). *War and Revolution: Lenin Collected Works*. Moscow: Progress Publishers, Vol. 24, pp. 398–421. https://www.marxists.org/archive/lenin/works/1917/may/14.htm and (accessed 23 October 2023).

Lenin, V.I. (1964). *Imperialism; The Highest Form of Capitalism: Selected Works: Vol. 1*. Moscow: Progress Publishers, pp. 667–766. https://www.marxists.org/archive/lenin/works/1916/imp-hsc/index.htm (accessed 23 October 2023).
Loft, P. and Brien, P. (2022) *Reducing the UK's aid spend in 2021 and 2022*. London: House of Commons Library https://researchbriefings.files.parliament.uk/documents/CBP-9224/CBP-9224.pdf (accessed 10 February 2023).
Lukes, S. (2005) *Power: A Radical View*. New York: Red Globe Press.
Mac Ginty, R. (2004) Looting in the context of violent conflict: a conceptualization and typology. *Third World Quarterly* 25(5): 857–70.
Mac Ginty, R. (2006) *No War, No Peace: The Rejuvenation of Stalled Peace Processes and Peace Accords*. London: Palgrave.
Mac Ginty, R. (2014) Everyday peace: bottom-up and local agency in conflict-affected societies. *Security Dialogue* 45(6): 548–64.
Maček, I. (2009) *Sarajevo Under Siege: Anthropology in Wartime*. Philadelphia: University of Pennsylvania Press.
Macharia, J. (2012) At least 32 police dead in cattle raid ambush. Reuters, 12 November. http://uk.reuters.com/article/2012/11/12/us-kenya-security-idUSBRE8AB0NL20121112 (accessed 8 September 2015).
Mandelbaum, M. (2002) *The Ideas that Conquered the World: Peace, Democracy and Free Markets in the Twenty-First Century*. New York: PublicAffairs.
McIlwaine, C. and Moser, C. (2004) *Encounters with Violence in Latin America: Urban Poor Perceptions from Colombia and Guatemala*. London: Routledge.
Miall, H. (2007) *Emergent Conflict and Peaceful Change*. Cambridge: Polity.
Milhench, C. NRGI (National Resource Governance Institute); Reuters (2017) A new breed of sovereign wealth fund – without the wealth. 17 March 2017. https://www.reuters.com/article/uk-emerging-swf-investment-idUSKBN16O22B (accessed 25 October 2023).
Mills, C.W. (1956) *The Power Elite*. Oxford: Oxford University Press.
Mitchell, C. (2014) *The Nature of Intractable Conflict*. London: Palgrave Macmillan.
Morefield, J. (2022) *Unsettling the World: Edward Said and Political Theory*. Lanham, MA: Rowman and Littlefield.
Münkler, H. (2005) *The New Wars*. Cambridge: Polity.
Murshed, S.M. and Tadjoeddin, M.Z. (2009) Revisiting the greed and grievance explanations for violent internal conflict. *Journal of International Development* 21(1): 87–111.
Nathan, L. (2005) The frightful inadequacy of most statistics: A critique of Collier and Hoeffler on the causes of civil war. LSE Crisis States Development Research Centre, Discussion paper 11, September.
Neate, R. (2014) Desmond Tutu tells G4S to stop supplying Israeli prisons. *The Guardian*, 4 June. http://www.theguardian.com/business/2014/jun/04/desmond-tutu-g4s-israeli-prisons-noam-chomsky (accessed 8 September 2015).
Newman, E. (2014) *Understanding Civil Wars: Continuity and Change in Intrastate Conflict*. London: Routledge.
Nordstrom, C. (2004) *Shadows of War: Violence, Power, and International Profiteering in the 21st Century*. University of California Press.
Nordstrom, C. (2007) *Global Outlaws: Crime, Money and Power in the Contemporary World* University of California Press.
Nordstrom, C. (2008) Casting Long Shadows: War, Peace and Extralegal Economies. In Darby, J. and Mac Ginty, R. (eds) *Contemporary Peacemaking: Conflict, Peace Processes and Post-war Reconstruction*. Basingstoke: Palgrave, pp. 289–99.

NRGI (2021) How Mozambique could strengthen its proposed sovereign wealth fund. 17 February 2021. https://resourcegovernance.org/articles/how-mozambique-could-strengthen-its-proposed-sovereign-wealth-fund (accessed 25 October 2023).

Nussbaum, M. (2006) *Frontiers of Justice*. Cambridge, MA: Belknap Press.

Obi, C. and Rustad, S.A. (eds) (2011) *Oil and Insurgency in the Niger Delta*. London: Zed.

Odhiambo, M.O. (2003) *The Karamoja Conflict: Origin, Impact and Solutions*. London.

Ofcom (2023) *Facts and figures*. Ofcom website. https://www.ofcom.org.uk/media-use-and-attitudes/media-habits-adults/adults-media-use-and-attitudes/ (accessed 15 August 2023).

OHCHR (2020) UN Human Rights: Office of the High Commissioner, 17 July 2020 'Israel's collective punishment of Palestinians illegal and an affront to justice: UN expert'. https://www.ohchr.org/en/press-releases/2020/07/israels-collective-punishment-palestinians-illegal-and-affront-justice-un (accessed 8 August 2020).

OHCR (2022) OHCHR assessment of human rights concerns in the Xinjiang Uyghur autonomous region, People's Republic of China. https://www.ohchr.org/sites/default/files/documents/countries/2022-08-31/22-08-31-final-assesment.pdf (accessed 31 August 2023).

Olorounbi, R. and Orjinmo, N. (2023) Nigeria seeks to tighten rules to curb raw mineral exports. *Bloomberg and Yahoo Finance*, 24 October 2023. https://www.bloomberg.com/news/articles/2023-10-24/nigeria-to-tighten-rules-to-curb-raw-mineral-exports?leadSource=uverify%20wall (accessed 24 October 2023).

Oxfam (2023) https://oxfamilibrary.openrepository.com/bitstream/handle/10546/121081/bk-karamoja-conflict-uganda-010103-en.pdf?sequence=5&isAllowed=y (accessed 7 February 2023).

Payne, A. (2005) *The Global Politics of Unequal Development*. London: Red Globe Press.

Phillips, T. (2023) My city is destroying itself: Juliaca under siege as death toll rises in uprising. *The Guardian*, 5 February 2023. https://www.theguardian.com/world/2023/feb/05/juliaca-under-siege-as-death-toll-rises-in-uprising-against-perus-government. (accessed 6 February 2023).

Pinker, S. (2012) *The Better Angels of our Nature: A History of Violence and Humanity*. London: Penguin.

Plaut, M. (ed.) (2021) *The Tigray War and Regional Implications*. Eritrea Focus/Oslo Analytica: https://eritreahub.org/the-tigray-war-and-regional-implications-full-report (accessed 16 February 2023).

Plaut, M. and Vaughan, S. (2023) *Understanding Ethiopia's Tigray War*. London; Hurst.

Posen, B. (1993) The security dilemma and ethnic conflicts. In Brown M. (ed.) *Ethnic Conflict and International Security*. Princeton, NJ: Princeton University Press, pp. 103–24.

Pugh, M., Cooper, C. and Turner, M. (eds) (2008) *Whose Peace? Critical Perspectives on the Political Economy of Peacebuilding*. London: Palgrave.

Pugh, M., Goodhand, J. and Cooper, N. (2004) *War Economies in a Regional Context: Challenges of Transformation*. Boulder, CO: Lynne Rienner

Rapier, R. (2022) The U.S. is still the world's top oil producer. *Forbes*, 22 July. https://www.forbes.com/sites/rrapier/2022/07/22/the-us-is-still-the-worlds-top-oil-producer/?sh=3dfe6e307b39 (accessed 7 February 2023).

Reno, W. (1997a) War, markets, and the reconfiguration of West Africa's weak states. *Comparative Politics* 29(4): 493–510.

Reno, W. (1997b) African weak states and commercial alliances. *African Affairs 96*: 165–85.

Rodney, W. (2018, 1981) *How Europe Underdeveloped Africa*. London: Verso.

Rodney, W. (2022) *Decolonial Marxism; Essays from the Pan-African Revolution*. London: Verso.

Rogers, P. (2000) *Losing Control: Global Security in the Twenty-First Century*. London: Pluto.

Said, E. (1978) *Orientalism*. London: Penguin.

Sen, A. (2006) *Identity and Violence*. New York: Norton.

Simmel, G. (1955) *Conflict: The Web of Group Affiliations*. New York: Free Press.

Slim, H. (2022) *Solferino 21: Warfare, Civilians and Humanitarians in the Twenty-First Century*. London: Hurst.

Snyder, T. (2011) *Bloodlands: Europe between Hitler and Stalin*. New York: Vintage.

Spear, J. (2012) Trade and Resources: A Security Perspective. In Spear J.D. Williams P.D. (eds) *Security and Development in Global Politics: A Critical Comparison*. Washington, DC: Georgetown University Press, pp. 229–46.

Tajfel, H. (1978) Social Categorization, Social Identity and Social Comparison. In Tajfel, H. (ed.) *Differentiation between Social Groups: Studies in the Social Psychology of Intergroup Relations*. London: Academic Press.

Tareke, G. (2009) *The Ethiopian Revolution: War in the Horn of Africa*. NJ: Yale University Press.

Terry, F. (2011) Myanmar: Playing Golf with the Generals. In Magone, C., Neuman, M. and Weissman, F. (eds) *Humanitarian Negotiations Revealed: The MSF Experience*. London: Hurst, pp. 109–26.

Tilly, C. (1985) War Making and State Making as Organized Crime. In Evans, P., Rueschemeyer, D. and Skocpol, T. (eds) *Bringing the State Back In*. Cambridge: Cambridge University Press, pp. 169–91.

UN Sustainable Development Goals Report, 2022. https://unstats.un.org/sdgs/report/2022/The-Sustainable-Development-Goals-Report-2022.pdf (accessed 27 October 2023).

United Nations (2023) UN Commission of Inquiry on Ukraine finds continued war crimes and human rights violations gravely impacting civilians. New York/Vienna, 20 October. https://ukraine.un.org/en/250021-un-commission-inquiry-ukraine-finds-continued-war-crimes-and-human-rights-violations-gravely#:~:text=NEW%20YORK%2FVIENNA%20(20%20October, the%20Russian%20Federation%2C%20according%20to (accessed 23 October 2023).

UNODC (2012) Corruption in Afghanistan: Recent patterns and trends. https://www.unodc.org/documents/lpo-brazil//Topics_corruption/Publicacoes/Corruption_in_Afghanistan_FINAL.pdf (accessed 24 October 2023).

UNODC [United Nations Office on Drugs and Crime, Vienna] (2012) New UNODC campaign highlights transnational organized crime as a US$870bn a year business. United Nations Office on Drugs and Crime, 16 July. http://www.unodc.org/unodc/en/frontpage/2012/July/new-unodc-campaign-highlights-transnational-organized-crime-as-an-us-870-billion-a-year-business.html (accessed 24 October 2023).

US Department of State, Bureau of Democracy, Human Rights and Labor (2021) *2020 Country Reports on Human Rights Practices: Eritrea*. Washington: US Department of State. https://www.state.gov/reports/2020-country-reports-on-human-rights-practices/eritrea/ (accessed 16 February 2023).

Van de Walle, N. (2001) *African Economies and the Politics of Permanent Crisis, 1979–1999*. Cambridge: Cambridge University Press.

Vittori, J. (Carnegie Endowment) (2021) Corruption and self-dealing in Afghanistan and other U.S.-backed security sectors. 9 September 2021. https://carnegieendowment.org/2021/09/09/corruption-and-self-dealing-in-afghanistan-and-other-u.s.-backed-security-sectors-pub-85303 (accessed 24 October 2023).

Walker, R. (2013) *Enduring Violence: Everyday Life and Conflict in Eastern Sri Lanka*. Manchester: Manchester University Press.

Wilmer, F. (2002) *The Social Construction of Man, the State, and War: Identity, Conflict, and Violence in the Former Yugoslavia*. London: Routledge.

World Bank (2021) Haiti. https://data.worldbank.org/country/HT (accessed 10 February 2023).

World Bank (2022) *The World Bank in Angola*. https://www.worldbank.org/en/country/angola/overview (accessed 7 February 2023).

World Green Economy Summit [WGES] (2023) Bridging the gap to the future: Advancing the global green economy, 28–29 November 2023. https://wges.ae/?utm_campaign=WGES_2023&utm_medium=CPC&utm_source=SEARCH&utm_content=Text&utm_term=EN&gclid=EAIaIQobChMI_qzW9KSRggMVhdrtCh1NzQnkEAAYBCAAEgKh2PD_BwE (accessed 25 October 2023).

Yergin, D. (1991) *The Prize: The Epic Quest for Oil, Money and Power*. New York: Simon and Schuster.

Young, C. (2003) Explaining the Conflict Potential of Ethnicity. In Darby, J. and Mac Ginty, R. (eds) *Contemporary Peacemaking: Conflict, Violence and Peace Processes*. Basingstoke: Palgrave Macmillan, pp. 9–18.

Zartman, I.W. (2005) Need, Greed and Creed in Intrastate War. In Arnson, C. and Zartman, I.W. (eds) *Rethinking the Economics of War: The Intersection of Need, Greed and Creed*. Baltimore, MD: Johns Hopkins University Press, pp. 256–84.

2 Institutions
Hardware and software

Introduction: US hegemony?

The twentieth century has often been termed 'the American century' and, until recently, the twenty-first century had seemingly continued the trend of US dominance in formal global institutions like the United Nations (UN) and ad-hoc institutions like the 'G7' economic grouping of developed countries. States like India, and particularly China, are growing in influence and economic might, but they have yet to eclipse the United States, and neither has the European Union (EU). One reason is that the only really conceivable economic or political blocs to emerge in recent times as competitors to the West have far too many internal disagreements. Jim O'Neil, the former Goldman Sachs economist and the inventor of the term 'BRICS' in 2001, Brazil, Russia, India, China and South Africa is dismissive of these countries ever being able to form a common currency or a central bank. He called it 'ridiculous … embarrassing almost' and added that they had 'never achieved anything since they first started meeting'. He did agree that 'the global financial system [is] not beneficial for emerging countries' because the US Federal Reserve works 'in the interests of the US' (O'Neil 2023). But many observers disagree. The narrative of global power and influence is visibly shifting from a unilateral to a multilateral one. Meetings where China, Russia and India, all BRIC nations, sit down to discuss world affairs are as followed in the international press as much as are G7 or G20 developed nations (Canada, France, Germany, Italy, Japan, the United States, with the EU in attendance). Russia was expelled in 2014 over its invasion of Crimea, a stance strengthened at the G7 Summit in Hiroshima, Japan, in 2023 (Al Jazeera 2023).

What we therefore have might be termed a global system that is failing but with another one struggling to replace it for both economic and political reasons. This is because the United States has withdrawn from its period of outright assertion of power; it withdrew from Afghanistan in the Summer of 2021, has largely withdrawn from its former role in Iraq, and has played an increasingly 'hands-off' role globally since at least the end of the Obama presidency in 2016. It is

DOI: 10.4324/9781003353485-3

worth remembering that the very term the 'American century' was only coined during the Second World War (by Henry Luce in *Life* magazine in 1941) and that before that, the United States was by no means seen in that light. As has often been said, '*tout empire périra*' (Duroselle 1981), all empires pass, and in the circumstances in which this edition was written, that warning from history has never been more accurate. So while the United States can still attempt to impose global sanctions, often through institutions like the UN, what is called in international relations 'dissatisfied powers' or those who oppose what John Foster Dulles suggested in 1939 was the 'status quo', are increasingly able to assert their influence. As Dulles pointed out, peace became identified with the 'status quo' and 'in such a world, both war and peace are alternating phases of an inevitable cycle' (Williams 2007a: 86). So we might suggest that while such states are shy of directly challenging the United States and have strong vested interests in stability under US stewardship, the challenge might not be long in coming, if only through surrogate states as in Ukraine.

As well as being dominated by large powers, the twentieth century, particularly from 1945 onwards, was a century of international institutions, even if they now represent a 'status quo' that is far more fragile than we might hope. Institutions such as the UN, NATO, African Union (AU) and EU have continued to play a key role in thwarting and facilitating conflict and development in the twenty-first century, but will they continue to be able to do so, and can we not already see a weakening of those abilities? A select group of international institutions comprises the primary international instruments dedicated to preventing and minimizing the impact of war, spearheading post-war reconstruction and promoting development. However, at the same time, this select group of international institutions is often blamed for underdevelopment and de-development, and by extension contributing to conflict. To complicate matters, the same institutions have been blamed for contributing both to the escalation and transformation of violent conflict. That process of blaming has often been associated with the feeling that institutional actions have ideological origins, which are themselves usually, if not always, assertions of power. Hence, what we call 'hardware' (the institutions themselves) and the 'software' (the ideas that underpin them) can be seen as intimately linked and needing interpretation accordingly.

And to further complicate matters, the essentially liberal institutions outlined earlier have become threatened in their very existence by the re-emergence of what can only be termed 'empires', thought after the 1960s to have been consigned to the dustbin of history. We must ask what role these (re-) emerging imperial powers, of which China and Russia are the most prominent, will have on a world that thought geopolitics was now regulated in some way from Washington and New York. There are even some who would categorize the emergence of the EU in these terms, such as its global and regional impact. The nation state is a relatively new structure; empires long pre-date them, and there is a widespread belief that 'neoliberalism' has imperial roots (especially in the Austro-Hungarian Empire). This is important, as the EU, that Empire's successor in part, has also been inspired by many liberal politicians and thinkers over centuries; European integration has nearly always made great steps forward in the aftermath of major wars (Slobodian 2018; Ghervas 2021). War itself, it might be concluded, is as much a powerful promoter of beneficial change as a horrific experience for all those involved in it. This is a paradox that we can further explore in Chapter 5 on 'reconstruction'.

This chapter sketches the principal international architecture that provides the context for contemporary conflict and development. Just as it is difficult to discuss computing hardware in isolation from software, it is difficult to gain a comprehensive understanding of international political structures without discussing their operating 'software' or their behaviour. The chapter begins by discussing 'software' and then moves on to hardware. Under software, we discuss the liberal peace or the overarching philosophy that shapes many international peace-support operations, post-war reconstruction programmes and development interventions. International institutions are given shape and purpose by the principles, behaviours and worldviews of their key members. It is argued here that the liberal peace has a peculiarly Western flavour that reinforces the dominance of existing elites and promotes highly specialized Western ideas, namely versions of liberalism, democracy and economics. As will be shown, the liberal peace has decidedly illiberal aspects. A crucial element of the liberal peace is hyper-globalization which supports multiple connections through multiple networks so that conflict and development on various parts of the planet are linked. The software section ends by considering alternatives to the liberal peace and finds that spaces for these are often marginal.

The chapter then moves on to discuss 'hardware' or the principal institutions that comprise the international political system. It begins with a brief discussion of the key constituent feature of the international system: the state. The important factor from our point of view is the political organization of the state, particularly in terms of its relationships with its citizens and market and its ability to resist or adapt to external pressure. Especially pertinent to the focus of this book is how the influences of 'established' and 'non-established' power of the state and non-state actors direct the purpose and practice of actors in the aid and development sectors. The chapter then reviews the roles and effectiveness of the primary collective security and development promotion organizations (the UN, EU, etc.) before examining the role of the international financial institutions (the World Bank, IMF, etc.). The chapter concludes by recommending that we adopt the lens of complex multilateralism when reviewing the role of international institutions in relation to conflict and development. In other words, we need to move away from a view in which we have compartmentalized entities such as states and international organizations that have formal and well-defined linkages between them. Instead, conflict and development operate in a much more complex environment in which multiple transnational and international actors cooperate and clash. This 'cast list' is much more extensive than the traditional list of states and international institutions and includes globalized multinational companies, non-governmental organizations (NGOs) and transnational social movements.

A final introductory point is to highlight the importance of resisting US bashing as the catch-all explanation for the state of international relations, development and conflict. Yes, the United States is the hegemon, or the world's most powerful state, and has global interests and almost global reach. This dominance has not always been so clear and might not be again in the future. After 1918, the institutions that emerged from the First World War reflected the 'prerogatives of a few European governments ... primarily those of the victorious allied powers, and above all Britain' (Martin 2022: 2). These were to evolve further after 1945 to reflect the changed balance of power in favour of the United States but with important input from the declining, previously hegemonic, powers, Britain and France, both attempting to build the UN in such a way as to protect imperial interests (Mazower 2009). That was, as Mazower points out, upset by the rise after 1945 of newly emerging countries like India and later Brazil. Now China is widely predicted to try and influence the policies and practices of the UN. The evolution of international organizations has thus always tended to follow changing patterns of power, and in the future, these may not be as 'liberal' as those envisaged by Britain and the United States. So rather than simply bashing the United States, or indeed any other hegemonic actor(s), it is important to ask why they act in the ways they do. Each state, or group of states, acts in usually rational ways to safeguard their own interests. Moreover, at present, the US top-dog position suits many others, too, and they are active supporters of the stability and creed of economic growth that it supports. This is often referred to as the 'Liberal International Order' (LIO). Challenges to it can be expected to be vocal and are clearly being demonstrated by the actions and pronouncements of China, India and Russia. They were thought until the 2010s to be broadly supportive of US aims, which is now in question. One of the most important questions for students of conflict and development is now whether China wishes to *replace* the US-backed LIO or merely modify it (Jones 2018; Mearsheimer 2021;

Allison 2017; Ogden 2022b). Might that even lead to war between the United States and China, as one hegemon refuses to give way to a new one? That did not happen in the case of the changeover between Britain and the United States, but they were both liberal states and had great cultural affinities. We might further ask whether liberalism is a necessary ingredient of the economic process we call 'globalization' or whether it is sufficient to have states within the global economic order that are economically liberal but not necessarily politically liberal (Hughes 1995). One thing that is for sure is that we are now experiencing geopolitical turmoil between the greatest powers on the planet since 1990. As the African proverb has it, when the elephants fight, the grass gets trampled, so such conflict, whether violent or not, has great effects on development policies and issues. That is already evident in rivalry over development in Africa and Asia with the emergence of the Chinese government's 'Belt and Road Initiative' (see Box 2.1) (Yu and Wallace 2021; Taylor 2006). Is the economic globalization encouraged by this initiative merely a form of 'autocratization' (Widmalm 2021)? The jury is out on all these questions.

Software

Without guidance, international institutions such as the UN, the Asian Development Bank or the AU are merely empty vessels. They do not have autonomous lives of their own. Instead, they reflect the positions of their most powerful members and lobbies. Crucially, international organizations are not neutral (despite professions otherwise). International organizations are the product of a political, economic and cultural reality that has been and is, heavily contested. At the heart of this struggle is the old-fashioned concept of 'power', a concept that once dominated political science and international relations and now tends to be overlooked. It is not over-simplistic to say that a basic struggle between power-holders and power-seekers defines the structure and operations of many international organizations. The power-holders, who are often rich Western states, international organizations, international financial institutions and corporate interests, wish to maintain their stranglehold over economic, political and cultural power over the power-seekers. The latter are often developing world and sometimes conflict-ridden states and their populations, minority communities, the dispossessed and those of a critical perspective.

The 'software' that directs international institutions to act in prescribed ways is often specifically designed to reinforce the position of power-holders and thwart attempts by power-seekers to achieve a more egalitarian share of resources. The power-holders, and particularly the hegemon in the form of the United States, have been remarkably successful in perpetuating their power. Tim Jacoby (2007: 523) notes how the hegemon is skilled at convincing states that its perpetual dominance is in their own interests, maintains enough military power to cow opposition, and guarantees its material superiority through a heavily biased distribution of resources and capital. This section discusses 'software' or the global computer program that helps run the dominant international political system and many of the international organizations within it.

The liberal peace

Different scholars use different lenses with which to interpret the world. Feminists, for example, might argue that the main software package that drives the international political system is the patriarchy or the male dominance that is infused into many aspects of life (Tickner 2001). Marxists might be tempted to interpret their known universe through an analysis of the means of production and patterns of ownership and consumption (Maclean 1988). This study finds that the liberal peace lens is particularly useful in explaining many of the meta-influences at work in situations of conflict and development. The liberal peace is a highly specialized form of peace and development intervention promoted by leading Western states, leading international organizations and international financial institutions in their attempts to shape the international political system and its constituent parts (Paris 2004). It uses the language of liberalism (democracy, freedom, free markets, human rights), hence the phrase 'liberal peace'. Whether it is actually liberal in its execution is a matter of great debate (Campbell et al. 2011; Roberts 2011; Tadjbakhsh 2011; Richmond and Mac Ginty 2014; Tom, 2017). The liberal peace is the 'ideology upon which life, culture, society, prosperity and politics are assumed to rest' (Mac Ginty and Richmond 2007: 493). It is capable of constructing an attractive rationale for its own promotion. Thus it speaks of 'responsibility', 'development', 'common interests' and above all, intervention (Williams 2007b: 543).

Sometimes called 'liberal interventionism' or 'liberal internationalism', the liberal peace is most visible in societies undergoing Western-backed peace-support interventions in the aftermath of civil war. But many of the tools of the liberal peace, particularly in disciplining societies, governments and economies, are also at work in developing states that have not experienced recent war. In non-post-war environments, these interventions are often covered by the terms 'good governance' and 'reform', and we find the same commitment to the market as a prerequisite for debt relief and poverty reduction strategy funding (Craig and Porter 2003; Abrahamsen 2004). Indeed, the 2013–15 disciplining of an indebted Greece by the European Commission, the European Central Bank and the International Monetary Fund is instructive of the power wielded by an elite group of states and institutions. Greece, a democratic sovereign country and a member of NATO and the EU, was bullied into accepting loan extension deals that were unpopular and clearly disadvantageous to the Greek people (through cuts to health care, pension and wages). Creditors used the language of responsibility and living within one's means, but it was clear that liberal economic internationalism was backed up with coercion and threats. Greek Finance Minister Yanis Varoufakis gave a startling insider account of the talks trying to avert Greece's exit from the Eurozone: 'You had might as well sung the Swedish national anthem, you'd have got the same reply.... You either sign on the dotted line or you are out' (cited in Lambert 2015).

An advantage of the liberal peace tool is that it allows us to make comparisons, sometimes across contexts that may not have obvious similarities or connections. It also allows us to make sense of the strategy employed by the leading political and economic actors in the international system. The liberal peace can be seen as a normatively neo-liberal system of compliance that is variously recommended, induced and enforced by leading states, leading international organizations and international financial institutions. Developing world states and states emerging from conflict often have little choice but to accept the liberal peace. If they do not conform, for example, Cuba, then they risk being locked out of access to funds and international platforms.

Many aspects of the liberal peace are deeply illiberal. It promotes a highly specialized form of liberalism that is often prescriptive and reflective of Western norms (Mac Ginty 2006: 33–57). Rajiv Chandrasekaran's (2007: 7) exposé of life inside Baghdad's Green Zone in the aftermath of the 2003 invasion provides a stark illustration of the crass ethnocentrism at the heart of the liberal peace as manifested in Iraq's Coalition Provisional Authority (CPA). He tells of how US administrators lived in a hermetically sealed compound complete with air-freighted fast food, US sports television channels and other home comforts. The security situation meant that many administrators rarely left the compound, while only a very few Iraqis could gain access to their new rulers. Many CPA staffers were woefully inexperienced college graduates whose sole work experience had been as an intern for a Republican member of Congress. This was often enough to justify their appointment to tasks such as helping to write the new Iraqi constitution or organizing the privatization of public services. Most US CPA personnel got their first passport to travel to Iraq, and six of the young 'gofers' were 'assigned to manage Iraq's $13 billion budget, even though they had no previous financial management experience' (Chandrasekaran 2007: 104–5).

The liberal peace promotes the individual as the primary unit of society. While such a viewpoint is unproblematic in Western societies, it clashes with many developing world and non-Western contexts in which the family or clan group may also have significant importance. By empowering individuals (for example, as consumers or as voters with free choice), the liberal peace introduces a cultural clash (sometimes characterized as traditionalism versus modernism) in many societies.

If one searches the liberal peace or the operating philosophy of leading states for 'red lines' or non-negotiable elements, then support for the United States seems to be only inviolable principles. Other possible red lines wilt under scrutiny. Commitments to democracy, international law, human rights or ideas of common humanity ebb according to circumstances. They are upheld in certain cases but conveniently overlooked in others. Even commitments to the inviolability of state sovereignty are abrogated when actors feel strong enough to do so. For example, US drone strikes in Pakistan occur without the support of the Pakistan government. Even the commitment to the free market can be overridden. Despite the rhetoric of austerity and slimming down the state, many cases of Western intervention have involved massive outlays in cash. The US wars of the 2000s were fought on borrowed money, and compliant regimes – like in Afghanistan before 2021 and Iraq – have been very costly allies (Bacevich 2010: 246–8). So the free market is not a sacred cow. The leader of Lebanon's Hezbollah, Hassan Nasrallah (2008), made an extremely perceptive observation about the red line being compliance with US political objectives:

> The US administration is not concerned with ideologies, beliefs, religions, races and nationalities. Frankly, I tell you, the US administration has no objections should an Islamic party or movement rule any Arab or Islamic country.... What counts is your political program.... The Americans and Israelis are not concerned with our prayers and fasting. Pray, fast and make pilgrimage to Mecca as much as you want, but leave control and great political interests to be schemed by America and Israel.
>
> (Nasrallah 2008)

The bottom line for Nasrallah was that the United States was engaged in simple power politics: protecting its interests and those of its allies. Talk of ideology and principles was erroneous.

If Nasrallah is correct, and the liberal peace has few real principles (despite the rhetoric), then it is worth asking what is the point of the international political and economic system operated by leading Western states. It also explains how radical movements, such as anti-globalization campaigners or even violent fundamentalists in the form of Al-Qaeda or Islamic State, are able to depict 'the West' as an ethical vacuum. Indeed, a number of the martyrdom videos of actual and failed suicide bombers in the United Kingdom mention the baselessness of Western popular culture. Perhaps there is merit to Huntington's (1993) much criticized 'clash of civilizations' thesis, though according to violent fundamentalists, the clash is between an empty mammon-obsessed West and the more sophisticated entities of non-Western identity, religion and customs.

It is very difficult to identify the unifying core behind many Western states. Other than the fact that large numbers of people live there, a rallying theme seems absent. The key point is that the combination of economic globalization and a hollow form of liberalism risks producing soulless societies with few centripetal forces. Those with particularist agendas, for example, promoting exclusive ethnic or religious views, have the advantage of telling a straightforward story that can capture the imagination and inspire (no matter how illusory it may actually be). In addition, illiberal leaders like Vladimir Putin can use simplistic rhetoric about US actions in Afghanistan and Iraq to claim that they are 'liberating' areas of Ukraine from Western forces, whereas in reality, they are doing the opposite, a stance that nevertheless appeals to many countries that are suspicious of previous Western actions. In other words, the liberal peace can end up encouraging distinctly illiberal actions. Maybe liberalism's nineteenth-century instinct for *non-intervention* has some sense to it, 'to let other nations alone' (Mill [1859] 2006).

As noted earlier, the manifestations of the liberal peace are most visible in situations of internationally supported peacebuilding and post-war reconstruction. Here the levers of the liberal peace include Western encouragement (or coercion) to reach a peace deal, direction in writing a new constitution, assistance in establishing and advising political parties, a donors' conference to fund and direct post-war reconstruction, help in holding electoral contests (and sometimes outright interference in the result), programmes of capacity building for the state and civil society, the introduction of 'good governance' targets and the attaching of economic reform conditions to any reconstruction assistance. In effect, the liberal peace has manifested itself in post-peace-accord societies as a conveyor belt of Western inputs. This has resulted in criticisms that it has been formulaic, using a template style of intervention that is unresponsive to the variations demanded by local circumstances (Mac Ginty 2006: 176; Gordon-Gibson 2021: 142). According to critics, it is as though IKEA made peace: flat-packed and with standardized parts.

The liberal peace is not only restricted to international interventions in societies emerging from civil war. International interventions can also be detected in a range of developing world societies. The main means of advance is through economic reform and leverage, and the 'good governance' agenda of bureaucratic reform. Such 'reform' can often sound innocuous, particularly in its promotion of concepts such as accountability, transparency, the regularization of bureaucracy and empowerment. Yet the cumulative effect of such interventions (often multiple interventions by multiple international organizations, bilateral relationships, international non-governmental organizations (INGOs) and NGOs) may have profound cultural and social impacts capable of influencing the relationships between states, their citizens and markets.

Fundamentally, liberal peace interventions are capable of altering the locus of power within a state. For example, citizens in a clientelist political system may have been used to transferring their allegiance to the ruling party at the municipal level in return for resources (Martz 1996). This was a rational transaction for both the citizen and the local political leader, and such relationships would have been deeply embedded into the socio-political culture. Politicians from the ruling party would be routinely invited to family weddings and in return for this public expression of loyalty, the family may expect a share of patronage, such as public-sector employment for a son (Hamieh 2007). According to the Western mindset, such behaviour may be regarded as 'corrupt' or 'nepotistic' (and indeed, it is often deeply inefficient and patriarchial).

'Good governance' interventions by Western states and international organizations in many developing and post-war contexts have attempted to reform political and economic cultures in order to promote transparency and accountability. In order to build the capacity of municipalities, the United Nations Development Programme (UNDP) may place one of its own personnel in the municipality offices to introduce new administrative procedures. This might involve the ending of discriminatory employment practices whereby loyal supporters received sinecures and protected employment according to who they knew rather than what they knew. The municipality may be unable to refuse the deployment of UNDP staff to its offices because of a directive from the central government or because the capacity-building scheme comes with financial inducements. Observers from a Western liberal perspective may applaud the introduction of meritocracy in the municipality. But they may be blind to the impact that such good governance 'reforms' have on altering power relationships within the target society. The citizen may no longer attach legitimacy to the municipality, a factor that is likely to have consequences for political participation, legitimacy and stability. So rather than 'statebuilding', capacity-building activities funded by the international community might actually help to undermine the state or make apparent state weakness. Chandler (2006: 478) refers to the process as 'the privileging of governance over government'. The key problem is that while people can vote for governments (and government opponents), their votes carry little weight in Washington, Paris, New York or Geneva, where important decisions that affect their lives are made.

While many Western liberal interventions are behind the scenes and take the form of bureaucratic meddling, others are much less subtle. Then US Vice President Joe Biden visited Beirut ahead of the 2009 Lebanese general election. He made clear that people should vote for the incumbent government or that US aid would cease (Daragahi 2009). Just to make sure that the message was clear, a US naval fleet was parked within sight of the Lebanese coast for much of the election campaign.

The impact of such capacity-building interventions is often quite subtle and sometimes seems incredibly minor, but their cumulative effect can transform the operating culture and orientation of state institutions (Abrahamsen 2000; Larmour 2005). Whereas a verbal promise between the local mayor and the head of household may have been sufficient in the traditional dispensation, good governance reforms may require the filling in of a standardized form. This apparently innocuous change goes to the heart of some core political and cultural relationships involving issues of trust and reciprocity upon which many societies operate.

It should not be assumed that everything connected with international intervention, the liberal peace and good governance is harmful. Often, for example, in the case of state collapse, it is only international actors who are empowered to organize national processes (such as elections) or have the capacity to provide the security necessary for the introduction of humanitarian or development assistance. Since the end of the Cold War, Western-backed interventions have saved and improved lives across war zones and development contexts. Ways of organizing and thinking about international interventions have been inflected with more humane and civilian-orientated sensibilities. The post–Cold War period has seen the development of the human security concept, and innovations in international human rights law (most notably the Responsibility to Protect (R2P) initiative). While these have not delivered on their potential, they have opened space in which the concept of civilian protection has been taken seriously in relation to some conflicts.

Defenders of the liberal peace point out that many underdevelopment and conflict situations would be worse off if there was no international intervention and that we should swallow objections to the imposition of Western values and look to the bigger picture of stability and the potential for economic growth (Quinn and Cox 2007: 518; Paris 2010). Moreover, we should not conceptualize the liberal peace as a dastardly plot solely perpetrated by the United States. Rather than being coerced into the liberal peace, many states, organizations and enterprises see it as a way to further their own ends. France, for example, protested loudly at the prospect of the 2003 US invasion of Iraq. It did not, however, sever its ties with the United States, impose sanctions or take any real steps to prevent the invasion. In short, France had much to gain from its generally good relations with the United States and its connections with the global economy. Other states, even apparent competitors of the United States like China, have an interest in stability. As a result, they are often complicit or silent on many foreign policy issues.

We should be under no illusions that international assistance is neutral. It reflects the worldview of those who fund and direct it. Although the liberal peace and the good governance agenda have standardized elements that are applied to different societies with minimal regard to local circumstances, it is important to note that Western interventions and reforms are applied with different levels of enthusiasm in different locations (Richmond 2005a: 217–18). Bosnia-Herzegovina received immense statebuilding and peacebuilding assistance (BBC News 2023a; Boulden 2001), but the Democratic Republic of Congo did not. International powers pursued Iran to reach a deal on the non-proliferation of nuclear weapons but have avoided similar activism over Israel and North Korea's nuclear weapons. There were US and UK boots on the ground in Afghanistan, Iraq, Sierra Leone and Bosnia, yet no such enthusiasm in the conflicts in Yemen, Syria and Ukraine. The selectivity in interference is often based on the political mood in the United States, the United Kingdom and other potential troop-contributing countries – not on needs on the ground in the conflict-affected country.

We must be careful not to represent the developing world and post-conflict states and communities as mute or powerless actors without agency. In fact, there are many cases of communities resisting, modifying and subverting the original intentions of Western policymakers (Richmond and Franks 2007; Franks and Richmond 2008). Whether in Bosnia, Sierra Leone or Timor-Leste, local political leaders, bureaucrats and communities have found ways to exploit the resources and intentions of international organizations and states involved in peacebuilding and development activities. Indeed, an often overlooked but absolutely crucial part of the story of international intervention concerns how local bureaucrats bend international edicts to suit their own interests and pace.

While this book is interested in contemporary conflict and development, the parameters of modern liberal peace are historical. We have had numerous 'defining moments' in which political leaders make 'never again' pronouncements and boldly set a 'new' course: Versailles, Yalta and the declaration of a new world order at the end of the Cold

War are just a few examples. Yet many of the apparently 'new' structures, institutions and modes of operation seem to resemble those of the previous era (Williams 2007a: 5–18). Alex Callinicos (2005: 596) finds that the key to explaining the 2003 Anglo-American invasion of Iraq lies not only in the contingencies of the Bush White House but also in long-maintained historical projects: 'The ideal of a global order in which free markets and democratic institutions promoted peace and prosperity was eloquently articulated by Woodrow Wilson during the First World War'. These continuities guide us towards examining international structures that can survive the tectonic shifts in the international economy and polity.

Alternatives to the liberal peace?

As explained in the previous section, the liberal peace is a complex system, with many actors complicit in its maintenance. As a result, the spaces for alternatives to the liberal peace often occur on the margins of societies – in spaces beyond the gaze of leading states and international financial institutions. Hezbollah's creation of a parallel state in Lebanon, along with its own social welfare system, can be regarded as an alternative to a Western-orientated system, but it is by no means a rival to the liberal peace. Similarly, attempts by some Latin American states (most notably Venezuela) to step out from beneath the US shadow may attract headlines, but they cannot be regarded as sustainable alternatives to the globalized economy.

This chapter has presented the rationale behind the liberal peace and some of the core concepts regarding its role in conflict resolution and development. These coalesce around notions of a shared responsibility and common interests. The problem is that the causes behind recent conflicts and the responses for their resolution point towards approaches and interests that are distant from the prescriptions of development and good governance promoted by the liberal peace.

What form might an alternative approach look like? One of the constraints already noted is the prescriptive nature of a system that reflects the norms and values of the Global North and foregrounds the individual as the primary unit in society. This is an assumption of a universal aspiration that is in tension with different customs and traditions. There needs to be a new approach that recognizes the differences in culture and the hierarchies in society and religion but which is able to legitimate and promote a shared sense of meaning. It needs to increase participation.

Key to participation in this reformulated space is a sense of solidarity and community: a social and political place of cooperation and contestation framed by a sense of common meaning. In her formulations of the Capability Approach, the philosopher Martha Nussbaum recognizes hierarchies and inequalities in common life but emphasizes a universal entitlement to dignity and respect as one which remains fixed and stable (Nussbaum 2006). The right to a life worth living is non-negotiable. This presents a humbler approach that is freed from the exceptionalism of the liberal peace discourse around humanitarian aid and development, where assumptions of a blueprint on principles and values are framed by its formulations of best practice and good governance.

An alternative will continue to promote human security, protection, shared responsibility and international humanitarian law (IHL), which are the core tenets of the liberal peace. But it must enable new approaches to the relationship with power that are less prescriptive, and which escape from the vagaries of the prevailing political mood. This new way of thinking disrupts comfortable patterns of engagement and engages the margins of society. It is a variation of the liberal peace approach that allows participation across cultures and religions – one that is more relevant and representative and which recognizes the messiness of humanity. This enables a diversity of practice but secures the stability of purpose centred around a universal common interest – that of shared dignity and respect.

A proto-alternative, as we said at the beginning of this chapter, remains the BRICS, led by China. They are all, and especially collectively, an economic powerhouse. They demand an end to a US-dominated global system for one based more on national self-determination and sovereignty, unexceptional if read with the 1945 UN Charter, but certainly not if read against the liberal peace as it has developed since the 1980s. China has invested heavily in infrastructure in developing world states as part of its 'Belt and Road Initiative'(Box 2.1) in return for access to natural resources and assertion of its own brand of 'soft power' in the geopolitical landscape. But it is often accused of loading up developing countries with huge debts, asking for military bases in. return for its 'help'. It is also mute on many issues, including human rights. It is, however, showing signs of assertiveness in global arenas beyond its own 'backyard', the South China Sea, positioning itself as an intermediary in peace talks between Russia and Ukraine, between Saudi Arabia and Iran and even between Saudi Arabia and Israel (an initiative now in some doubt as we write given the October 7 events in Gaza and Israel: Aboudouh/Chatham House 2023). Some commentators are fearful that even a misunderstanding, say between Iran and the United States, could lead to war. President Xi Jinping announced a more forceful foreign policy stance 'to stand guard over world order' alongside Vladimir Putin in Moscow in March 2023 and further defined this as a 'dare to fight' doctrine (Leahy et al. 2023).

Box 2.1 The One Belt, One Road Initiative

The BRI is an ambitious plan to develop two new trade routes connecting China with the rest of the world. Initially referred to as two separate projects, it was combined first as the 'One Belt, One Road' initiative, then as the 'Belt and Road Initiative'. The idea was first proposed by Chinese president Xi Jinping during a visit to Kazakhstan in 2013 when plans for a revitalized series of ancient overland trading routes connecting Europe and Asia were to be built largely with Chinese expertise. Central Asia is seen as the most vital region for the 'Belt' element. In 2014, Xi Jinping outlined plans to establish new sea trade infrastructure along the old Marco Polo route – a maritime silk road connecting China, Southeast Asia, Africa, and Europe: the 'Road'. The initiative is about far more than infrastructure. It is an effort to develop an expanded, interdependent market for China, grow China's economic and political power, and create the right conditions for China to build a high-technology economy. In this respect, it has similarities with the Marshall Plan that followed the Second World War (Yu Jie and Wallace, 2021). Three main motivations have been identified with the BRI: (1) China's rivalry with the United States. The vast majority of Chinese international trade passes by sea through the Malacca Strait off the coast of Singapore, which is a major US ally. The initiative is integral to China's efforts to create its own more secure trade routes; (2) to make participating nations interdependent with the Chinese economy, and thereby build economic and political influence for China; and (3) to stimulate economies of China's central provinces, which historically lag behind richer coastal areas.

Source: Yu Jie and Jon Wallace (2021). 'What Is China's Belt and Road Initiative (BRI)?' Chatham House. Available at https://www.chathamhouse.org/2021/09/what-chinas-belt-and-road-initiative-bri

In order to understand 'structural liberalism', we now turn to the 'hardware' or international institutions that define contemporary conflict and development (Deudney and Ikenberry 1999: 180).

Hardware

States

Rather than provide a 'political science 101' conceptualization of the ideal state, this section will content itself with making three points about the state or the principal unit in the international political system. In lieu of an elementary exposition of the concept of the state, readers may wish to consult classics on the subject, such as Waltz's (2001) *Man, the State and War*; Migdal's (1988) *Strong Societies,*

Weak States; or Tilly's (1990) *Coercion, Capital and European States*. The first point is an obvious one: there are enormous variations in the resources, power, legitimacy and capabilities of states and, thus in their conflict and development status. Some states are 'lucky' in that they are in a stable region with dependable neighbours, though (as we saw in Chapter 1) states 'fortunate' enough to have natural resources are often cursed with conflict over the distribution of these resources. The massive variance in state capability leads to uneven international institutions, with some states (and corporate bodies and lobby groups) wielding disproportionate influence. In stark terms, while some states and lobby groups can afford to maintain extensive bureaucracies at international organizations, other states cannot (Robbins 2003: 105). Indeed, Japan is reported to pay the International Whaling Commission membership fees for some states in return for support for the right to hunt whales (*Sunday Times* 2010).

State capacity is a crucial factor in facilitating, developing and constraining conflict. Adrian Leftwich's (1996: 284) concept of the 'developmental state' is useful in illustrating how states need to 'concentrate sufficient power, authority, autonomy, competence and capacity at the centre to shape, pursue and encourage the achievement of explicit developmental objectives'. As Leftwich (1996) points out, states have pursued this capacity to develop in very different ways, with Taiwan and South Korea achieving very high rates of growth under authoritarian political systems. Other states simply did not have the capacity or inclination to engage in national mobilization for economic growth.

This leads us to the second point: the type of state matters, particularly in terms of its internal political organization. Some forms of state organization may be more conducive than others to promoting development and constraining conflict. Democracy and development may not always be compatible. The state is often the central clearing house for societal conflict, providing rules and mechanisms to allow individuals and groups to coexist. But it may also be inept or incapable or may have little interest in maintaining its conflict regulating responsibilities. William Reno (1997) paints a dystopian picture of West African states in the late 1990s, in which ruling elites regarded citizens as an expensive encumbrance and thus made no pretence at offering public services. Instead, they fixed their energies on the extraction of precious minerals, usually in concert with external commercial interests. In some cases, especially societies with identity fissures, the state may privilege some groups and discriminate against others. An enormous amount of literature posits a link between democracy and peaceful relations between states, but the relationship is by no means simple, and this 'democratic peace' literature tends to overlook the locus of most violent conflict within the state itself (Doyle 1980; Henderson 2002: 2).

A key point to bear in mind is that in many societies, the state is the only political and economic prize worth having. To be excluded from the state means to be cut off from virtually all public resources. Indeed, one can see this quite literally in parts of Africa where the road and electricity pylons stop abruptly because a district did not support the ruling elite. Ian Taylor (2005) observes that

> control of the state serves the twin purposes of lubricating patronage networks *and* satisfies the selfish desire of elites to enrich themselves, in many cases in quite spectacular fashion. That is what lies at the heart of the profound reluctance by African presidents to hand over power voluntarily and why many African regimes end messily, often in coups.
>
> (Taylor 2005: 4)

He goes on to note how neo-patrimonialism or clientelistic 'big man' politics is alive and well across the region. This is despite enormous democratization and 'good governance' interventions over many years. In other words, a political culture may be surprisingly resilient regardless of externally promoted institutional engineering. Democratization is a difficult policy and aspiration to pin down precisely because it always has to exist in a particular cultural context. Pakistan (see Box 2.2) considers itself democratic, as does India because it (rightly) asserts that it has a long, if not uninterrupted, history of elections

Box 2.2 Pakistan: strong or weak state?

Ostensibly, Pakistan is a strong state. It covers a huge area (over twice the size of Germany), is rich in resources (including natural gas) and is strategically located (it borders China, India, Iran and Afghanistan). It has an enormous army, is a nuclear power and its economy, though underdeveloped, has experienced strong levels of growth in recent years. Yet the state is chronically weak. It does not control all of its territory (with insurgent groups regularly defeating the national army in the Waziristan region). The bureaucracy is unable to fulfil its stated aims; thus, a large proportion of the economy is corrupt and unregulated and – unable to conduct a census – the state can only estimate the population. Democratic institutions are weak, with the result that military coups, states of emergency, mass riots, human rights abuses and political assassinations have been commonplace. Long-running boundary disputes with India and the Kashmir conflict have been a drain on state coffers and attentions. So is Pakistan a strong or weak state? The answer, despite its formidable nuclear arsenal, must be that it is weak. The state faces too many alternative sources of power to be able to assert itself.

resulting in changes of government and that the rule of law tends to be respected. India is the oldest democracy in Asia and has a lively civil society that contests government decisions. But for many Western observers, the caste system (in India), abuse of women and religious intolerance, often aided and abetted by the state under the guise of nationalism, disqualifies either India or Pakistan from a full status as 'democratic' (Chowdhury 2003). What unites them (and China, even Russia) is their drive for modernization and development and a desire to be taken seriously as global powers in a post-colonial era. They all have elements of authoritarian rule, usually very personalized (China, Russia and India share this trait) (Bunce et al. 2020).

This is reflected in the current coalition of agreement over the invasion of Ukraine, in spite of their obvious difficulties with each other. For what divides China and India are major geopolitical tensions along the borders they share in the Himalayas (Ogden 2022a). In Central Asia, China and Russia have long-standing differences over borders, with Russia having annexed a huge area of its Eastern provinces from China in the nineteenth century. They have been as much in conflict as in harmony with each other, and the current relationship is deeply asymmetrical in China's favour (Snow 2023). If these states mostly claim to be (somewhat) democratic and (some) have liberal aspirations with national characteristics, they seemingly do not fulfil the notion of the liberal peace as a 'natural' restraint on warlike behaviour.

The third point to make in relation to the state is to note the international community's near addiction to statebuilding as the standard response to civil war or political transition. It seems as though the only tool in the toolbox is state rebuilding (as manifested through the already mentioned capacity-building programmes, good governance, public-sector reform, etc.). In development contexts, Kenny (2003) notes,

> The state is either the solution, the only way to combat structural weaknesses that hold back growth, or it is the problem, tying down the invisible hand; or it is the facilitator, vital for the efficient functioning of the free market.
>
> (Kenny 2003: 413)

Either way, the state is a key ally or enemy in international efforts to promote development or reduce conflict.

Failed states

In some parts of the world, the state is patently a dysfunctional political and economic model. Terms like 'failed' or 'fragile' state are laden with baggage, but there are those who believe they can be 'fixed'

(Ghani and Lockhart 2008). We can often see the paradigm of foreign agencies doing the fixing, with a nod being made to local agency. Strong leadership is seen as vital, with President Paul Kagame of Rwanda being seen as a role model (*ibid.*: 191–2). Eritrea was seen as a role model for Africa throughout the 1990s, and Ethiopia was similarly praised. John Garang was South Sudan's only hope until he was killed in a helicopter accident only a few months after leading his country to independence in 2005. All three of these states have ended up even more 'un-fixed' after a huge investment by intergovernmental organizations (IGOs) and Western leaders. *Some* have even argued that the United States is a failed state, given its permanent budget deficit and massive prison population (Whitmeyer 2013). But the United States can wield immense power in a controlled way. Other states, like Yemen, Iraq, Libya, Syria, Nigeria or Ukraine, do not control all of their territory. In entire regions, like the Horn of Africa or Central and West Africa, the state has not taken root. In fact, for many people, the state is irrelevant in daily life (Kabamba 2010). People are left to get on with their own survival and prosperity while the state does little in terms of providing public goods like health care or security. The state may manifest itself occasionally in the shape of a police car visiting the village every few weeks, but otherwise, it might be absent, occasional, irrelevant or incompetent. Indeed, the state is often the body that most citizens fear the most, as evidenced by South Sudan since its formal independence from 2013 to the present, one fought along ethnic lines (Martell 2018).

How can resilience be built into small, vulnerable or 'weak' states?

It is relatively easy for people who live in developed and rich states to denounce other states in developing areas as beyond help. It is true that many states are 'vulnerable'; maybe all were when they were first established, like Kosovo in 1999 or the Irish State in 1922 (it became the Republic of Ireland in 1948). But most people prefer to live in a state that not only functions well, but also they identify as their own, the principle known as 'self-determination'. Many scholars have pointed to both the benefits of being 'born' as a state and the dangers. They likely emerge in a hostile climate, resented by their neighbours, or in a 'bad neighbourhood' (Kenny 2003), or feared by those who see their possible success as a stimulus to other secessions. Kosovo has been portrayed in this light, for example, as has South Sudan (Martell 2018; Ker-Lindsay 2011; Ker-Lindsay and Fabry 2023). They may be faced with climate change, unruly or violent neighbours and a myriad of other challenges. One example of how these dilemmas have been approached through local agency is illustrated in Box 2.3, the

Box 2.3 The Bridgetown Initiative

The world is faced with a series of interlocking crises, including climate change, pandemics, conflict and a rising cost of living. While this is a global phenomenon, the impacts are felt most acutely in the developing countries and small island states of the Global South, where natural catastrophes are made more acute by international loans affected by high-interest rates and growing government debts. At the COP 27 meeting held in Egypt in November 2022, the government of Barbados, led by its Prime Minister Mia Mottley, promoted an initiative to transform how lending is made to developing nations in a climate crisis. Named the *Bridgetown Initiative*, after the island's capital city, the project is compared to the Marshall Plan of 1948, when the United States provided more than $13 billion of foreign aid to help Western Europe recover after World War II. It proposes to reform the processes of international development finance, setting out three key steps: (1) changing some of the terms around how funding is loaned and repaid with an aim to stop developing nations spiralling into a debt crisis when their borrowing is forced up by successive disasters like floods, droughts and storms. This might include suspending or cancelling interest payments while a country is tackling a pandemic or a natural disaster. (2) International development banks guarantee funding to developing nations for climate resilience. (3) new private-sector mechanisms to fund climate mitigation and reconstruction after a climate disaster.

Sources: Government of Barbados https://pmo.gov.bb/wp-content/uploads/2022/10/The-2022-Bridgetown-Initiative.pdf; https://www.weforum.org/agenda/2023/01/barbados-bridgetown-initiative-climate-change/

'Bridgetown Initiative'. But they hope for aid from global development banks. We live in a world that is globalized beyond the dreams (or nightmares!) of our forbears, so if a person comes from a kind of state that is very common, indeed they need to take account of the immediate environment and their links to, often, former colonial overlords, the demands of foreign capital and so forth. These are the small island states which make up most of the membership of the 'Commonwealth', all of whom were members of the former British Empire, with one or two outliers like Madagascar.

The question is how to transform their evident vulnerability into some sort of resilience. This term has become very fashionable in recent years, and for good reasons (Chandler 2013; Duffield 2012; Hussain 2013). With global networks of crime and the problems of climate change, vulnerability can come at any time and in unexpected ways. So how can a Caribbean island make itself more resilient?

One way has been to make itself attractive to foreign tourists and investors, as in the issuing of 'golden passports' that allow citizens of non-EU countries to have better access to the EU, as well as the chance of a house in the sun.

But not all vulnerable states have the leisure to try and rely on such ideas. Despite abundant evidence that some states (like South Sudan) just don't work, the international community remains addicted to the state as the primary unit of political organization. Thus immense international effort is invested in statebuilding and state reform. The phrase attributed to the car manufacturer Henry Ford, that a customer can have a car painted any colour – as long as it is black, can be remodelled to read that any political unit is tolerable – as long as it is a state. The fear of statelessness (regarded by many as a deviant form of political organization) is very grave indeed. According to one observer, 'Failed or failing states are often Petri dishes for transnational criminal activity such as money laundering, arms smuggling, drug trafficking, people trafficking, and terrorism' (Wainwright 2003: 486).

While huge international resources are poured into shoring up and reconstructing states (Mayall and Soares de Oliveira 2011), it is worth noting that the international system is selectively tolerant of different types of states. This tolerance often depends on three factors: the ability of a state to resist or subvert Western influence (for example, a powerful China), the strategic importance or unimportance of the state (for example, oil-rich Iraq or strategically marginal Haiti), and the points of economic and geopolitical confluence between the state and leading states in the international political system (for example, Israel's ability to organize a strong lobby in the United States). Good examples of 'bad' but tolerated states come in the form of the oil-rich Gulf monarchies that can, at best, be described as authoritarian and controlling and, at worst, as despotic and tyrannical. Although Saudi Arabia announced important reforms in 2020 and 2021, Human Rights Watch reports that ongoing repression and contempt for basic rights, with the near-total repression of independent civil society, are major barriers to progress (Human Rights Watch 2023). Accusations of serious mistreatment (including shootings) of refugees crossing the border from Yemen have appeared in the international media, as well as reports from human rights monitors (BBC News 2023b; Human Rights Watch 2023).

Since the kidnapping and murder of Jamal Khashoggi, a Turkish journalist (murdered in the Saudi Consulate in Istanbul) in 2018, relations between Riyad and the West have not been good. Accusations that the Crown Prince of Saudi Arabia was allegedly involved in some way have made bilateral relations with the United States difficult ever since.

It is clear that the geo-strategic concerns of leading Western states outweigh the desire to upset regional power-holders (Ehteshami and

Wright 2007) or did so until the Khashoggi murder. The worldviews of the ruling elites in the Gulf region were congruent with Western political and economic interests, and so their unsavoury political backyard was spared scrutiny and intervention. Gulf states invest their oil wealth in Western economic markets, buy prodigious quantities of Western arms and were on the 'right side' in the War on Terror and the Sunni versus Shiite struggle. The recent (2022–3) closeness between Saudi Arabia and China, on the one hand, which helped mediate a ceasefire between itself and Yemeni Houthi rebels, and Iran with Russia over oil and the Ukraine war has clearly poisoned relations with the West. The results of this have not yet become clear.

International organizations

That international organizations exist at all is remarkable. The jealousies of national interests, as well as the fallout from international crises, have meant that the international institutions of previous eras often resembled short-lived tactical alliances rather than permanent forums for the regulation of international society. Even more remarkable has been the existence of some international institutions for so many decades (the UN, World Bank and International Monetary Fund for over seven decades, and the EU – albeit in a greatly modified form – for over six decades).

Some neo-realist commentators are scathing about them. John Mearsheimer and Susan Strange have both asserted that such 'institutions are basically a reflection of the distribution of power in the world ... they are based on the self-interested calculations of the great powers, and they have no independent effect on state behavior'. Strange made the maybe more telling argument that neither states nor IGOs can control global financial flows and that multinational corporations wield far more power than economic IGOs (Mearsheimer 1995, 2021; Strange 1991, 2009). Again, we come to the fundamental question of where power and authority in the international system really lie. In a situation unparalleled since 1990 and 'dissatisfied' states claiming their own version of a 'new world order' to replace that of the USA, we have to accept that the answer to these questions is unclear.

The United Nations

As the planet's premier collective security organization, the UN is the target of immense criticism. Common chants are that it's too inefficient, bureaucratic, slow, corrupt, underfunded and unwieldy (MacFarlane and Khong 2006; Urquhart and Weiss 2012). Yet, the organization is merely the sum of its parts (member states) and is a largely accurate reflection of states' relationships with one another and their attitudes towards pooling sovereignty for the collective good. That it tends to be reactive rather than proactive, is highly selective in its interventions and is inconsistent in its attitude towards state sovereignty is not the fault of international mandarins working in a vacuum in New York and Geneva. Instead, it is the fault of the national governments who have created and maintained the system.

While it is easy to criticize the UN, it is also easy to overlook the immense (and often unsung) development, conflict-amelioration and humanitarian work it undertakes. Often, this work is conducted through specialist agencies, has long-term impacts and makes a qualitative difference to the lives of millions in underdeveloped and war-torn states. Thus UN agencies are responsible for the physical security, legal protection, nourishment, shelter and repatriation of substantial numbers of individuals and communities, a fact that is routinely overlooked by those who use broad-brush criticisms against the organization.

The UN has developed as a result of disjointed incrementalism. Rather than planned strategic growth, UN capabilities have developed reactively in the face of crises, with the end of the Cold War simultaneously lifting an immense constraint on its ability to operate and presenting it with a vastly increased workload. It responded in the 1990s by having ever more ambitious operations in an increased number of theatres (Bellamy and Williams 2010: 93–120). The growth in peacekeeping operations occurred despite the UN Charter not mentioning the term 'peacekeeping'. Nor does the UN have its own army; it relies on member states to contribute troops. In the post–Cold War period, there has been a trend towards 'outsourcing' or 'subcontracting', with the employment of regional organizations, INGOs and NGOs to carry out peace-support and development activities on behalf of the UN (Richmond and Carey 2005). A privatization of UN functions has also been visible, with – among other trends – a growth in the use of private security companies to provide security for UN personnel (Pingeot 2012: 7).

On the positive side of the ledger, the UN's post-1990 operations showed that the organization was adopting a more sophisticated understanding of conflict, especially with regard to the complexities of the relationship between conflict and development. At the end of December 2022, the organization had 76,000 personnel serving in 12 peacekeeping missions – most of which are in Africa (UN Peacekeeping; Williams 2011: 185–92). On the negative side of the ledger, there was discomfort at the more robust aspects of some peacekeeping interventions in which peace 'keeping' became peace 'making' and peace 'enforcing' (Boulden 2001). Such qualms echoed those over the notion of 'humanitarian war' and concerns that humanitarian interventions could prolong and intensify war (Janzekovic 2006; Belloni 2007; Bellamy and Williams 2010: 229).

There were also accusations of 'mission creep', whereby originally modest UN interventions became victim to ever-broadening mandates. This was especially the case in missions that required extensive nation and statebuilding, activities that are necessarily long term and expensive (Pugh 1997). The term 'mission creep' came to prominence in relation to an originally modest UN effort in Somalia in the early 1990s. What was originally an attempt to secure supply routes for humanitarian aid convoys led to direct conflict between UN contingents and Somali warlords.

The UN has undergone significant reform and embarked on major initiatives since the 1990s. Its eight Millennium Development Goals helped inject focus into development interventions undertaken by it and its member states and have been replaced by 17 Sustainable Development Goals, with 169 targets to be met by 2030. The 2001 Brahimi Report, although not fully implemented, recognized many of the UN's organizational shortcomings. The 2003–6 High-Level Panel on Threats, Challenge and Change considered the new range of transnational threats faced by states and the utility of UN structures and practices to deal with them (Hannay 2005; Stedman 2007). The 2005 establishment of a Peacebuilding Commission signalled a willingness among member states to take on board the lessons from its previous and ongoing peace-support operations. A 2015 review sought to 'propose ways to strengthen the performance and impact of the Peacebuilding Architecture, with a view to realizing its full potential' (United Nations 2014).

Plate 5 UN HQ New York.

Source: Shutterstock/Osugi.

Prior to his reappointment in January 2022, the current United Nations Secretary-General presented a vision statement entitled 'Restoring Trust and Inspiring Hope', focusing on how to make the UN more relevant and representative in an interconnected and interdependent world. The aim is to advance the 2030 Agenda for sustainable development and achievement of the SDGs. Amongst directions towards alleviation of the physical and political threats posed by conflict, climate change and natural disaster, the Secretary-General proposed a set of internal UN reforms promoting innovation and adaptation in a system that needed to free itself from unnecessary bureaucratic processes and fostered a collaborative work culture. These reforms target improvements in five core areas of technical and strategic performance (see Box 2.4). Key elements of the UN Reform include a new generation of UN Country Teams (UNCTs) and a UN Development Nations Assistance Framework (UNDAF), supported by a reinvigorated Resident Coordinators system with enhanced capacity, leadership, accountability and impartiality. Work will be rationalized through a common business service and back-offices function with a system-wide approach to partnerships (for details of reforms, see https://reform.un.org/content/development-reform).

Box 2.4 UN reform

On 31 May 2018, the UN General Assembly unanimously adopted the resolution 'Repositioning of the United Nations development system in the context of the quadrennial comprehensive policy review of operational activities for the Development of the United Nations system'. The resolution contains measures to enhance the capacity of the United Nations Development System (UNDS) to better support countries in implementing the 2030 Agenda for Sustainable Development (2030 Agenda) and provides guidance to the Secretary-General and the UNDS on their implementation. The UNSG's Vision Statement for the five years 2022–6 specifies transformation through a quintet of change: (1) Data, analysis and communications – building on the UN data strategy to provide the best possible global expertise in these areas of technology and strategy; (2) innovation and digital transformation; (3) strategic foresight – global collaboration to provide anticipatory action and preparedness; (4) performance and results orientation – delivery of services measured by impact; and (5) work culture – simplifying and reducing unnecessary bureaucratic processes and fostering a work culture of collaboration.

Source: https://www.europarl.europa.eu/RegData/etudes/BRIE/2019/635517/EPRS_BRI(2019)635517_EN.pdf; https://reform.un.org/content/development-reform

The principal problem facing the UN is the perennial struggle between state sovereignty and the collective good. On many occasions, the two are simply incompatible. Neo-conservative elements in the United States are at least honest when they voice their suspicions about multilateralism or the potential of any national or multilateral power source to rival its hegemonic position (Callinicos 2005: 598). As the 2003 Anglo-American invasion of Iraq and its bloody aftermath showed, leading states are willing to override the UN when it suits but not above appealing for UN assistance when they find themselves in a mess. Important decisions on restructuring the Security Council to make it more representative have been dodged, and it is likely that the pattern of disjointed incrementalism will continue. The permanent five members of the UN Security Council – United States, China, Russia, United Kingdom and France – reflect the world in 1945. There are few credible reasons for excluding powerful populous states like Brazil or India from the Security Council other than the concentration of power in the hands of those that already have it.

Ultimately, the UN relies on its moral authority – a sense that a collective security organization represents world opinion. This moral authority, to the extent that it ever existed, is under severe pressure. Failures to prevent or resolve the protracted conflicts which have marked the past decade and more, together with accusations of abuse and corruption (BBC 2022), have seriously dented respect for the Security Council and UN Offices in the Secretariat. States are wary of the power they are willing to cede to the UN. They are content for relatively anonymous figures to be appointed Secretary-General. A good indication of the importance placed on the UN in the policing of world events is the number of troops and police states are willing to provide for UN peacekeeping missions. Once seen as a reluctant supporter of UN peacekeeping operations (PKOs), China has voted in favour of all new PKOs since 2000 and is the second largest funder of PKOs after the USA. This has been attributed to shifts in China's foreign and security policy, as well as to reforms of peacekeeping missions after the Brahimi Report in 2000 (Stähle 2008). Whatever the reasons, China is currently the largest contributor of peacekeeping troops amongst the UNSC Permanent 5 and was the tenth largest contributor worldwide at the end of 2022, in a list dominated by low-middle-income countries. In December 2022, the Russian Federation, with an armed force of over 1.4 million, had 87 personnel serving on UN peacekeeping missions. The figure for the United States (with an armed force of over 1.3 million) was 34 (United Nations Peacekeeping 2023). Peacekeeping has become a developing world activity, with countries like Bangladesh (7,233 personnel serving with the UN), Nepal (6,251 personnel) and India (6,043 personnel) taking the strain of providing troops for UN missions (United Nations Peacekeeping 2023).

Other international organizations

Regional security and economic organizations such as the Association of South-East Asian Nations (ASEAN) and the AU have shown themselves to be increasingly capable of pursuing conflict-amelioration and development agendas (see Box 2.5). The AU is composed of all African states. It was established to replace the ineffective Organisation for African Unity and was partly a response to the 1994 Rwandan Genocide and the need for an African response to sudden-onset emergencies (Plant 2014). It has steadily expanded its operations to include peacekeeping, election observation and humanitarian assessments, but it is hobbled by capacity and political-will issues (Renwick 2015). It is not clear that the AU's promise of adopting 'African solutions for African problems' has been met. Its definitions of peacebuilding, for example, seem to be cut and pasted from Western sources and overlook indigenous and traditional forms of dispute resolution that might be found closer to home (Mac Ginty 2008). An indicator of continent-wide commitment to the AU comes from the fact that its new headquarters in Addis Ababa was a gift from China, and only about 40 per cent of the organization's costs are met by African states. The rest is made up of donations by the UN, EU and major donors (Al Jazeera 2013).

Box 2.5 ASEAN Cyclone Nargis 2008

On 2 May 2008, Cyclone Nargis crossed the coastline of Myanmar from the Bay of Bengal and caused massive loss of life and damage to large areas of the central part of the country, principally the Ayeyarwady Division. Over the 2nd and 3rd of May, the cyclone moved inland across the delta towards Yangon, the country's most populated city and its commercial hub. It caused a devastating loss of life, with an estimated 140,000 people killed or missing. On 5 May 2008, the ASEAN Secretary-General called on its member states to urgently provide relief assistance to Myanmar as part of the ASEAN Agreement on Disaster Management and Emergency Response. A special meeting of ASEAN Foreign Ministers on 19th May led to an agreement to form the Tripartite Core Group (TCG), comprised of the government of Myanmar (Chair), ASEAN and the UN, 'to act as an ASEAN-led mechanism to facilitate trust, confidence and cooperation between Myanmar and the international community in the urgent humanitarian relief and recovery work after Cyclone Nargis hit Myanmar' (PONJA 2008: 178). Independent evaluations were consistent in commending these regional structures for its facilitation of the humanitarian response to Cyclone Nargis, and helping to streamline government interactions and decision-making.

Sources: Telford et al., 2006; Bennett, 2006; Turner et al., 2008

NATO and the EU are probably examples of the most capable regional security organizations. In part, this is because of the relative wealth of their members and their alignment with major states with a record of intervention. The EU has beefed up its foreign policy capabilities with its External Action Service to give weight to the commitments of a Common Foreign and Security Policy. The EU has engaged in approximately 30 overseas military and civilian operations, usually in cooperation with NATO, the UN or AU. One mission worth mentioning is the EU Monitoring Mission in Georgia. A response to the 2008 Russia–Georgia conflict, it is small, unarmed, mainly civilian and was placed in the field very quickly. Like other international organizations, member states are wary of giving up power to the EU, but it is a major player in peace, development and emergency response.

After looking as though it had no role in the post–Cold War world, NATO's future as a security provider is secure. In part, this is because of concerns over the Russian Federation and its displays of aggression in countries neighbouring NATO, such as Ukraine and Moldova. But it is also because NATO has redefined itself as an organization able to operate 'out of area' – that is, outside of the European theatre of operations it was designed for. NATO has deployed to Afghanistan, Iraq, Somalia and counter-insurgency operations off the Horn of Africa.

An interesting development has been the 2010 establishment of the G7+ group of fragile and conflict-affected states. Composed of 20 states (among them South Sudan, Haiti and Yemen), it was a reaction to the top-down nature of much development and peacebuilding interventions. The states involved realized that they could benefit from sharing experiences among themselves and lobbying as a group. Their main initiative thus far has been agreeing on the New Deal for Engagement with Fragile States, initially with the Organisation for Economic Co-operation and Development (OECD) in 2011 (G7+ 2011), a manifesto that seeks to shape peacebuilding and statebuilding activities from within conflict-affected countries. Members committed to 'focus on new ways of engaging, to support inclusive country-led and country-owned transitions out of fragility'. It is too early to tell if the G7+ and the New Deal constitute what the World Bank optimistically declared as a 'paradigm shift' in how peace and development were managed (World Bank 2011; Woollard 2013). Sarah Hearn (Hearn 2016) was less optimistic than these early promoters. While much work lies ahead in its implementation, with an agenda that suffers from critical gaps between policy and practice, the agreement is seen as a major accomplishment and has changed the debate on fragility by defining it in their own terms, based on understandings of their own context, and ensuring that conversations about development in fragile states focus on the importance of security, justice and legitimate political processes. It has not stopped wars breaking out on some of the signatories' territories and has also not noticeably improved the life chances of many of the inhabitants. It has not stopped countries from being devastated by internal conflict or appalling climate events. But it is a start.

International non-governmental organizations

INGOs have transformed international responses to conflict and development since the 1970s. Organizations such as the International Red Cross and Red Crescent Movement (see Box 2.6), Médecins Sans Frontières or Oxfam have been important for at least three reasons. First, they have allowed official development assistance, conflict prevention and peacebuilding activities to have a much further reach than traditional bilateral or international organization activity. INGOs are often cheaper and more flexible than intergovernmental mechanisms of aid delivery, particularly in cases where they subcontract to local NGOs (Richmond and Carey 2005). Second, INGOs have prompted more cases of development intervention by leading states and international organizations. This has been particularly the case in their ability to 'bear witness' and engage in advocacy, thus bringing humanitarian emergencies and chronic development and conflict situations to the attention of the public and polities. The third and most significant point has been the ability of INGOs to play a role in shaping the development, conflict prevention and development assistance agenda of leading states, leading international organizations and international financial institutions. This influence has been by no means total, nor has the process been one way, yet INGOs have been incredibly important in broadening governmental understanding of conflict and making obvious the multiple connections between conflict and development.

Box 2.6 International Red Cross and Red Crescent Movement

The International Red Cross and Red Crescent Movement is comprised of three distinct organizations that are legally independent of each other but are united within the movement through common basic principles, objectives, symbols, statutes and governing organizations: The International Committee of the Red Cross (ICRC), founded in 1863, the International Federation of Red Cross and Red Crescent Societies (IFRC), founded in 1919 and (currently) 191 National Societies worldwide. The Movement works to protect human life and health, to ensure respect for all human beings, and to prevent and alleviate human suffering.

The **Seven Fundamental Principles** which guide the Movement are as follows:

Humanity: The International Red Cross and Red Crescent Movement, born of a desire to bring assistance without discrimination to the wounded on the battlefield, endeavours, in its international and national capacity, to prevent and alleviate human suffering wherever it may be found. Its purpose is to protect life and health and to ensure respect for the human being. It promotes mutual understanding, friendship, cooperation and lasting peace amongst all peoples.

Impartiality: It makes no discrimination as to nationality, race, religious beliefs, class or political opinions. It endeavours to relieve the suffering of individuals, to be guided solely by their needs and to give priority to the most urgent cases of distress.

Neutrality: In order to continue to enjoy the confidence of all, the Movement may not take sides in hostilities or engage at any time in controversies of a political, racial, religious or ideological nature.

Independence: The Movement is independent. The National Societies, while auxiliaries in the humanitarian services of their governments and subject to the laws of their respective countries, must always maintain their autonomy so that they may be able at all times to act in accordance with the principles of the Movement.

Voluntary service: The International Red Cross and Red Crescent Movement is a voluntary relief movement not prompted in any manner by a desire for gain.

Unity: There can be only one Red Cross or one Red Crescent Society in any one country. It must be open to all. It must carry on its humanitarian work throughout its territory.

Universality: The International Red Cross and Red Crescent Movement, in which all Societies have equal status and share equal responsibilities and duties in helping each other, is worldwide.

Sources: https://www.icrc.org/en/who-we-are/history; https://www.ifrc.org/who-we-are/about-ifrc/our-history-and-archives

Some INGOs have been co-opted into the liberal peace project. Rather than acting as bulwarks against the ambitions of rich Western states, some INGOs are the principal transmission agents of the liberal peace through their promotion of good governance and reform agendas (Richmond 2005b). The privatization of development and humanitarianism (for example, through tender processes for development contracts) leaves many INGOs with little choice. To stay in business (and to stay relevant and to fund their other advocacy work), they must compete for contracts released by governments and international organizations. Yet all contracts come with conditions, and some of these conditions may constrain the original development and conflict-amelioration objectives of the INGO. Increasing privatization in the humanitarian and development sectors has raised many practical and ethical questions.

Commercial relationships between humanitarian organizations and multinational companies have been a growing reality since the end of the Cold War. The brokerage of public-private partnerships (PPPs) featured at the World Economic Forum in Davos, 2008, when WEF member companies and UN agencies launched the Joint Guiding Principles for private-public collaboration in humanitarian response (Andonova and Carbonnier 2014). The growth of business–humanitarian partnerships focusing on a strategic provision of technical expertise that goes beyond traditional philanthropy has shown positive benefits. Logistics and the provision of specialized equipment were critical in support of emergency responses to natural disasters such as the Indian Ocean tsunami in 2004 and response to the Ebola crisis in West Africa in 2013. The provision of digital finance assistance through commercial banking partnerships with the UN and INGOs has provided highly successful assistance to refugees and displaced populations in contemporary conflict and complex emergencies. Beyond the provision of technical expertise, studies have shown that the private sector could replace more traditional humanitarian actors where governments in low- and middle-income countries play an increasingly strong role in responding to crises and rebalance power dynamics between the national and international in emergency responses (Andonova and Carbonnier 2014; Zyck and Kent 2014). Privatization does not mean monopolization. There are no recorded instances of the private sector given sole or primary responsibility for mainstream ('classical') humanitarian activity.

International financial institutions

UN and INGO personnel are often highly visible in conflict and development environments through their blue helmets, white 4×4s and prominent banners advertising their projects (Mills 2006). Yet one set of highly influential international actors is frequently less visible: the international financial institutions (IFIs). The World Bank (originally the International Bank for Reconstruction and Development) and the International Monetary Fund date from the era when Adolf Hitler was still in power. The framers of both organizations were conditioned by their experiences of the 1930s, with the contraction of world trade and state attempts to 'beat' economic depression through tariffs. Although the IMF and World Bank are often discussed in the same breath, and although both share the same aim of facilitating free trade, they are 'not identical twins' (Hanlon 1996: 25). The IMF's chief role is to smooth international trade through short-term balance of payments assistance to states. Its main focus is on macro-economic matters, and it regards low inflation (achieved by tight government spending regimes) as the most important weapon in the regularization of trade and the stabilization of currencies (Pollard 1997: 79). The World Bank has a broader brief and is focused on development and longer-term interventions. It aims to encourage states to change the structure of their economies so that growth and trade can operate unhindered. A welter of associated bodies and permanent conferences (the General Agreement on Tariffs and Trade, the World Trade Organization, G8, the UN Conference on Trade and Development and the Uruguay and Doha Rounds of negotiations, etc.) assist in the overall aim of promoting free trade.

As was made clear in the Introduction to this book, the international financial architecture has been heavily criticized as being designed to perpetuate the poverty and the disadvantaged positions of developing world states (Stiglitz and Charlton 2005). Thus, the IMF and World Bank are routinely demonized as being 'the high priests of neo-classical orthodoxy' (Shutt 1998: 160) or progenitors of 'predatory globalization' (Falk 1999). The reasons for this derision are clear: the international financial architecture has facilitated the post–Second World War growth in the disparity between rich and poor states and the income disparities in developing states (Seligson 2003: 2). Falk (1999: 13) characterizes the international economy as structurally racist: it is an 'economic apartheid' that reinforces a rich white world and is comfortable with the impoverishment of the non-white world. Joseph Stiglitz (2003), for many years an economist with the IMF, pursues a similar theme: the IFIs prioritize the interests of global capital above those of the global community.

Given our interest in the international financial architecture is in how it facilitates and thwarts conflict and development, three points can be made. The first point is that the IFIs are here to stay. Although many commentators paint them as pantomime villains (with considerable justification), there are no alternative sources of international economic regulation waiting in the wings. The IFIs have undergone considerable change throughout their history and have presided over many economic downturns (for example, the 1973 oil and dollar shocks, the 1994–5 Mexican peso crisis, the 1997–8 Asian financial crisis and the 2007–9 market instability and credit crunch). Yet, for all of this mismanagement and the disjointed incrementalism of their institutional evolution, there is no political appetite among leading states for root and branch reform (Harris 1999; Underhill 2001: 291). Despite the iniquities of the situation, societies emerging from civil war or suffering from underdevelopment simply have to work with the IFIs. States emerging from civil wars are unable to avail themselves of reconstruction loans or to conduct international trade unless they sign up for IFI strictures. These strictures may include honouring the external debts of the previous regime, even if that regime lacked a democratic mandate.

There is a new rival in the form of the Asian Infrastructure Investment Bank (AIIB), a Chinese-backed initiative that also has the support of the other BRICS (Brazil, Russia, India, China and South Africa). It is too early to say whether it will succeed, but it is likely to become a way of allowing China to expand its soft power by lending developing countries sums to build infrastructure (Perlez 2014).

The second point concerns the worrying absence of democracy and transparency in the affairs of the IFIs. There is a certain irony here in that while democracy and democratization are often pushed on states emerging from conflict as the route to guarantee them stability, these states must also enter into deeply undemocratic relationships with the IFIs. Joseph Hanlon (1996: 24) noted how, in the aftermath of a civil war, Bretton Woods officials in Washington had 'more power in Mozambique than any Mozambican, up to and including President Chissano'. Unelected IFI officials have been present at many civil war peace negotiations: they are difficult guests to refuse in that reconstruction assistance is often routed through them. While the World Bank, IMF and other international agents encourage transparency among recipient states, the World Bank's own lack of transparency was illustrated by the controversy surrounding a massive pay rise awarded to its president's girlfriend in 2005. Decision-making power in the IMF is based purely on financial clout, with states enjoying voting rights commensurate with their contributions to the Fund. As a result, the poorest states have little say in how the Bank operates.

The third point is that IFI strictures can have profound impacts on social harmony in the aftermath of a civil war and the ability of a post-civil-war government to fulfil its social provision pledges. This can have serious consequences since the post–civil war political dispensation will require a broad base of stakeholders if the peace accord is to hold. IMF and World Bank mantras of 'small government' can severely restrict the ability of governments to honour the 'peace dividend' pledges that accompany a peace accord. As Addison (2003: 264) notes, 'Decisions made in the early phase of the war-to-peace transition alter the distribution of productive assets and thereby the benefits of growth to the rich and the poor for years to come'. The 2006 Nepalese peace agreement, for example, contains a highly ambitious list of social pledges that would make a 1970s Scandinavian socialist government proud. Yet the Nepalese state does not have autonomy over how it disburses the contents of the public coffers. While the IFIs do not have peacekeepers or storm troopers on the battlefields, they are not entirely removed from the dynamics that encourage conflict and development.

Conclusions

To return to the software and hardware metaphor, it would seem that an incredibly sophisticated software prevails in the form of the liberal peace. Through a mixture of subtle and not-so-subtle inducements and compliance mechanisms, states are encouraged to conform to an order that reinforces existing power-holders. Its chief aims are to stabilize dysfunctional war-torn states, to promote liberal economics and to perpetuate the control of leading states (Duffield 2001: 10). The liberal peace regime allows some wriggle room: it has been softened by the human security perspective, will face increasing challenges from a more internationally active China and – in certain circumstances – can be subverted by states and communities. While the dominant software is constantly upgraded, the hardware (in the form of international institutions) is clunky, slow and often fails to work. In a sense, the software has found ways in which it can work without hardware or certainly without the original hardware.

The institutional factors that shape contemporary conflict and development are undergoing simultaneous processes of continuity and change. The forces of continuity are to be found in the rigidity of a UN Security Council that reflects the 1945 balance of power and in the inflexibility of state sovereignty as interpreted by some states. But there are many signs of change, all of which impact the ebb and flow of conflict and development and the ability of international institutions to react. The range of factors deemed capable of causing conflict (or instability) and thwarting economic and human development is deemed to have grown in the post–Cold War era. The transnational nature of many of these threats (for example, legal and illegal labour flows or environmental degradation) means that traditional state-centric institutions are often poorly equipped to deal with them. While many international institutions have been changing to anticipate or react to a new range of threats, it is not at all clear that the current suite of collective security organizations and development organizations is the best equipped to deal with these threats. It is an incontrovertible fact that none of the current international or regional security organizations in existence were explicitly formed to deal with contemporary civil war or underdevelopment. Quite simply, the international political system is stuck with the wrong tools for the job.

Having said that, many international organizations have undergone processes of reform, and there are growing interconnections between the traditional international units (such as states and collective security organizations) and more modern units in the form of INGOs and organized global capital. This 'complex multilateralism' allows numerous multilateral associations to come together, often on specific issues and for limited time periods. Thus campaigns to address blood diamonds, child soldiers or landmines will see 'mixed-actor coalitions' of states, international organizations, pressure groups and private-sector bodies cooperate on a single issue. There are ongoing initiatives within the areas of policy and research that address the need for reform. Amongst these is the research project Everyday Peace Indicators that uses community-generated indicators to inform peacebuilding and social justice initiatives in conflict-affected areas (EPI https://www.everydaypeaceindicators.org/). Urwin et al. (2023) urge a new approach to conventional monitoring and evaluation practices that they see as restricting the agency and participation of communities affected by crisis which supports the call for a new articulation of the discourse around conflict and development urged in this book. The 'new way of working' called by the UN Secretary-General at the World Humanitarian Summit in 2016 must be complemented by a 'new way of thinking' that engages the hierarchies of power and influence in the traditional system that shapes the spaces of participation. These are alternative articulations of peace and conflict that go beyond the quantifiable expressions and log-frames with which the current architecture is comfortable (Mac Ginty and Firchow 2014. See also Further reading). They are not seen to replace the tested methods of gauging impact and accountability but complement these by offering different sources of information and more relevant places of representation.

There seems to be no end in sight to the perennial problem of reconciling the national interests of states and the need for states to put aside narrow national interests to deal with pressing collective problems such as global warming or poverty. International organizations are the clearing house for many of these tensions. In some cases, such as agreement on the Sustainable Development Goals, states have been able to cooperate on joint aspirations. Logjams in international organizations, however, have tempted some states and interests to act either unilaterally or multilaterally, though not through an international organization. Some commentators have suggested the need for a 'league of democracies' or an alliance of Western democratic states who could act without the encumbrance of an unwieldy international organization (Kagan 2008). The advantage would be swiftness of action and unity of purpose in the face of a humanitarian or complex political emergency. Critics, however, have pointed out that such a league would amount to a charter for interventionism and would conveniently free Western states of the need to persuade non-Western states of the merits of their case. Changes in the geopolitical landscape, together with UN reforms and emerging new partnerships described in this chapter, direct towards changes in the institutional responses to conflict and development. Recent scholarship has urged a new way of thinking to complement the new way of working promoted at the World Humanitarian Summit in 2016. A cultural shift that will take time.

Summary

- Rather than simply look at international institutions, it is also important to look at the principles and values behind the institutions.
- The concept of 'the liberal peace' has been used to describe the peace- and development-support interventions by leading states, leading international organizations and IFIs in war-affected and developing societies. According to critics, the liberal peace is guilty of Western bias and can be a form of neocolonialism.
- While good governance reforms might seem wholly sensible to Western eyes, they can have a profound effect on important relationships in target societies. For example, they might alter how citizens interact with and perceive the state and fellow citizens.
- The IFIs wield immense economic and political power, but it is difficult to think of alternatives to them.
- The international political system increasingly operates according to 'complex multilateralism' whereby a constantly shifting mix of states, international organizations, INGOs, NGOs and private-sector interests coalesce and recoalesce at different times on different issues.

Discussion questions

- If the UN collapsed, what would you replace it with? What principles should underpin any new collective security organization?
- Is it necessary to have a powerful 'hegemonic' state to have a stable institutional international system and approach to development issues?
- Is there any way around the tension between the national interests of states and the collective needs of a community of states?
- Are there any serious alternatives to the state as the basic political unit in the international system?
- What advantages can private sector firms bring to security, humanitarian and development interventions?
- Has R2P worked?

Further reading

M. Douglas (1986) *How Institutions Think*, Syracuse, NY: Syracuse University Press provides a seminal account on the inner workings of large organisations. Insider's guides to the role of the IMF and US capital in the developing world can be found in J. Stiglitz (2003) *Globalization and Its Discontents*, New York: Norton, and J. Perkins (2004) *Confessions of an Economic Hit Man*, San Francisco, CA: Berret-Koehler. An extremely good survey of some of the problems facing INGOs and

NGOs is H. Yanacopulos and J. Hanlon (eds) (2005) *Civil War, Civil Peace*, Oxford: James Currey. Revealing insights into the nature of Western interventions can be found in S. Autessere (2014) *Peaceland: Conflict resolution and the everyday politics of international intervention*, Cambridge: Cambridge University Press, R. Chandrasekaran (2007) *Imperial Life in the Emerald City: Inside Iraq's Green Zone*, London: Bloomsbury, and T. Rick (2006) *Fiasco: The American military adventure in Iraq*, London: Penguin. A wonderfully honest (and funny) reminder of how Western preconceptions sometimes travel poorly can be found in N. Barley (1983) *The Innocent Anthropologist: Notes from a Mud Hut*, London: British Museum Publications.

Useful websites

Most international organisations, IFIs and INGOs have content-rich websites. Major international organisations include the United Nations (www.un.org), the European Union (http://europa.eu/), the African Union (www.Africa-union.org), the Organization of American States (www.oas.org), the League of Arab States (www.arableagueonline.org) and the South Asian Association for Regional Cooperation (www.saarc-sec.org). Some of these organisations have agencies or sections with a specific remit for conflict and/or development. For example, the European Commission's Humanitarian Aid Office (http://ec.europa.eu/echo) and the UN's Peacebuilding Commission (www.un.org/peace/peacebuilding) or Department of Peacekeeping Operations (www.un.org/Depts/dpko). The World Bank (www.worldbank.org), IMF (www.imf.org), Asian Developm ent Bank (www.adb.org) and European Bank for Reconstruction and Development (www.ebrd.com) are all represented online. There is much data available at the individual state level. See, for example, the Fragile States Index (https://fragilestatesindex.org/), or the Global Peace Index (visionofhumanity.org) and reports by Freedom House (https://freedomhouse.org/) and Amnesty International (www.amnesty.org). Brettonwoodsproject.org keeps an eye on the IMF and World Bank. Many INGOs have very useful resources in terms of reports and primary documents. See, for example, sipri.org, international-alert.org, accord.org.za and c-r.org. For a study of bottom-up community-sourced indicators of peace, safety and social change, see *Everyday Peace Indicators* https://www.everydaypeaceindicators.org/

Bibliography

Aboudouh, A. Chatham House (2023) 25 October: China's approach to the war in Gaza is not anti-Israel. It is designed to contain the US https://www.chathamhouse.org/2023/10/chinas-approach-war-gaza-not-anti-israel-its-designed-contain-us (Accessed 27 October 2023).

Abrahamsen, R. (2000) *Disciplining Democracy: Development Discourse and Good Governance in Africa*. London: Zed.
Abrahamsen, R. (2004) Poverty reduction or adjustment by another name? *Review of African Political Economy 31*(99): 184–87.
Addison, T. (2003) Communities, Private Sectors, and States. In Addison, T. (ed) *From Conflict to Recovery in Africa*. Oxford: Oxford University Press, pp. 263–87.
Al Jazeera (2013) African Union seeks financial independence. *Al Jazeera*. 24 May. Available at: http://www.aljazeera.com/news/africa/2013/05/2013524 12928567270.html (accessed 28 September 2023).
Al Jazeera (2023) G7 tightens sanctions over Russia's war on Ukraine, 19 May. https://www.aljazeera.com/news/2023/5/19/g7-tightens-sanctions-over-russia-war-ukraine (accessed 27 October 2023).
Allison, G. (2017) *Destined for War: Can America and China Escape Thucydides's Trap?* New York: Houghton Mifflin Harcourt.
Andonova, L.B. and Carbonnier, G. (2014) Business–humanitarian partnerships: Processes of normative legitimation. *Globalizations 11*(3), 349–67
Bacevich, A.J. (2010) *Washington Rules: America's Path to Permanent War*. New York: Metropolitan Books.
BBC News (2022). UN sexual abuse claims 'must be investigated'. *BBC News Online*. 21 June. Available at: https://www.bbc.co.uk/news/world-61826551 (accessed 28 October 2023).
BBC News (2023a) Country profile: Bosnia-Herzegovina. *BBC News Online*. 1 January. Available at: http://www.bbc.co.uk/news/world-europe-17211415 (accessed 28 October 2023).
BBC News (2023b). Hundreds of migrants killed by Saudi border guards – Report. 21 August) https://www.bbc.co.uk/news/world-middle-east-66545787 (accessed 31 August 2023).
Bellamy, A. and Williams, P. (2010) *Understanding Peacekeeping*, 2nd edn. Cambridge: Polity.
Belloni, R. (2007) The trouble with humanitarianism. *Review of International Studies 33*: 451–71.
Bennett, Jon (principal author) (2006) *Coordination of International Humanitarian Assistance in Tsunami-Affected Countries*. London: Tsunami Evaluation Committee.
Boulden, J. (2001) *Peace Enforcement: The United Nations experience in Congo, Somalia, and Bosnia*. Westport, CT: Praeger.
Bunce, V., Koesel, K. and Weiss, J. (2020) *Citizens and the State in Authoritarian Regimes: Comparing China and Russia*. New York: Oxford University Press.
Callinicos, A. (2005) Iraq: Fulcrum of world politics. *Third World Quarterly 26*(4–5): 593–608.
Campbell, S., Chandler, D. and Sabaratnam, M. (eds) (2011) *A Liberal Peace? The Problems and Practices of Peacebuilding*. London: Zed.
Chandler, D. (2006) Back to the future? The limits of neo-Wilsonian ideals of exporting democracy. *Review of International Studies 32*(3): 475–94.
Chandler, D. (2007) EU Statebuilding: Securing the liberal peace through EU enlargement. *Global Society 21*(4): 593–607.
Chandler, D. (2013) Resilience and the autotelic subject: Toward a critique of the societalization of security. *International Political Sociology* 7(2): 210–26.
Chandrasekaran, R. (2007) *Imperial Life in the Emerald City: Inside Iraq's Green Zone*. London: Bloomsbury.
Chowdhury, M.H. (2003). *Democratization in South Asia: Lessons from American Institutions*. Farnham: Ashgate.

Craig, D. and Porter, P. (2003) Poverty reduction strategy papers: A new convergence. *World Development 31*(1): 53–69.
Daragahi, B. (2009) Joe Biden, in Lebanon, hints that US might cut aid if Hezbollah wins election. *Los Angeles Times*, 23 May. Available at: http://articles.latimes.com/2009/may/23/world/fg-biden-lebanon23 (accessed 8 September 2015).
Deudney, D. and Ikenberry, G.J. (1999) The nature and sources of liberal international order. *Review of International Studies 25*(2): 179–96.
Doyle, M. (1980) Liberalism and world politics. *American Political Science Review 80*(4): 1151–69.
Duffield, M. (2001) *Global Governance and the New Wars: The Merging of Development and Security*. London: Zed.
Duffield, M. (2007) *Development, Security and Unending War: Governing the World of Peoples*. Cambridge: Polity.
Duffield, M. (2012) Challenging environments: Danger, resilience and the aid industry. *Security Dialogue 43*(5): 475–92.
Duroselle, J.B. (1981) *Tout empire périra: Théorie des relations internationales*. Paris: Editions de la Sorbonne.
Ehteshami A. and Wright, S. (2007) Political change in the Arab oil monarchies: From liberalization to enfranchisement. *International Affairs 83*(5): 913–32.
Falk, R. (1999) *Predatory Globalization: A Critique*. Cambridge: Cambridge University Press.
Franks, J. and Richmond, O. (2008) Coopting liberal peace-building: Untying the Gordian knot in Kosovo. *Cooperation and Conflict 43*: 81–103.
G7+ (2011, 2023 update) New deal for engagement in fragile states. 11 December. https://www.g7plus.org/new-deal-implementation/ (accessed 27 October 2023).
Ghani, A. and Lockhart, C. (2008) *Fixing Failed States: A Framework for Rebuilding A Fractured World*. Oxford: Oxford University Press.
Ghervas, S. (2021) *Conquering Peace: From the Enlightenment to the European Union*. Cambridge MA: Harvard University Press.
Gordon-Gibson, A. (2021) *Humanitarians on the Frontier: Identity and Access along the Borders of Power*. Lanham: Rowman and Littlefield.
Hamieh, C. (2007) Intra-party competition in a divided society: Amal versus Hezbollah in Lebanon, 1998–2003. Unpublished doctoral thesis, University of York.
Hanlon, J. (1996) *Peace Without Profit: How the IMF blocks rebuilding in Mozambique*. Dublin: Irish Mozambique Solidarity and the International African Institute in association with James Currey and Heinemann.
Hannay, D. (2005) Reforming the United Nations: The use of force to safeguard international security and human rights. A member's perspective from the Secretary-General's high level panel on threats, challenges and change. *Conflict, Security and Development 5*(1): 109–17.
Harris, L. (1999) Will the Real IMF Please Stand Up? In Grieve, Michie J. and Smith, J. (eds) *Global Instability: The Political Economy of World Economic Governance*. London: Routledge, pp. 198–211.
Hearn, S. (2016) *Independent Review of the New Deal for Engagement in Fragile States for the International Dialogue on Peacebuilding and Statebuilding*. New York: Center on International Cooperation, New York University.
Henderson, E.A. (2002) *Democracy and War: The End of an Illusion?* Boulder, CO: Lynne Rienner.
Hughes, C.R. (1995) China and liberalism globalised. *Millennium: Journal of International Studies 24*(3): 425–45.

Human Rights Watch (2023) Saudi Arabia: Mass killings of migrants at Yemen border. Available at: https://www.hrw.org/news/2023/08/21/saudi-arabia-mass-killings-migrants-yemen-border (accessed 31 August 2023)
Huntington, S. (1993) The clash of civilizations? *Foreign Affairs* 72(3): 22–49.
Hussain, M. (2013) Resilience: Meaningless jargon or development solution?, 5 March, https://www.theguardian.com/global-development-professionals-network/2013/mar/05/resilience-development-buzzwords (accessed 28 October 2023).
Jacoby, T. (2007) Hegemony, modernisation and post-war reconstruction. *Global Society: Journal of Interdisciplinary International Relations* 21(4): 521–37.
Janzekovic, J. (2006) *The Use of Force in Humanitarian Intervention: Morality and Practicalities*. Aldershot: Ashgate.
Jones, C. (2018) *China's Challenge to Liberal Norms* London: Palgrave Macmillan.
Kabamba, P. (2010) Heart of Darkness: Current images of the DRC and their theoretical underpinnings. *Anthropological Theory* 10(3): 265–301.
Kagan, R. (2008) The case for a league of democracies. *Financial Times*, 14 May.
Kenny, C. (2003) Why Aren't Countries Rich? Weak States and Bad Neighbourhoods. In Seligson, M. and Passé-Smith, J. (eds) *Development and Under-Development: The Political Economy of Global Inequality*, 3rd edn. Boulder, CO: Lynne Rienner, pp. 413–25.
Ker-Lindsay, J. (2011) *Kosovo: The Path to Contested Statehood in the Balkans*. London: I.B. Tauris.
Ker-Lindsay, J. and Fabry, M. (2023) *Secession and State Creation: What Everyone Needs to Know*. Oxford: Oxford University Press.
Lambert, H. (2015) Yanis Varoufakis – Full transcript: Our battle to save Greece. *New Statesman*, 13 July. Available at: http://www.newstatesman.com/world-affairs/2015/07/yanis-varoufakis-full-transcript-our-battle-save-greece (accessed 28 October 2023).
Larmour, P. (2005) *Foreign Flowers: Institutional Transfer and Good Governance in the Pacific Islands*. Honolulu: University of Hawai'i Press.
Leahy, J. et al. (2023). Dare to Fight: Xi Jinping unveils China's new world order, *Financial Times*, 31 March, https://www.ft.com/content/0f0b558b-3ca8-4156-82c8-e1825539ee20 (accessed 5 April 2023).
Leftwich, A. (1996) Two Cheers for Democracy? Democracy and the Developmental State. In Leftwich, A. (ed) *Democracy and Development*. Cambridge: Polity, pp. 279–95.
Leigh, D. and Evans R. (2010) BAE admits guilt over corrupt arms deals. *Guardian*, 6 February. Available at: http://www.theguardian.com/world/2010/feb/05/bae-systems-arms-deal-corruption (accessed 28 October 2023).
Mac Ginty, R. (2006) *No War, No Peace: The Rejuvenation of Stalled Peace Processes and Peace Accords*. London: Palgrave.
Mac Ginty, R. (2008) Indigenous peacemaking versus the liberal peace. *Cooperation and Conflict* 43: 139–63.
Mac Ginty, R. and Firchow, P. (2014) 'Everyday peace indicators: Capturing local voices through surveys'. *Shared Space* 18, 33–40.
Mac Ginty, R. and Firchow, P. (2016) Top-down and bottom-up narratives of peace and conflict. *Politics* 36(3): 308–23.
Mac Ginty, R. and Richmond, O. (2007) Myth or reality: Opposing views on the liberal peace and post-war reconstruction. *Global Society* 21(4): 491–7.

MacFarlane, N. and Khong, K.F. (2006) *Human Security and the United Nations: A Critical History*. Bloomington, IN: Indiana University Press.

Maclean, J. (1988) Marxism and international relations: A strange case of mutual neglect. *Millennium* 17(2): 295–320.

Martell, P. (2018) *First Raise a Flag: How South Sudan Won the War and Lost the Peace*. Oxford: Oxford University Press.

Martin, J. (2022) *The Meddlers: Sovereignty, Empire, and the Birth of Global Economic Governance*. Harvard MA: Harvard University Press.

Martz, J. (1996) *The Politics of Clientelism*. New Brunswick, NJ: Transaction.

Mayall, J. and Soares de Oliveira, R. (eds) (2011) *The New Protectorates: International Tutelage and the Making of Liberal States*. London: Hurst.

Mazower, M. (2009) *No Enchanted Palace: The End of Empire and the Ideological Origins of the United Nations*. Princeton NJ: Princeton University Press.

Mearsheimer, J. (1995) The false promise of international organizations. *International Security* 19(3): 5–49.

Mearsheimer, J. (2021) The inevitable rivalry: America, China and the tragedy of great-power politics. *Foreign Affairs*, November/December.

Migdal, J. (1988) *Strong Societies, Weak States: State–Society Relations and State Capabilities in the Third World*. Princeton, NJ: Princeton University Press.

Mill, J.S. (1859, 2006) A few words on non-intervention. *New England Review* 27(3): 252–64 (Reprint of the original].

Mills, K. (2006) The postmodern tank of the humanitarian international. *Journal of Social Justice* 18, 261–67.

Mills, K. (2015) *International Responses to Mass Atrocities in Africa* Philadelphia: University of Pennsylvania Press.

Nasrallah, H. (2008) Speech in solidarity with Gaza, 28 December. Formerly. http://almusawwir.org. No longer available.

Nussbaum, M. (2006) *Frontiers of Justice*. Cambridge, MA: Belknap Press.

O'Neil, J. (2023) BRICS creator slams 'ridiculous' idea for common currency. *Financial Times*, 15 August: https://www.ft.com/content/d8347bd0-cc4f-4c3b-9225-0ccd272330a6 (accessed 15 August 2023).

Ogden, C. (2022a) The Double-Edged sword: Reviewing India–China relations. *India Quarterly* 78(2): 210–28.

Ogden, C. (2022b) *The Authoritarian Century: China's Rise and the Demise of the Liberal International Order*. Bristol: Bristol University Press.

Paris, R. (2004) *At War's End: Building Peace after Civil Conflict*. Cambridge: Cambridge University Press.

Paris, R. (2010) Saving liberal peacebuilding. *Review of International Studies* 36(2): 337–65.

Perlez, J. (2014) US opposing China's answer to World Bank. *New York Times*. 9 October.

Pingeot, L. (2012) *Dangerous Partnership: Private Military and Security Companies and the UN*. New York: Global Policy Forum.

Plant, M. (2014) African Union missing in action in conflicts from Mali to South Sudan. *Guardian*. 6 January.

Pollard, S. (1997) *The International Economy since 1945*. London: Routledge.

Post-Nargis Joint Assessment (PONJA) report (2008) https://reliefweb.int/report/myanmar/myanmar-post-nargis-joint-assessment

Pugh, M. (1997) *The UN, Peace, and Force*. London: Routledge.

Quinn, A. and Cox, M. (2007) For better, for worse: How America's foreign policy became wedded to liberal universalism. *Global Society* 21(4): 499–519.

Reno, W. (1997) African weak states and commercial alliances. *African Affairs* 96: 165–85.
Renwick, D. (2015) Peace operations in Africa. Council on foreign relations backgrounder, 15 May. Available at: http://www.cfr.org/peacekeeping/peace-operations-africa/p9333 (accessed 8 September 2015).
Richmond, O. (2005a) *The Transformation of Peace*. London: Palgrave.
Richmond, O. (2005b) The Dilemmas of Subcontracting the Liberal Peace. In Richmond, O. and Carey, H. (eds) *Subcontracting Peace: The challenges of NGO peacebuilding*. Aldershot: Ashgate, pp. 19–35.
Richmond, O. and Carey, H. (eds) (2005), *Subcontracting Peace: The challenges of NGO peacebuilding*. Aldershot: Ashgate.
Richmond, O. and Franks, J. (2007) Liberal hubris? Virtual peace in Cambodia. *Security Dialogue 38*(1): 27–48.
Richmond, O. and Mac Ginty, R. (2014) Where now for the critique of the liberal peace? *Cooperation and Conflict 50*(2): 171–89.
Robbins, P. (2003) *Stolen Fruit: The Tropical Commodities Disaster*. London: Zed.
Roberts, D. (2011) *Liberal Peacebuilding and Global Governance: Beyond the Metropolis*. London: Routledge.
Seligson, M. (2003) The Dual Gaps: An Overview of Theory and Research. In Seligson, M. and Passé-Smith J. (eds) *Development and Under-development: The Political Economy of Global Inequality*, 3rd edn. Boulder, CO: Lynne Rienner, pp. 1–6.
Shutt, H. (1998) *The Trouble with Capitalism: An Enquiry into the Causes of Global Economic Failure*. London: Zed.
Slobodian, Q. (2018) *Globalists: The End of Empire and the Birth of Neoliberalism*. Cambridge MA: Harvard University Press.
Snow, P. (2023) *China and Russia: Four Centuries of Conflict*. New Haven, NJ: Yale University Press.
Spearin, C. (2008) Private, armed and humanitarian? States, NGOs, international private security companies and shifting humanitarianism. *Conflict, Security and Development 39*(4): 363–82.
Stähle, S. (2008). 'China's shifting attitude towards United Nations peacekeeping operations' *The China Quarterly 195*: 631–55.
Stedman, S. (2007) UN transformation in an era of soft balancing. *International Affairs 83*(5): 933–44.
Stiglitz, J. (2003) *Globalization and Its Discontents*. New York: Norton.
Stiglitz, J. and Charlton, A. (2005) *Fair Trade for All: How Trade Can Promote Development*. Oxford: Oxford University Press.
Strange, S. (1991, 2009) *Rival States, Rival Firms*. Cambridge: Cambridge University Press.
Sunday Times (2010) Japan's bribes on whaling exposed. 13 June. No longer available.
Tadjbakhsh, S. (ed) (2011) *Rethinking the Liberal Peace: External Models and Local Alternatives*. London: Routledge.
Taylor, I. (2005) *NEPAD: Towards Africa's Development or Another False Start?* Boulder, CO: Lynne Rienner.
Taylor, I. (2006) *China and Africa: Engagement and Compromise*. London, Routledge.
Telford, J. Cosgrave, J. and Houghton, R. (2006) *Joint Evaluation of the International Response to the Indian Ocean Tsunami: Synthesis Report*. London: Tsunami Evaluation Commission.
Tickner, A. (2001) *Gendering World Politics: Issues and approaches in the post-Cold War era*. Columbia, NY: Columbia University Press.

Tilly, C. (1990) *Coercion, Capital and European States*. Oxford: Blackwell.

Tom, P. (2017) *Liberal Peace and Post-Conflict Peace-Building*. London: Palgrave.

Turner, R., Baker, J., Oo, Z.M. and Aye, N.S. (2008) Inter-Agency Real Time Evaluation of the Response to Cyclone Nargis. United Nations Office for the Coordination of Humanitarian Affairs on behalf of the Inter Agency Standing Committee.

Underhill, G. (2001) The Public Good versus Private Interests and the Global Financial and Monetary System. In Drache, D. (ed) *The Market or the Public Domain: Global Governance and the Asymmetry of Power*. London: Routledge, pp. 274–95.

United Nations (2014) Letter from President of the General Assembly to President of the Security Council. 14 December. A/69/674-S/2014/911.

United Nations Peacekeeping (2015) Troop and police contributions. http://www.un.org/en/peacekeeping/resources/statistics/contributors.shtml (accessed 8 September 2015).

Urquhart, B. and Weiss, T.G. (2012) *What's Wrong with the United Nations and How to Fix It?* Cambridge: Polity.

Urwin, E., Botoeva, A., Arias, R., Vargas, O. and Firchow, P. (2023) Flipping the power dynamics in measurement and evaluation: International aid and the potential for a grounded accountability model. *Negotiation Journal 39*(4): 401–26.

Wainwright, E. (2003) Responding to state failure – The case of Australia and the Solomon Islands. *Australian Journal of International Affairs 57*(3): 485–93.

Waltz, K. (2001) *Man, the State and War*. Columbia, NY: Columbia University Press.

Whitmeyer, A. (2013) Good news: The US still isn't a failed state. *Foreign Policy*. 2 October. Available at: http://foreignpolicy.com/2013/10/02/good-news-the-united-states-still-isnt-a-failed-state/ (accessed 30 October 2023).

Widmalm, S. (ed) (2021) *Routledge Handbook of Autocratization in South Asia* London: Routledge.

Williams, A. (2007a) *Failed Imagination? New World Orders of the Twentieth Century*. Manchester: Manchester University Press.

Williams, A. (2007b) Reconstruction: The bringing of peace and plenty or occult imperialism? *Global Society: Journal of Interdisciplinary International Relations 21*(4): 539–51.

Williams, P. (2011) *War and Conflict in Africa*. Cambridge: Polity.

Woollard, C. (2013) Peace in the post-2015 development goals. *Journal of Peacebuilding and Development 8*(1): 84–9.

World Bank (2011) *World Development Report: Conflict, Security and Development*. Washington, DC: World Bank.

World Bank (2015) Armed forces personnel, total. http://data.worldbank.org/indicator/MS.MIL.TOTL.P1 (accessed 30 October 2023).

Yu, J. and Wallace, J. (2021) *What Is China's Belt and Road Initiative (BRI)?* London: Chatham House. https://www.chathamhouse.org/2021/09/what-chinas-belt-and-road-initiative-bri (accessed 3 November 2022).

Zaum, D. (2006) The authority of international administrations in international society. *Review of International Studies 32*(3): 455–73.

Zyck, S. A. and Kent, R. (2014) *Humanitarian crises, emergency preparedness and response: The role of business and the private sector Final report*. HPG/ODI [web page no longer available]

3 People
Participation, civil society and gender

Introduction

While the previous chapter examined the institutions and structures that shape conflict and development, this chapter looks at the makers, users, bystanders and victims of these institutions and structures: people. It is interested in the concept and practice of participation and how particular groups (such as women or civil society) are included or excluded from processes of conflict and development. The notion of 'civil society' is important because it is often said to be missing in many developing (and some developed) countries that have autocratic governments and an understanding of 'democracy', which is at odds with what Western policymakers believe it to be. Robert Putnam makes the point that for democracy to flourish, it must have non-state institutions that are guardians of societal trust and knowledge independent of the state. Popularized in conflict studies by Mary Kaldor in the 2000s (Kaldor 2003a, 2003b, 2012, 2018), it has been theorized by other thinkers like John Keane and Jurgen Habermas, who see civil society as the part of society that is not the state or (historically) the domain of state-controlled religion, but rather the 'public sphere' (Keane 1988; Putnam 2000). The hoped-for resulting public participation is widely regarded as the silver bullet of Western democracy and peacebuilding: through participation comes legitimacy, and with legitimacy comes a discourse to justify a particular course of action.

In recent years, 'public participation', 'local participation' and 'local ownership', 'community partnership' and 'customary wisdom' have become prominent themes in relation to the discourses on development and peacebuilding processes (Mac Ginty and Richmond 2013). But to what extent do leading states, international organizations and INGOs deploy these terms as rhetorical devices, and to what extent have they a genuine interest in local empowerment in areas that are undergoing peacebuilding and development? This chapter examines the issue of participation in relation to conflict and development. One of the overall themes of this book is a consideration of the extent to which people are included or excluded from peace and development processes. So-called ordinary people are usually excluded from the

DOI: 10.4324/9781003353485-4

major decisions that surround war, peace and development, and they are often written out of the histories of these events. The chapter will illustrate how this 'writing out' occurs and consider the steps that have been taken to include or 'write people in' to political and economic processes connected with war and peace.

Participation

As Chapter 2 discussed, many of the structures and institutions connected with conflict and development are elite-led and non-participatory. In fact, parts of the international financial architecture have been deliberately designed in order to maintain an oligopoly (Glenn 2008). Similarly, many political structures are maintained to limit participation or channel it in particular directions. Against a backdrop of the essentially conservative nature of many political and economic structures, there has been a growing realization by some international organizations, governments, INGOs and others that popular participation is the key to legitimacy and thus sustainability. Interestingly, this realization has tended to be projected from the Western developed world towards the developing world and has not been reflected back into developed world polities, many of which also suffer from democratic deficits. Across many development and peacebuilding spheres, there has been an increased emphasis on 'local participation', 'ownership' and 'partnership' (see Further reading).

Before examining participation in development and peacebuilding, it is worth noting that popular participation (voluntary and involuntary) is required in conflict (Conteth-Morgan 2004: 112–13). There is a startling disparity between the war-fighting practised by a handful of Western electronically advanced states and that practised by combatants on the rest of the planet. Techno-war, as practised by the United States or the United Kingdom, requires relatively few direct protagonists. The actual number of troops deployed on the ground is often quite small due to technology and a division of labour. Even operations in Afghanistan, Iraq or Sierra Leone, in which large numbers of US and UK troops were deployed, benefited from tremendous 'back office' support in the form of communications and coordination bases thousands of miles from the combat theatre (US military operations are directed from Tampa, Florida), while many support functions (for example, catering and logistics) may be conducted by private contractors (Boot 2007). By late 2003 – just months after a full-scale invasion – there were more private security contractors in Iraq than British soldiers (Traynor 2003). Indeed, the combination of professional militaries and 'outsourcing' to private firms means that many citizens in the United States and the United Kingdom are insulated from the fact that their states are engaged in permanent war.

This position differs enormously from most other violent conflict environments. In such cases, the impact and costs of conflict are often felt closer to home and are more difficult to avoid. For many people, participation in violent conflict is not a choice, and there are few opportunities to escape.

Consider, for example, a family in Syria as the anti-Assad movement escalated into a civil war involving President Assad's regime and a host of anti-regime militant organizations, many of which were engaged in feuding. Family members would have to make tough and risky choices, often based on sketchy information and rapidly changing circumstances. Some family members may be lucky enough to escape to Türkiye, Lebanon or Jordan, but this might depend on luck and having access to money. For others, particularly the elderly and infirm (and some women who, for customary reasons, could not travel unaccompanied), escape would be very difficult. Of those left behind, some might depend on the state and outright displays of loyalty to the Assad regime for their survival. Young males conscripted into the military may have to show enthusiasm while looking for a way out. Others may be able to support, and help, the rebels. But the rebels are a mixed bunch who range from relatively secular opponents of Assad to religious extremists. Again, family members may be faced with uncomfortable choices depending on who held the territory where they lived. The key point is that conflict, war and, indeed, many development processes impose invidious decisions on people. Participation in the conflict becomes mandatory because the conflict overwhelms the society and leaves little or no space that can be identified as somehow neutral or beyond the reach of the war.

One of the defining characteristics of the so-called new wars in the post–Cold War period has been the deliberate targeting of civilian populations through 'ethnic cleansing' or the calculated displacement of civilians to cause refugee flows (Kaldor 2012). Conflicts ongoing at the time of writing – Libya, Syria, Ukraine, Iraq, Yemen, Gaza, Sudan, South Sudan and Nigeria – have all involved the displacement of large numbers of people. Once displaced, these populations are vulnerable to a variety of problems ranging from public health emergencies to predation by human traffickers. While this is not unique to the post–Cold War period, the displacement of populations often leads to serious knock-on effects in terms of humanitarian emergencies. All organized violence involves 'violence specialists' or actors who take on specific roles during conflicts (Tilly 2002: 20–1). Yet many conflicts involve the mass mobilization of people, whether through rallies, the formation of popular movements or recruitment into militant or militant support groups (Horowitz 1985: 443–59). While Western armed forces may be defined by their high-tech methodologies, they also work hard to mobilize populations. The veneration of the US military in the United States, for example, is remarkable. While many US institutions have suffered a decline in legitimacy (the presidency, a gridlocked Congress, the police, Wall Street, and churches), the military evokes a romanticized loyalty and honour that few dare to criticize. The notion that the military 'can do no wrong' is a powerful mobilizing tool in times of conflict.

Many non-Western military forces (state and non-state) are decidedly informal and low-tech, and rely on large numbers of lightly armed combatants. In some cases, this may involve child soldiers, militias (who farm their land by day but serve as a 'home guard' by night), 'dollar soldiers' (whose loyalty is bought) or clan groups (as in Afghanistan) (Marriage 2007). The presence of young unemployed or underemployed males ('a disposable population' in neo-Malthusian parlance) may aid this mass involvement in violent conflict. The chief point is that in such cases, war may be a mass-participation activity rather than an outsourced and compartmentalized activity, as it is in Western countries.

The case study of how Sri Lanka decisively won its war with the separatist Liberation Tigers of Tamil Eelam (LTTE) in 2009 is instructive about popular mobilization for conflict. The insurgency had been ongoing since the early 1980s and had racked up high costs, including about 100,000 dead, many thousands displaced, the retardation of economic development in Sri Lanka and the assassination of the Sri Lankan and Indian presidents. A peace process in the mid-2000s fizzled out, and a new Sri Lankan president in 2005 decided upon an all-out effort to defeat the LTTE. In order to do this, President Rajapaksa invested in preparatory groundwork. Human rights organizations were harassed and booted out of the country (Barnes 2013), loans and military expertise and hardware were sourced from countries with little concern for human rights (Israel, China and Pakistan), government-backed death squads 'disappeared' hundreds of Tamils and some civil society activists so as to minimize voices of opposition (Mac Ginty 2010). But crucially, the Rajapaksa government invested heavily in mobilizing the majority Sinhalese population in favour of a sustained military offensive. This involved much propaganda on behalf of the military and addressing public perceptions of an institution with a mixed reputation. The war against the LTTE became a popular project and, ultimately, a successful one in the sense that the LTTE were defeated.

In addition to actual combatants, mobilizations require political support. Pro- and anti-Syrian forces in Lebanon sought to demonstrate their strength through rallies and counter-rallies in 2006–8. A key form of mobilization comes through political and 'ethnic entrepreneurs' who may seek to mobilize large crowds in support of their cause, often using public demonstrations as a way of indicating their popular support and thus their legitimacy (Kaufman 2001: 5–7). Ukraine, Lebanon, Azerbaijan, Iran, Egypt and Venezuela have seen popular demonstrations and counter-demonstrations, sometimes accompanied by violence and repression.

Political leaders, aided by symbolism and references to a historic past (real or imagined), may be incredibly successful in mobilizing large numbers of people (Mac Ginty 2003: 235–44). War and conflict become, in many cases, mass events involving large numbers of people as participants or active supporters. Over 50,000 people were estimated to have taken part in rioting in Kosovo in March 2004 (United Nations Mission in Kosovo [UNMIK] 2004). Importantly, mass-participation war and conflict may contrast sharply with the situation that prevails during peace negotiations and subsequent peacebuilding in which the opportunities for popular participation are severely constrained. Populations that felt involved in a 'people's war' – as in Colombia or Nepal – may find peace or peace negotiations a much less inclusive process. Box 3.1 illustrates the secrecy that surrounded the origins of the 'Oslo Process' between Palestinians and Israelis.

Box 3.1 The Oslo Peace Process: a closed system

The 1990s 'Oslo Process' between Israel and the Palestine Liberation Organisation (PLO) began as clandestine negotiations in a secluded house owned by the Norwegian Ministry of Foreign Affairs on the outskirts of Oslo (Pruitt et al. 1997: 177–82). Staff at the house were told that the occupants were academics working to a strict deadline imposed by an exasperated publisher. At the time of the negotiations, it was illegal for Israeli citizens to meet with members of the PLO. The talks were so secret that many senior members of the PLO and the Israeli government did not know of their existence. The talks were a useful way for each side to test the seriousness of their adversary should more formal and detailed negotiations develop. Problems arose when news of the talks was made public (Wolfsfeld 2003: 92–3). Both the Israeli government and the PLO had invested heavily into demonizing the other side, so they now had to prepare their respective constituencies for the prospect that it might be possible to make a deal with their historic enemies. The seriousness of this task was made clear with the 1995 assassination of one of the architects of the Oslo Process, Israeli Prime Minister Yitzhak Rabin, by a right-wing Israeli who thought that the peace process would lead to ruin. Those involved in the early phase of the Oslo Process faced a real conundrum: do they try to make advances away from the glare and turbulence of publicity, or do they open up the process to public scrutiny before it has had a chance to take root in the minds of the negotiators? Ultimately, the Oslo Process was not sustained. Crucial in its demise was the differential perception of gains and losses held by Israelis and Palestinians. 'Each side to the conflict has continued to see the "other" as being the sole recipient of all the benefits, while not offering anything in return' (Hermann and Newman 2000: 110–11). It is, of course, overly simplistic to say that this perceptual differential was due solely to a failure by each side to prepare their constituencies for the give and take of a peace process, but the interface between the public and private aspects of a peacemaking process requires careful management.

Sources: Pruitt et al. (1997); Hermann and Newman (2000); Wolfsfeld (2003)

The example of peace negotiations in the aftermath of civil war shows how many political networks are 'closed systems' that have firm boundaries of inclusion and exclusion. In the case of peace negotiations, this may be due to understandable security reasons: spoiler groups may be tempted to use violence outside the conference room to influence events inside it (Darby 2001). In practical terms, complex negotiations are often best left to small and dedicated teams who have mastery over their brief. In many cases (for example, peace processes in Sri Lanka, Colombia or Ukraine), the talks have taken place away from the locus of the conflict to allow negotiators to concentrate on wider issues without being blown off course by the immediacy of political violence and the attendant headlines. Problems arise when the existence of talks are made public. Antagonists often invest considerable energy in demonizing their enemies, so to turn around and admit to having talks with 'the enemy' requires a delicate massaging of the message. Similarly, the contents of the talks and an admission that compromises may have to be made, also require the preparing of constituencies that are more used to being sold messages of complete victory (Darby and Mac Ginty 2003: 267–8).

Opportunities for public involvement in peace negotiations may be limited to public demonstrations to signify a desire for peace (or at least an end to violence), a referendum on a peace accord or new constitution or the election of a post-peace-accord government (Reilly 2003). Given the circumstances, these opportunities may be valuable (for example, if elections had not been held before or for a very long time), yet they can be best described as one-off events rather than sustained processes that allow for continued and meaningful relationships between citizens and wider political processes. Electoral turnout in some post-conflict and post-authoritarian countries shows that, after initial enthusiasm, many voters realize that elections change little. Post-civil-war parliamentary elections in Mozambique in 1994 had a turnout of 88 per cent. By 2014, the figure was less than 50 per cent. Bosnia-Herzegovina, Cambodia, Timor-Leste, Iraq, Liberia and South Africa have all seen declining voter engagement after initial euphoria (IDEA 2015).

One reason that elections and subsequent constitutions have limited appeal for societies that are recovering from war is that the local population has learned not to put as much faith in them as foreign governments and IGOs, who want to prove 'success' in the disbursement of what are often eye-watering amounts of money on election promotion, constitution-building and associated aid and development programmes. Liberal governments need 'good news' stories. In the post-1990 period, one such was provided by both Eritrea and Ethiopia, who were emerging from a long and vicious war of separation. IGOs were anxious that the fragile peace brought about by the end of that war was not squandered. The leaders of the now two countries were feted, but another war erupted in 1998, and a subsequent even worse conflict has been seen since 2020 between the membership of the initial post-1993 government of Ethiopia and their successors in Tigray with the new Ethiopian government, the leader of which, Abiy Ahmed, was awarded the Nobel Peace Prize in 2019 and is now accused, along with Eritrea, of terrible war crimes.

One of the key among these 'good news' stories was Myanmar, where a military junta was persuaded after decades of refusal and civil war to allow the release of political prisoners, a freer media, and serious peace talks that led to democratic elections in 2015. These led to a massive victory (of 86 per cent of the voting electorate) for the opposition leader Aung San Suu Kyi. But the war continued in parts of the country, notably in the North (Kachin State and the Shan State) and most notably in Rakhine State. Democracy merely brought to the surface long-repressed animosities, notably between the Rakhine and the "Rohingya" peoples who had been singled out by Burman-dominated military juntas for decades, based partly on their Muslim identity. Inter-communal violence resumed with terrible ferocity against this 'alien population' with Rohingya militants fighting with Myanmar military units. Rohingya are considered to be stateless, and many were killed, subject to sexual violence and expelled from the country, mainly to neighbouring Bangladesh, which also hardly recognizes them (Ware and Laoutides 2018). Elections proposed for 2023 were described by

Crisis Group as 'a road to nowhere' that would 'trigger escalated violence' (Crisis Group 2023).

What possible solutions have been suggested to such dilemmas? Specialist post-conflict programmes may require or invite the direct participation of particular sectors: for example, members of the security forces may be compelled to take part in a truth and reconciliation process or combatants may be required to disarm and take part in retraining programmes. These formal aspects of a peace process may be described as being overwhelmingly male, limited to specific demographic and political groups, and time-limited. For many other people, however, a peace process may be something that occurs elsewhere (in a capital city) or their only connection with it may be through the media or word of mouth. Thus, the peace process may seem to be 'a creature of the international community and their co-opted national elites and [have] limited connection with the bulk of citizens in the war-affected state' (Darby and Mac Ginty 2008: 5). There may be few opportunities to affect what Harold Saunders (1999) calls 'a public peace process', in which the citizenry can have a substantive input and broaden out formal political processes to involve a wider constituency.

A growing interest in participation among international organizations, INGOs and NGOs is partially a recognition of the dominance of closed systems and the failings that accrue from excluding key constituencies. The pro-participation consensus is based on the belief that people can become 'stakeholders' in projects and feel that as a result of their investment, they have ownership in a process. The theory continues that such locally 'owned' processes are more legitimate and more likely to succeed because local constituencies will be able to shape them to suit local needs and aspirations. In such circumstances, external assistance and direction will not be as necessary, and local actors will be able to sustain the process. International organizations, INGOs and NGOs now routinely pay attention to issues of participation, regarding it as 'a fundamental human right', a way of enhancing 'the sustainability of the settlement' and a means for 'managing inclusion' (ACCORD 2003; Herbolzheimer 2015). Major policy documents from leading international organizations that work on issues of peacebuilding are now full of references to local participation, partnership and legitimacy. Two decades ago, that was not the case. The 1992 *Agenda for Peace* document published by the UN is often seen as the seminal explanation of peacebuilding (Boutros-Ghali 1992). The word 'local' is not mentioned in it. By contrast, the World Bank's 2011 *World Development Report* (World Bank 2011) and the 2011 United Nations Development Programme's *Governance for Peace* document (UNDP 2011) mention the word 'local' 382 and

197 times, respectively. Closer support to national and local actors was the core message given at the World Humanitarian Summit (2016: see Chapter 6).

There has, however, been criticism that many participatory schemes are superficial and less empowering than their advocates would suggest. Cooke and Kothari (2002) point to the 'tyranny of participation', whereby communities in developing-world and post-conflict environments are shoehorned into superficial participation mechanisms that suit the requirements of donors (allowing them to tick a box attesting to 'local participation') but do not necessarily involve local communities in meaningful and sustainable ways. Similarly, Alternative Dispute Resolution techniques have been criticized for their 'coercive harmony' whereby individuals and groups are encouraged to participate but find themselves pressured to reach an accommodation, even if it leaves grievances unaddressed and unbalanced power structures in place (Nader 1997). Rather than being truly inclusive, the criticism made in many development and post-conflict environments is that participation mechanisms are just another part of a suite of essentially Western mechanisms that are top-down and conceived and funded in the West (for example, Mac Ginty and Firchow 2016). According to this view, many participation mechanisms are actually part of a process of disempowerment. Different cultural explanations of 'participation' may be in operation. Western notions of participation may prioritize institutional and technocratic mechanisms such as quantifiable, one-off events like elections or consultation exercises. In other societies, participation may be interpreted in more people-centric and relationship-orientated ways, in which participation is a process rather than an event (Lederach 1995: 26). It is important, though, not to romanticize non-Western and indigenous means of popular involvement in political or economic decision-making since many of these processes can reinforce social order and conservatism.

Certain phases of development and peacebuilding processes may be more open to public participation than others. Necessarily, for example, the security and 'stabilization' phases in the aftermath of a violent conflict might actively exclude people and concentrate power in the hands of military forces and a limited number of 'institution builders'. 'Institutionalism before Liberalisation' (Paris 2004) or the belief that stability and functioning bureaucracy must be achieved before democratization has been accepted by most states and international organizations in post-war reconstruction contexts. Yet, at some stage, all development and peacebuilding processes require legitimacy. The ways in which political leaders (international and national) seek to build and maintain legitimacy are crucial. In some societies, guerrilla movements have transformed very quickly into political parties (the morphing of the Kosovo Liberation Army into the Kosovo Democratic Party provides one example). While the transition from protagonist to pragmatist is obviously crucial in any post–civil war transformation, there is a danger that some political parties are too narrowly based and merely continue the civil war by peaceful means (Gormley-Heenan 2001). The case of Robert Mugabe's Zanu-PF in Zimbabwe provides an example of a party reluctant to leave behind its civil war roots. When under pressure, it has repeatedly used the language of liberation and evoked the independence struggle against white rule: a good strategy for mobilizing supporters but of little help for Zimbabwe's economic woes. Many post-civil-war societies have struggled to find a political process that sustains public interest and participation.

Civil society

Civil society can be regarded as both an institution and a space and is often regarded as a bulwark against arbitrary control by government. It is defined by the Centre for Civil Society as

> the arena of uncoerced collective action around shared interests, purposes and values. In theory, its institutional forms are distinct from those of the state, family and market, though in practice, the boundaries between state, civil society, family and market are often complex, blurred and negotiated. Civil society commonly embraces a diversity of spaces, actors and institutional forms, varying in their degree of formality, autonomy and power.
>
> (Centre for Civil Society 2008)

Through the media, trade unions, businesses, professional associations, voluntary and church groups, and online campaigns, citizens can become organized, voice their concerns and engage in activities that may support or challenge a government (Burnell and Calvert 2004). In the Western political mind, an unimpeded civil society is regarded as a key indicator of political freedom. American sociologist Robert Putnam (2000), whose classic study *Bowling Alone* captured the decline of community in the United States, listed the ways in which US citizens could become politically involved:

> contacting local and national officials, working for political parties and other political organizations, discussing politics with our neighbors, attending public meetings, joining in election campaigns, wearing buttons, signing petitions, speaking out on talk radio, and many more.
>
> (Putnam 2000: 31)

Plate 6 Mine warning sign, Tyre, Lebanon. Copyright ICRC. Photographer: Marko Kokic. 27/3/2007.

To this, we can add the Internet, which has opened up multiple ways through which political views can be aired. For many of us, civil society is non-contentious, has regular and structured access to the corridors of power and adds colour to the dull routine of politics.

In authoritarian, deeply divided or post-war societies, however, civil society can play a more delicate role and may face significant difficulties in organizing and expressing itself. China and Russia, for example, introduced laws to control civil society and limit the activities of internationally sponsored NGOs (Denyer 2015). In Israel, those who speak out against the occupation of Palestinian lands and human rights abuses against Palestinians are often targeted by right-wing mobs who seem unhindered by the police (Zonszein 2014). In Cambodia, protestors against land-grabbing (or the transfer of subsistence agricultural land to commercial landowners and agribusinesses) have faced intimidation and have received little protection from the state (Pilorge 2012). General Morsi's counter-revolution in Egypt saw the imprisonment of three Al Jazeera journalists and a clampdown on the media (see Box 3.2). The key point is that civil society is often on the front line in development processes, the prevention of conflict, and peacebuilding and reconciliation in the aftermath of conflict. Civil society actors are often most prominent in 'bearing witness' or publicizing injustice. However, many actors in conflict and developing contexts make it their business to sideline civil society actors. For example, most peace negotiations include armed groups (state and non-state) but tend to exclude civil society actors. The 2014–15 Havana peace talks held to end the long-running civil war in Colombia were unusual in that victims' groups were present at the peace talks.

Box 3.2 Press freedom

By May 2015, the NGO Reporters Without Borders reported that 25 journalists had been killed and 158 had been imprisoned so far that year in Eritrea. While all states and organizations will attempt to influence the media so that they are reflected in a flattering light, some states and organizations attempt to control or suppress the media. The WikiLeaks episode and the harassment and jailing of whistleblowers like Edward Snowden and Bradley Manning showed that Western states could be intolerant of those who publicize the wrongdoings of the state. The World Press Freedom Index, however, shows that the greatest press freedom 'offenders' are developing-world states or authoritarian Arab states like Saudi Arabia and Bahrain. Eritrea props up the Index. The UK government reported, 'The government of Eritrea controls all domestic media outlets and requires all publications to be approved. It is reported that at least 28 journalists are behind bars. Eleven journalists are reported to have been detained since 2001, seven of whom are reported to have died [in captivity]' (FCO 2015). There is no privately owned media, and journalists are not permitted to enter the country (Greenslade 2014).

Recognizing the utility of civil society, international organizations and INGOs often regard it as a key transmission agent for development, conflict prevention and peacebuilding. As a result, immense energy and resources are invested into creating or shoring up civil society. It is here that problems can emerge. Western interveners may have a particular concept of civil society in mind that may not always fit with the political, social and cultural norms of the host society. Western actors may attempt to create a civil society in the image of their home civil society, oblivious to the fact that a vibrant civil society exists in the post-war or developing-world context. A classic example of this comes in the form of the US response to the looting that followed the 2003 American-led invasion of Iraq (Mac Ginty 2004). According to US military commanders, the problem was that there was no civil society that could control the civilian population and coordinate attempts to repatriate looted goods with their original owners (BBC News 2003). The view that Iraq was bereft of civil society was widespread. According to Toby Dodge, 'Before the liberation of Baghdad it was impossible to talk about civil society in Iraq … autonomous collective societal structures beyond the control of the Ba'athist state did not survive' (cited in Pirouz and Nautré 2005: 3–4). Yet, notwithstanding the brutalities of the Saddam Hussein regime, there was some sort of civil society in place: mosques, family groups, professional associations, local Red Crescent branches and other networks. Certainly, these networks were very far removed from the plural and voluntary forms of civil society found in Western liberal democracies. Yet, to assert that there is 'no civil society' may overlook various forms of social capital that may not be immediately visible to Western eyes.

In many post-war societies, Western interveners have taken it upon themselves to create civil society, often in the image of Western models (Prusher 2003). Through 'local capacity building' schemes and the direct funding of NGOs, international actors can create a civil society that is recognizable to them but may not necessarily have deep and effective roots in the host society (Adamson 2002: 200). One may be forced to ask: what is the essential purpose of this new civil society? Its key role is often to provide a recognizable interface with which international organizations and INGOs can engage. The connection between civil society and the host society may be thin, limited to a particular stratum of the host country that has been artificially created by international actors. The new civil society may be unsustainable because of its dependence on Western funding streams. A Western-created and -inspired civil society, or the 'democracy sector', is often highly visible in post-war societies. It is often urban, metropolitan and English-speaking (Mac Ginty 2006: 52). Indeed, in many cities, it often inhabits the same physical space as the INGO sector that was responsible for spawning it and fuels the same micro-economy of vehicle hire, Western-style coffee shops and bars, and a conference support sector. The key issues with civil society in any context are the extent to which it can be inclusive, voluntary and act as a real counterweight to state institutions. These issues are particularly acute in deeply divided societies in which society may be structured in a way that excludes or discriminates against particular groups. Elaborate structures (for example, human rights commissions) or legal codes (that outlaw discrimination) may be created in the wake of a civil war, but it does not always follow that institutional engineering will bring with it changes in societal behaviour and thought. India has had legislation outlawing the caste system for decades, but that system of apartheid is still firmly rooted in everyday life. Indeed, some would claim that India has become a more intolerant society since the rise to power of President Narendra Modi (though others disagree: Singh 2023).

It is worth stressing that the 'creation' of civil society by external agents and consequent problems between the new civil society and the host society are by no means restricted to post-conflict settings. The phenomenon of the artificial civil society, especially the democracy-promotion sector, is common in many non-conflict developing contexts.

Importantly, civil society need not be restricted to one polity or location. Civil societies can be transnational, especially so in a globalized and digitally connected age. Indeed, given that so many conflict and development-related problems are transnational, it makes sense that civil society mirrors this. Moreover, some locations are so unwelcoming to dissenting civil society organizations that the only meaningful opposition can come from outside the state boundaries in the form of diaspora organizations.

The concluding point of this section is to differentiate between civil society as a set of organizations and civil society as a behaviour. There is a danger that civil society is reduced to being civil society organizations, such as NGOs with offices, staff, programmes and projects. This type of civil society (while being valuable in many cases) risks being shallow. Its agendas might be donor led; it exists only so long as overseas bodies fund programmes; it may be incestuously urban, with the same faces turning up to a narrow range of events in the same city and saying the same thing. Travel to Belfast, Sarajevo, Pristina, Beirut and other 'post-conflict cities', and one is likely to realize that there is a small world of civil society actors who speak the same professional language, attend the same events and often socialize together. A broader view of civil society can be more fruitful, however. This view regards civil society as a phenomenon not just restricted to civil society organizations. Instead, it regards civil society as a much more organic and evolutionary space composed of individuals, groups and organizations. The role, profile and public recognition of the volunteer have had an increasing profile since the COVID-19 pandemic, though voluntary service has played a crucial role in civil society since human civilisation began. The Red Cross and Red Crescent Movement has 16.5 million active volunteers living and working in at-risk communities around the World (IFRC Annual Report 2022).

This space will be a complex and dynamic tapestry of relations. As befits a deeply divided society, it may be riven with fractures and reflect major sectarian or identity divisions. Much civil society operates at the intra-group level. Here it may be divisive and contribute to the intensification of division in society. But, alongside this negative purpose, it might also be positive at the intra-group level: providing social services, restraint and guidance.

In this view, civil society is complex and dynamic. It can simultaneously restrain and inflame conflict. It is not restricted to artificially created organizations and instead inhabits the realm of everyday life. It is made and remade on a daily basis through intra- and inter-group contact. It occurs in the stairwells of apartment buildings, on the street in the city neighbourhood and in the village market. This everyday civil society might be difficult for outsiders to spot, but it can contribute to the glue that prevents a society from moving from chronic tension to all-out war (Mac Ginty 2014).

Gender

As our colleague in this series, Janet Henshall Momsen, writes, '[G]ender is a development issue', which she explains in detail in the three editions of her work on the topic. She deals at length with issues such as the 'after-effects of colonialism' and the often pernicious effects of modernization that has often, and maybe paradoxically, 'increased the gender-based disadvantages' of everyday life in 'subsistence and centrally planned economies'. The modernization of agriculture is singled out for altering age-old divisions of labour to the detriment of women's traditional roles, which, though often involving harsh labour, gave them a certain degree of autonomy and respect (Momsen 2004, 2019: 1–2). We cannot hope to fully reflect these themes, but what we can attempt to do here is to try and relate them to how they have fed conflict in developing societies, often in ways entirely unintended by well-meaning liberal development specialists.

The 1997 United Nations Human Development Report noted starkly that 'no society treats its women as well as its men' (UNDP 1997: 39). There is every reason to believe that this judgement still holds true. Even in apparently fair and peaceful societies, the gap between male and female experiences is startling. According to the Office for National Statistics (ONS), median hourly pay for full-time employees was 8.3 per cent less for women in the United Kingdom than for men in April 2022 (ONS 2023). Four out of five firms in Britain pay women less than men (Duncan 2023). These gender differentials may be stretched and distorted during conflict and development. And in the UN itself, there is a real disparity between both the roles and ranks of women in the organization compared with those of men. Kirsten Haack's contribution to the debate is important in that she shows that not only are women underrepresented as leaders in wider society, their only recent ability to influence development policy, but that scholarship has not yet caught up with how women have made a difference. This has

> meant that scholars have been unable to systematically account for women's contributions to policy-making in international organisations—specifically how gender norms and gender equality agendas are advanced alongside advocacy by the women's movement or, notably, in areas where the global women's movement is less likely to reach.
>
> (Haack 2022:5)

Discussion of gender issues in relation to conflict is made difficult by a potential minefield of stereotyping in which women are characterized as life-givers, nurturers and generally pacific, while men are depicted in the warrior mode (Jacoby 2008: 92). Such stereotypes are best avoided. There is also a tendency in some literature and practice to lump women together as 'victims' or an undifferentiated category without agency. Often, 'women and children' can be found as an apparent afterthought in the later stages of official reports. Having opened with a caveat, we can then say that development, conflict and peace impact men and women in different ways, with the burdens and benefits shared unevenly. But we cannot make blanket statements that have universal applicability; clearly circumstances and context matter.

Box 3.3 A clash of cultures?

A postgraduate class was having a discussion on democratization in post-war societies. The author asked a female Afghan student if she was looking forward to voting for the first time when she returned to Afghanistan. She said that she was not. At this point, the backs of the Western female students visibly straightened. They had been brought up in societies where votes for women were a right and a norm. The idea of an Afghan woman not wanting to vote when that right had been denied to them for so long was shocking. But the Afghan student continued, 'My father will decide who I should vote for. So this is not a vote for me. It's an extra vote for him.' She said that she wanted access to decent health care, education and other basic social services more than participating in an election. Moreover, she did not see a direct link between participating in an electoral process and the delivery of such services. 'Afghan society is clientelistic,' she explained. 'It's all about who you know and not how you vote. For me, voting isn't a "gift" from the West. At best it's an inconvenience.' The Western students were mollified by this explanation, but the incident illustrated a clash of cultures. The staging of elections in post-Taliban Afghanistan was more a priority for Western governments anxious to legitimize the rule of their appointed leader than a priority for many Afghans.

The principal gender distinction between men and women in times of conflict is that men are often (but not always) the main combatants, while women often perform (but not always) ancillary or support roles (Turshen and Twagiramariya 1998; Rehn and Johnson Sirleaf 2002). But as has been pointed out frequently in this book, contemporary civil wars and contexts of political suppression are very far removed from the classical notion of warfare with professional militaries and clear lines of distinction between combatants and non-combatants. This blurring of the demarcation between combatants and non-combatants has implications for how war impacts men and women. Often, entire ethnonational groups are targeted or mobilized, regardless of gender. Within these groups, men and women may be affected in different ways (women more prone to sexual violence or men more prone to arrest and torture), but the group may suffer the same overall experience. Women may not be the main frontline targets of groups like Islamic State or Boko Haram, but once these organizations seize territory, women become the object of social policing.

Development processes and the ending of wars provide 'transition moments' in which it is possible to make significant societal changes – for example, in relation to land reform, public access to power or the treatment of women. It is not always clear that these opportunities are taken. Certainly, in peace negotiations, issues of territory, security and constitutional design are often prioritized before 'secondary' issues such as gender, truth recovery or social inclusion. If, at best, peace processes can be labelled as 'gender blind', then they bring with them the danger that they reinforce the existing patriarchy. It may be the case that external actors, such as foreign governments and INGOs, favour the inclusion of gender reform in peace negotiations or the formulation of a new constitution. Local actors may not agree: either they may not see the issue as urgent, or they may see it as undesirable as it confronts dominant cultural norms. Or local civil society may advocate gender reforms that mark a significant advance for the country in question but are not as radical as some Western actors would favour. This raises an interesting series of questions: whose definition of gender equality is to be followed? What are the main components of any gender equality programme? Will the promotion of gender equality place women in danger? Such 'clashes of culture' (see Box 3.3) between Western norms and indigenous practices are common in virtually every aspect of development and post-conflict reconstruction. It is important to note that 'Western' and 'indigenous' do not comprise discrete categories: they vary enormously, and all social organizations and practices are – to some extent – hybrids. But the key struggle between Western and indigenous norms persists. It manifests itself in Western notions that there are 'proper' forms of social organization. Conflict may arise as different elements in developing and post-conflict societies may attempt to resist, accept or subvert external development interventions and the cultural baggage associated with them.

It is one thing to outlaw gender discrimination, but it is another thing to change the cultural norms and thought processes that support it. As Antonia Potter (2008: 105) notes, '[R]eality lags far behind rhetoric on women's involvement in peace processes'. Certainly, an impressive legal and normative framework has been erected internationally and in many national contexts over the past few decades to protect the rights of women and encourage their full participation in political life. The 1979 Convention on the Elimination of All Forms of Discrimination against Women, the 1995 Beijing Declaration and Platform of Action and the 2000 UN Security Council Resolution 1325 on Women, Peace and Security all count as international landmark documents for the inclusion of women. These and other initiatives are testaments to the incredible energy of women's movements that often operate in unwelcoming environments. Crouch (2004) calls women's mobilization 'a great democratic phenomenon' that has included

> extreme radicals; sober reformist policy-makers; cunning reactionaries taking the movement's messages and reinterpreting them; both elite and popular cultural manifestations of many kinds; the gradual suffusion into the conversation of ordinary people of elements of the language of an initially esoteric movement.
>
> (Crouch 2004: 62)

Yet, as Potter (2008) politely observes,

> it might not be unreasonable to worry that the slow pace of change implies that misogynist or bigoted views continue to hold a certain amount of sway in the most progressive and liberal societies, even if unconsciously.... A review of the literature and case studies on post-conflict situations, reveals ... a depressing paucity of examples of implementation.
>
> (Potter 2008: 107)

The very fact that global activism on gender issues is organized around 'Beijing +20' (that is, the Beijing Summit 20 years on) suggests that many of the original goals of equality have not been met. It is sobering to think of the gender differentials that exist despite the enormous political, economic and social change experienced by many societies since the 1950s and the massive development and peace-support interventions since the end of the Cold War. Janet Hunt (2004) sums up the situation thus:

> Women are said to be 70 per cent of the world's poor; have only about 10 per cent of all parliamentary seats in most countries; and in developing countries, on average, they earn only 73 per cent of male earnings. There is not a single developing country in which women and men enjoy equal rights under the law. In particular, women are discriminated against in areas such as their right to land and property, and their right to conduct business independently. The result is that women are more vulnerable to poverty than men, especially as a result of widowhood, separation or divorce, and the consequent loss of access to productive assets.
>
> (Hunt 2004: 243)

Many development interventions have sought ways in which to include women in economic growth. Thus women's participation is built into many development projects or programmes (see Box 3.4 on micro-finance).

Box 3.4 Women's empowerment through micro-finance initiatives

Many micro-credit or micro-finance initiatives have had a deliberate gender dimension. Indeed, in 2006, seven out of ten micro-finance clients were believed to be women (Armendáriz and Broome 2008). Small loans have been targeted at women in the hope that the resulting small businesses will first help alleviate poverty for them and their families, and second act as a means of empowerment. Women have been given loans to establish themselves as small-scale chicken farmers or to come together as a cooperative to market their handicrafts (Bigsten 2003: 115). Mayoux (2002) notes,

> Microfinance programmes have significant potential for contributing to women's economic, social and political empowerment. Access to savings and credit can initiate or strengthen a series of interlinked and mutually reinforcing 'virtual spirals' of empowerment. Women can use savings and credit for economic activity, thus increasing incomes and assets and control over income and assets.
>
> (Mayoux 2002: 76)

There has been something of a bandwagon championing micro-credit. The UN declared 2005 the 'Year of Micro-Credit', and the Grameen Bank (a Bangladeshi initiative that has subsequently spread to 28 countries) was awarded a Nobel Prize. There is no doubt that micro-credit schemes have economically empowered some women, and there is evidence that there have been additional one-off benefits in terms of social capital and public perceptions of women (Sanyal 2006). Critics, however, argue that not only are the economic benefits of micro-credit schemes often overblown but also the schemes may not be as liberating as their advocates suggest. For example, it is suggested that if economic empowerment is not linked to other empowerment (for example, greater legal powers for women such as the right to buy and sell land in certain countries), then stand alone micro-credit schemes will have limited impact (SIDA (Swedish International Development Cooperation Agency) 2006). Others have noted that indebtedness loads women with an additional burden and may bring them into conflict with men in the home. Just as importantly, without structural change to the organization of the economy (almost impossible given the power of international economic forces), any changes linked with micro-credit schemes or pro-poor growth strategies are likely to be sporadic and dependent on local circumstances.

Sources: Mayoux (2002); Bigsten (2003); Sanyal (2006); SIDA (2006); Armendáriz and Broome (2008)

The participation of women is seen as a crucial pillar in the delivery of humanitarian assistance in vulnerable communities. The hope is that multiple small schemes will have a cumulative impact, not only on the lives of women but also on the societal and structural factors that dictate women's position in society. Recent edicts banning the participation of women in public service in Afghanistan, as well as revelations of systemic misogyny among some leading public institutions in the United Kingdom and America, show there is still considerable progress to be made.

A final issue to consider in relation to gender and conflict is the role of equality. For some, full equality means the ability of females to serve in frontline positions. Thus, for example, some news sources have trumpeted Pakistan's first female fighter pilot as 'a role model for millions of girls' (Crilly 2013). Yet, much feminist literature would contest the notion that the equality to kill is equality at all. For critical feminists, it is merely the acceptance of a female into a male-dominated structure and does not alter the basic male dominance of systems of government and militarism (Enloe 2000, 2007; Sjoberg 2013). Most feminist theorists prioritize structural change of the system over piecemeal 'equality' that implicates women in a system that is unjust, violent and fundamentally exploitative of women.

Conclusions

A fundamental issue running through this book is culture and the potential of Western norms and practices connected to conflict, peacebuilding and development to clash with local, indigenous or traditional norms and practices. We should, of course, be careful not to build a strict dichotomy in which everything from 'the West' (by no means a discrete category) can be described as top-down, legalistic and rational, and everything found *in situ* in developing and post-conflict environments as bottom-up, organic and traditional. There is a very real danger of romanticizing the local and the traditional. At heart is the question of the appropriateness of external intervention and the appropriateness of the type of intervention. A case-by-case approach is best. In some cases, external states and actors are the only ones with the capability to effect the scale of change required to make a difference (for example, UNICEF and Médecins Sans Frontières). In other cases, the actions of external (often Western) states or actors undermine the capability of local states and actors. Many development and peacebuilding contexts are the scene of a cultural conflict between what might be described as variations of Western liberalism and local belief systems and practices of social and political organization. These conflicts revolve around the extent to which liberalism is adopted, enforced or resisted.

It is worth ending a chapter on participation by restating the agency of local populations. It is tempting to depict local communities as passive actors who receive or consume conflict, peace and development that is made elsewhere. Certainly, the international and transnational nature of conflict and development means that there are few hermetically sealed contexts on the planet. All societies are penetrated by external forces to some degree, and the sheer scale of development and peace-support interventions is awesome in some contexts. But we should not overlook the local. Keesing (1992: 2) stresses the importance of seeing 'peripheral populations as active agents in shaping and controlling their engagement with the outside world, giving local meaning to alien ideas, institutions and things, in various ways resisting them'. Local communities have indeed power to absorb, renegotiate, subvert and resist external pressure. Clearly, this power will differ from context to context, but one of the most remarkable aspects of globalization and liberal peace interventions is the extent of local variation. If nothing else, this local variation is a sign of participation in development, conflict and pacific processes.

Every international intervention in another country has unanticipated consequences. Whether it is the US invasion of Iraq, the Saudi Arabian bombing of Yemen or the presence of Nigerian peacekeepers in Darfur, local people react in ways that military and humanitarian planners did not foresee. The more farsighted of the international interveners are able to find ways to work with local actors, to calibrate their intervention and to know when to leave.

Summary

- Different types of conflict, peacebuilding and development have different ways of excluding and including people.
- Peacebuilders and those engaged in development processes often struggle to establish wide levels of popular participation in their projects.
- Civil society is often regarded as an important bulwark against the unrestrained actions of governments or corporations, but there is a danger that Western notions of civil society are regarded as best.
- It is important to see civil society as more than just civil society organizations. It is also a space that can be organic and evolutionary.
- Development and peacebuilding processes consistently sideline women.
- There is a serious clash between Western notions of gender rights and notions of gender rights in non-Western contexts.
- At the centre of many conflicts and tensions between Western and non-Western actors is culture: an issue that Western observers often overlook.

Discussion questions

- How should development projects attempt to gain popular legitimacy and acceptance in the context of a developing country?
- What avenues of political participation are available to you in your own country? Would you recommend them to people in other countries?
- Must civil society be open to all groups, especially in a deeply divided society?
- Should women in Afghanistan enjoy the same rights as women in America?

Further reading

There is an extensive literature on the role of civil society in peacebuilding, with Thania Paffenholz's (2012) *Civil Society and Peacebuilding: A Critical Assessment*, Boulder, CO: Lynne Rienner, probably the most comprehensive and relevant. Issues of culture (and how we interpret other cultures) are very important to inclusion and exclusion in development and conflict. Good starting points are P. Chabal and J.-P. Daloz (2006) *Culture Troubles: Politics and the Interpretation of Meaning*, London: Hurst, P. Chabal (2012) *The End of Conceit: Western rationality after postcolonialism*, London: Zed and Pankaj Mishra (2012) *From the Ruins of Empire: The Revolt against the West and the Remaking of Asia*, London: Allen Lane. Three modern classics are also worth reading to help us reflect about culture ('theirs' and 'ours'): S. Huntington, The clash of civilizations?, in *Foreign Affairs* 1993, 72(3); R. D. Kaplan, The coming anarchy, *Atlantic Monthly*, February 1994; and E. Said (1979) *Orientalism*, London: Vintage. Anthropology, journalism and personal testimony often give the best perspectives on the impact of conflict on women. See, for example, S. Drakulic (1994) *Balkan Express: Fragments from the Other Side of War*, London: Perennial or Halima Bashir (2012) *Tears in the Desert: One Woman's True Story of Surviving the War in Darfur*, London: Hodder and Stoughton. ALNAP has published an online collection of resources on local responses to the Covid-19 pandemic https://covid19.alnap.org/. A series of studies by HERE look at the role and profile of local actors in humanitarian response https://here-geneva.org/publications/

Useful websites

Many international organizations and INGOs have information on their websites labelled 'participation' or 'local partners', though it is useful to consider where the power lies in relationships between local and international actors. See, for example, the Asian Development Bank (www.adb.org) or ECHO – the European Commission's Humanitarian Aid Office (ec.europa.eu/echo/). United Nations Women (http://www.unwomen.org/en) contains very useful information on gender, development and conflict. Other development-related organizations, such as the World Bank (www.worldbank.org) or USAID (www.usaid.gov), contain material on the gender dimension of their programmes. You may have to use the search function on these organizations' websites. See also the gender-related information maintained by Human Rights Watch (www.hrw.org), the Hunt Alternatives Fund (www.huntalternatives.org), International Alert (www.international-alert.org) and Amnesty International (www.amnesty.org).

Bibliography

ACCORD (2003) Owning the Process: Public Participation in Peacemaking – Principles to Guide Policy and Practice. Presentation by Conciliation Resources ACCORD Programme at an International Peace Academy Conference. New York. 12 February. [No longer available online]

Adamson, F. (2002) International Democracy Assistance in Uzbekistan and Kyrgyztan. In Mendelson, S. Glenn, J. (eds) *The Power and Limits of NGOs: A Critical Look at Building Democracy in Eastern Europe and Eurasia*. New York: Columbia University Press, pp. 177–206.

Armendáriz, B. and Broome, N. (2008) *Gender Empowerment in Microfinance*. Cambridge, MA: Harvard University Press.

Barnes, J. (2013) Making torture possible: The Sri Lankan conflict 2006–2009. *Journal of South Asian Development* 8(3): 333–58.

BBC News (2003) Baghdad protests over looting. *BBC News Online*. 12 April. http://news.bbc.co.uk/1/hi/world/middle_east/2941733.stm (accessed on 19 March 2008).

Bigsten, A. (2003) Selecting Priorities for Poverty Reduction and Human Development in Ethiopia. In Addison, T. (ed) *From Conflict to Recovery in Africa*. Oxford: Oxford University Press, pp. 106–22.

Boot, M. (2007) In defence of Blackwater. *The Press* (Christchurch). 10 October.

Boutros-Ghali, B. (1992) *An Agenda for Peace, Preventive Diplomacy, Peacemaking and Peace-keeping*. New York: United Nations.

Burnell, P. and Calvert, P. (eds) (2004) *Civil Society in Democratisation*. London: Frank Cass.

Centre for Civil Society (2008) *What Is Civil Society?* CCS website. Available at: http://www.lse.ac.uk/collections/CCS/what_is_civil_society.htm (accessed 24 July 2015).

Conteth-Morgan, E. (2004) *Collective Political Violence: An Introduction to the Theories and Cases of Violent Conflicts*. New York: Routledge.

Cooke, W. and Kothari, U. (eds) (2002) *Participation: The New Tyranny?* London: Zed.

Crilly, R. (2013) Pakistan's only female fighter pilot becomes role model for millions of girls. *Daily Telegraph*, 1 September. http://www.telegraph.co.uk/news/worldnews/asia/pakistan/10279119/Pakistans-only-female-fighter-pilot-becomes-role-model-for-millions-of-girls.html (accessed 30 October 2023).

Crisis Group (2023) A Road to Nowhere: The Myanmar Regime's Stage-managed Elections. *Briefing No. 175/Asia*. 28 March. https://www.crisisgroup.org/asia/south-east-asia/myanmar/b175-road-nowhere-myanmar-regimes-stage-managed-elections (accessed 9 June 2023).

Crouch, C. (2004) *Post-democracy*. Cambridge: Polity.

Darby, J. (2001) *The Effects of Violence on Peace Processes*. Washington, DC: United States Institute of Peace.

Darby, J. and Mac Ginty, R. (2003) Conclusion: Peace Processes, Present and Future. In Darby, J. and Mac Ginty, R. (eds) *Contemporary Peacemaking: Conflict, Violence and Peace Processes*. Basingstoke: Palgrave, pp. 256–74.

Darby, J. and Mac Ginty, R. (2008) Introduction: What Peace? What Process? In Darby, J. and Mac Ginty, R. (eds) *Contemporary Peacemaking: Conflict, Peace Processes And Post-War Reconstruction*, 2nd edn. Basingstoke: Palgrave, pp. 1–9.

Denyer, S. (2015) NGOs in China fear clampdown. *Guardian*. 30 March.

Duncan, P. (2023) Women still paid less than men at four out of five employers in Great Britain. *The Guardian*, 5 April https://www.theguardian.com/world/2023/apr/05/women-paid-less-than-men-four-out-of-five-employers-uk-gender-pay-gap (accessed 30 October 2023).

Enloe, C. (2000) *Maneuvers: The International Politics of Militarizing Women's Lives*. Berkeley, CA: University of California Press.

Enloe, C. (2007) *Globalization and Militarism: Feminists Make the Link*. New York: Rowman and Littlefield.

Foreign and Commonwealth Office (2015) Eritrea: Country of Concern. *FCO*, 12 March. Available at: https://www.gov.uk/government/publications/eritrea-country-of-concern--2/eritrea-country-of-concern (accessed 30 October 2023).

Glenn, J. (2008) Global governance and the democratic deficit: Stifling the voice of the South. *Third World Quarterly 29*(2): 217–38.

Gormley-Heenan, C. (2001) From Protagonist to Pragmatist: Political leadership in societies in transition. Report published by INCORE, University of Ulster/United Nations University.

Greenslade, R. (2014) Press freedom body highlights plight of Eritrea's jailed journalists. *Guardian*, 8 December.

Haack, K. (2022) *Women's Access, Representation and Leadership in the United Nations*. London: Palgrave Macmillan.

Herbolzheimer, K. (2015) Multiple Paths to Peace: Public Participation for Transformative and Sustainable Peace Processes. *Kult-Ur 2*(3): 139–56.

Hermann, T. and Newman, D. (2000) A Path Strewn with Thorns: Along the Difficult Road of Israeli–Palestinian Peacemaking. In Darby, J. and Mac Ginty, R. (eds) *The Management of Peace Processes*. Basingstoke: Macmillan, pp. 107–53.

Horowitz, D. (1985) *Ethnic Groups in Conflict*. Berkley, CA: University of California Press.

Hunt, J. (2004) Gender and Development. In Kingsbury, D., Remenyi, J., McKay, J. and Hunt, J. (eds) *Key Issues in Development*. Basingstoke: Palgrave Macmillan, pp. 242–65.

IDEA (2015) Voter Turnout. International IDEA website. http://www.idea.int/vt/ (accessed 3 June 2015).

IFRC, Annual Report 2022 (published 2023) Available at https://www.ifrc.org/document/annual-report-2022 (accessed 17 November 2023)

Jacoby, T. (2008) *Understanding Conflict and Violence: Theoretical and Interdisciplinary Approaches*. London: Routledge.

Kaldor, M., (2003a) The Idea of Global Civil Society. *International Affairs* (Royal Institute of International Affairs), *79*(3): 583–93.

Kaldor, M. (2003b) *Global Civil Society: An Answer to War*. Cambridge: Polity.
Kaldor, M. (2012) *New and Old Wars: Organized Violence in a Global Era*, 2nd and 3rd edn. Cambridge: Polity.
Kaldor, M. (2018) *Global Security Cultures*. Oxford: Wiley.
Kaufman, S.J. (2001) *Modern Hatreds: The Symbolic Politics of Ethnic War*. Ithaca, NY: Cornell University Press.
Keane, John. (1988) 'Introduction' to *Civil Society and the State: New European Perspectives*, London: Verso.
Keesing, R. (1992) *Custom and Confrontation: The Kwaio Struggle for Cultural Autonomy*. Chicago, IL: University of Chicago Press.
Lederach, J.P. (1995) *Preparing for Peace: Conflict Transformation across Cultures*. Syracuse, NY: Syracuse University Press.
Mac Ginty, R. (2003) The Role of Symbols in Peacemaking. In Darby, J. and Mac Ginty, R. (eds) *Contemporary Peacemaking: Conflict, Violence and Peace Processes*. Basingstoke: Palgrave, pp. 235–44.
Mac Ginty, R. (2004) Looting in the context of violent conflict: A conceptualization and typology. *Third World Quarterly* 25(5): 857–70.
Mac Ginty, R. (2006) *No War, No Peace: The Rejuvenation of Stalled Peace Processes and Peace Accords*. London: Palgrave.
Mac Ginty, R. (2010) Social network analysis and counterinsurgency: A counterproductive strategy? *Critical Studies on Terrorism* 3(2): 209–27.
Mac Ginty, R. (2014) Everyday peace: Bottom-up and local agency in conflict-affected societies. *Security Dialogue* 45(6): 548–64.
Mac Ginty, R. and Firchow, P. (2016) Top-down and bottom-up narratives of peace and conflict. *Politics* 36(3): 308–23.
Mac Ginty, R. and Richmond, O. (2013) The local turn in peace building: A critical agenda for peace. *Third World Quarterly* 34(5): 763–83.
Marriage, Z. (2007) Flip-flop rebel, dollar soldier: Demobilisation in the Democratic Republic of Congo. *Conflict, Security and Development* 7(2): 281–309.
Mayoux, L. (2002) Microfinance and women's empowerment: Rethinking best practice. *Development Bulletin* 57: 76–81.
Momsen, J.H. (2019) *Gender and Development*. London: Routledge, 1st ed 2004, 3rd ed.
Momsen, J.H. (2004, 2019, 2020) *Gender and Development* London: Routledge.
Nader, L. (1997) Controlling processes: Tracing the dynamic components of power. *Current Anthropology* 38(5): 711–38.
ONS (2023) Gender pay gap in the UK: 2022 https://www.ons.gov.uk/employmentandlabourmarket/peopleinwork/earningsandworkinghours/bulletins/genderpaygapintheuk/2022 (accessed 31 August 2023).
Paris, R. (2004) *At War's End: Building Peace after Civil Conflict*. Cambridge: Cambridge University Press.
Pilorge, N. (2012) Conflict over land in Cambodia is taking a dangerous turn. *Guardian*. 25 September.
Pirouz, R. and Nautré, Z. (2005) *An Action Plan for Iraq: The Perspective of Iraqi Civil Society*. London: Foreign Policy Centre Report.
Potter, A. (2008) Women, Gender and Peacemaking in Civil Wars. In Darby, J. and Mac Ginty, R. (eds) *Contemporary Peacemaking: Conflict, Peace Processes and Post-war Reconstruction*, 2nd edn. Basingstoke: Palgrave, pp. 105–19.
Pruitt, D.G., Bercovitch, J. and Zartman, I.W. (1997) A brief history of the Oslo talks. *International Negotiation* 2(2): 177–82.

Prusher, I. (2003) Iraq's new challenge: Civil society. *Christian Science Monitor*. 8 October.

Putnam, R. (2000) *Bowling Alone: The Collapse and Revival of American Community*. New York: Simon & Schuster.

Rehn, E. and Johnson Sirleaf, E. (2002) *Women, War, Peace: The Independent Experts' Assessment of the Impact of Armed Conflict on Women and Women's Role in Peace Building*. New York: UNIFEM.

Reilly, B. (2003) Democratic Validation. In Darby, J. Mac and Ginty, R. (eds) *Contemporary Peacemaking: Conflict, Violence and Peace Processes*. Basingstoke: Palgrave, pp. 174–83.

Sanyal, P. (2006) *Credit, Capital and Collective Action: Microfinance and pathways to women's empowerment*. Paper presented at The Annual Meeting of the American Sociological Association, Montreal Convention Center, Montreal, Quebec, Canada.

Saunders, H. (1999) *A Public Peace Process: Sustained Dialogue to Transform Racial and Ethnic Conflict*. Basingstoke: Palgrave.

SIDA (2006) *Microfinance and Women's Empowerment: Evidence from the Self Help Group Bank Linkage Programme in India*. Stockholm: SIDA.

Singh, T. (2023) The authoritarian roots of democracy. *Journal of Democracy*, 34(3), July.

Sjoberg, L. (2013) *Gendering Global Conflict: Towards a Feminist Theory of Conflict*. Columbia, NY: Columbia University Press.

Smith, A. and Rigby, E. (2015) Women make inroads on UK boards but top slots remain elusive. *Financial Times*. 15 March. Available at: http://www.ft.com/cms/s/0/ab49c1e4-d215-11e4-b66d-00144feab7de.html#axzz3gPw1vtqQ (accessed 30 October 2023).

Tilly, C. (2002) Violent and Non-violent Trajectories in Contentious Politics. In Ungar, M., Bermanzohn, S. and Worcester, K. (eds) *Violence and Politics: Globalization's Paradox*. New York: Routledge, pp. 13–31.

Traynor, I. (2003) The privatisation of war. *Observer*. 10 December.

Turshen, M. and Twagiramariya, C. (1998) *What Women Do in Wartime: Gender and Conflict in Africa*. London: Zed.

United Nations Development Programme (1997) *Human Development Report 1997: Human Development to Eradicate Poverty*. New York: United Nations.

United Nations Development Programme (2011) *Governance for Peace: Securing the Social Contract*. New York: UNDP.

UNMIK (2004) Condemning violence in Kosovo, Security Council demands return to rule of law. *UNMIK Online*. 18 March. www.unmikonline.org (accessed 18 March 2008).

Ware, A., and Laoutides, C. (2018) *Myanmar's 'Rohingya Conflict*. London: Hurst.

Wolfsfeld, G. (2003) The Role of the News Media in Peace Negotiations: Variations Over Time and Circumstance. In Darby, J. and Mac Ginty, R. (eds) *Contemporary Peacemaking: Conflict, Violence and Peace Processes*. Basingstoke: Palgrave, pp. 87–99.

World Bank (2011) *World Development Report: Conflict, Security and Development*. Washington: World Bank.

World Humanitarian Summit (2016) https://ucdp.uu.se/ (accessed 7 November 2023).

Zonszein, M. (2014) How Israel silences dissent. *New York Times*. 26 September.

4 Conflict transformation and development

Introduction

The next three chapters are essentially about attempts to effect 'transitions' from conflict and war to a kind of 'peace'. First, they will look at the main strategies employed by the international community, individual powers or groups to try and do something about the conflicts and wars in developing (and some more developed) countries of which we have so far described the genesis. This can be by way of gentle or more muscled intervention, often referred to generically as 'peacebuilding'. Roland Paris, the author of a very influential book (Paris 2004) on the subject, describes it as follows:

> I define peace building as an activity that takes place in a post-civil war environment, the purpose of which is to create the conditions for a stable and lasting peace and to prevent the recurrence of large-scale violence.... Since the late 1990s, the term 'peace building' has been used to describe a much broader range of local, regional, national, and international initiatives intended to promote peace, including conflict prevention.
>
> (Paris 2004)

As Vivienne Jabri puts it, this catch-all phrase has become a 'norm' and one 'that informs the normative underpinning and legitimization of particular kinds of intervention'. It is nearly always supervised and implemented by the UN or another IGO, like NATO (Jabri 2013: 3–4). However, as Paris indicates, there has been a revival of interest in the idea of such peacebuilding needing very 'local' agents to make it work, or in David Chandler's words, a 'discovery of the local' (Chandler 2013: 22; Mac Ginty and Richmond, 2013). Underpinning nearly all such attempts is the idea that they should aim to create societies that are liberal economically and politically.

So, second, these chapters will investigate how successful such strategies have or have not been, and third, they will ask why such strategies have failed or succeeded and how we might try other tactics, up to and

DOI: 10.4324/9781003353485-5

including the possibility of total non-intervention. We have stopped our analysis in the Autumn of 2023 when we were finalizing the book.

In all these considerations, we have to take into account that the overarching categories we use are shorthand for much more complex and integrated issue areas. Hence the subject matter of Chapter 5, reconstruction, and Chapter 6, the delivery of aid or other peacebuilding attempts, cannot be separated from attempts at conflict resolution or 'reconciliation' after wars. They may take place sequentially, or simultaneously, while the violence continues at varying levels of intensity (see the debate on 'contingency' in Keashley and Fisher 1990). Equally, each of these concepts is a *matrioshka* doll of other concepts – disarmament, demobilization and reintegration (DDR) of ex-combatants being one important embedded 'doll' that will be explored in Chapter 5.

We will also look at other techniques and ideas that have been developed to try and heal communities after wars, and in particular techniques of 'reconciliation' and 'retributive justice' through truth and reconciliation commissions and war crimes tribunals. The overarching philosophical questions that underlie this enquiry and that we will engage with will be: how can we arrive at a shared understanding of the past, and of 'truth', of 'history' indeed, and how can we envisage the 'reintegration' of a society damaged by war, or as Rigby puts it, the 'envisioning of a common future together' (Rigby 2001: 12)?

Conflict 'transformation'

This chapter will first look at the concepts that can be used for understanding the practices of conflict resolution (or as we prefer, 'transformation'). Conflicts are unpredictable in their course, so trying to end one can be fraught with difficulty. One reason for this is that they aim to change the relationship between the parties to the conflict, all of which have 'constituencies' (Mitchell 1996) whose interests are more or less invested in the conflict and who may even wish it to continue, as they may be able to benefit from the opportunities for criminal or other activity that it provides. They nearly always start with a negotiated agreement to end the conflict, either through a treaty binding in international law (of which a good example is the Treaty of Versailles, 1919, which bought about the end of the First World War) or through bilateral or multilateral informal talks, some of which are conducted in secret, often over a lengthy period, and some of which are proclaimed as break-throughs in public announcements. These can be seen as 'hybrid' attempts at conflict transformation.

One such example of such hybrid agreements were the Oslo Accords of 1993 (for the official title, see Oslo Accords 1993), which were hailed optimistically at the time as a key breakthrough in the Israeli-Palestine conflict. This optimism has now faded, and they are generally seen as a failure, replaced by a feeling that such conflicts are 'intractable'. This feeling has been accentuated by the events of 7 October 2023 when Hamas, an Islamist group based in Gaza, attacked Israel, with subsequent loss of life in both Israel and Gaza. Such religious conflicts now seem endemic in the area, in Israel and Palestinian areas alike. Oslo has nonetheless been seen as the classic paradigm for conflict transformation since the end of the Cold War. As Scott Smith of the United States Institute for Peace (USIP) wrote, '[T]here are very few examples of a significant peace agreement that has been reached without some sort of third-party facilitation or mediation' (Smith 2021).

Might we suggest that such agreements are often fatally flawed? The Oslo Accords were co-signed by the Russian foreign minister and the American secretary of state, as well as the Palestinian Authority and the Israeli government. Could we ever, for example, trust a potential agreement between Putin's Russia and Ukraine? Were such attempts more a question of 'seeming to do something', futile gestures that left the most intractable issues unresolved?

Dealing with conflict – from 'settlement' to 'sustainability'

The period in the 1990s, after the end of the Cold War, has seen a swing from optimism about the possibilities of terminating conflict in a satisfactory way through third-party intervention of various kinds to a deep pessimism in the 2000s as 'intractability' has seemingly increased. We could even now speak of a 'nostalgia' for the 'conflict resolution' hopes of the 1990s, termed by Richmond as 'third generation' theories. In the 1980s, the Australian former diplomat and academic John Burton believed that to bring about a successful resolution of a conflict, there had to be the satisfaction of all the parties' 'human needs' or what Johan Galtung saw as 'incremental structural change to remove oppression' (Richmond 2007: 87–89). Linked to this was the ideal of 'an equal recognition of the stakeholders involved' (Jabri 2013: 16), the underlying structure of the Oslo Accords. The perils of such an approach may be said to have become clear in Afghanistan, where such 'equality' has been nothing of the sort and has in fact meant giving equal weight to the views of the Taliban and the Afghan government, with the results we have seen. Since 2020, such idealism has been replaced by a belief in the necessity of the use of force in the service of the 'international community', by which it is usually meant the UN in collaboration, or not, with powerful interventionary states. Ukraine is a current example of this new approach. In Niger, where a coup d'état was under way as this book was in the manuscript stage, the UN Peacekeeping Force, MINUSMA, which has been in the country since 2012, is to be withdrawn (Brooke-Holland 2023). The French government has expressed the view that force may be necessary to re-install the elected government. Crisis Group talks darkly of 'What Future for [the] UN', given the unceremonious ejection of the Blue Helmets (Crisis Group 2023). Meanwhile, the Russian Wagner Group (now 'Africa Corps') is suspected of having helped engineer the coup. The Russian government has predictably suggested that the Bamako government must be allowed to invite whoever it wishes to 'support' it and 'renounce their own "neo-colonial" habits (United Nations 2023). In short, global power politics are once again deciding the future of a developing country.

Equally, and as we have noted, war has certainly changed, from the predominantly international wars of the Cold War to the intra-state ones after it, though that statement perhaps needs qualification after the war in Ukraine since 2022. In the 1990s, there were approximately five intra-state conflicts for every one international war. The think tank PIOOM reported that the 1990s had one of the lowest numbers of battle deaths of the twentieth century ('World Conflict and Human Rights Map' by PIOOM, atf_world_conf_map.pdf). This was notwithstanding the highly publicized wars in the former Yugoslavia (1991–9), where an authoritative figure of 130,000 deaths has been recorded (Humanitarian Law Center). The outbreak of civil wars in the Middle East and Eastern Europe since the 'Arab Spring' of 2011 and the events in Ukraine since 2014 has seen a very disturbing upward trend in the most recent 2014 figures for deaths in armed conflicts (175,881, up 62,975 from 2013 (International Institute for Strategic Studies 2015a). The war in Ukraine has cost at least another 200,000 military deaths and many civilian casualties, at least 7,199 deaths and 11,756 injured as of 17 October 2022 (sources: UNHCR 2023; Milley 2022).[1] They have been climbing higher ever since. Equally, deaths in the regional war in Ethiopia, 2020–1, between the province of Tigray and the central government, but also involving several other provinces and neighbouring states, Eritrea and Somalia, have swelled the numbers considerably (Plaut and Vaughan 2023).

The nature of warfare has been changing since the 1990s, both in what is often referred to as 'hybrid warfare' (the combination of conventional and unconventional means of warfare, as in the Ukraine since 2014) and also in the way that many wars now take place in civilian areas, leading to very high numbers of civilian as opposed to military casualties in towns and cities (International Institute for Strategic Studies 2015b). Second, the sheer *number* of intra-state conflicts and their longevity is a new phenomenon when compared with the Cold War period, with one estimate being that more than one-third of countries have experienced an intra-state conflict, with two-thirds of these suffering from such conflicts over seven years or more (Gawerc 2006: 436). According to the Uppsala Conflict Data Programme (UCDP) there were

83 intra-state conflicts in 2022 (down from 87 in 2016 but up from 57 in 2012) (UCDP 2022a). And third, as noted earlier, 'geopolitics is centre stage' (International Institute for Strategic Studies 2022).

In addition, the overwhelming evidence that these 'civil' conflicts have had a habit of becoming regional problems can be seen in the contagion effect in many parts of the world (Long and Brecke 2003: 5). This was evident in West Africa, where Liberia, Sierra Leone and, arguably, Côte d'Ivoire were sucked into a regional conflict in the 1990s; Central and East Africa (where Rwanda, Burundi, the Democratic Republic of Congo, Uganda, Zimbabwe – among others – became, and still are, part of an arc of conflict after the Rwandan genocide of 1994); in the Horn of Africa, with Eritrea, Ethiopia, Somalia, Sudan and Chad forming varying geometries of inter- and intra-state conflict since the end of the Cold War, and in many other parts of the world, as with the archipelago of Indonesia, where Timor-Leste is an integral part of a network of conflicts. The most obvious evidence of regional contagion saw wars in Afghanistan and Iraq between 2001 and 2009 break out more generally over the Middle East in Libya, Egypt and Syria (all 2011) and Iraq (which redescended into civil war with the withdrawal of American troops in the same year), in particular.

The previous, very selective list of civil conflicts that have erupted since about 1990 gives a clue as to why war has become much less devastating on one level while simultaneously being seen as more vicious. Arguably, however, contemporary civil wars are not nearly as terrible as during the Cold War, the Indonesian civil war of 1965 (over 500,000 deaths), the Biafran war of 1967–70 (over a million deaths) and Cambodia 1975–9 (up to three million deaths). This is not to mention the appalling displacements and death tolls of the World Wars and the Chinese, Russian or Spanish revolutions (Kalyvas 2001: 110; Bellamy 2012). The worst casualty rates since 2011 have been in Syria, where well over 407,392 people had died as of the end of 2022. Yemen has seen 63,799 deaths since 2010 and Ethiopia 102,918 in 2022 alone (300,009 from 1989 to 2022) and Iraq 126,918, from 1989 to 2022. There was a 97 per cent increase in fatalities from organized violence from 2012 to 2022, from 120,000 to 237,000 (UCDP 2022b, various pages). In part, better media coverage has made those wars seem more terrible than wars in the past, as have the huge refugee flows to Europe and elsewhere (one and a half million to Lebanon alone and 3.6 million in Türkiye/Turkiye). But the death toll in 2022 was the worst since the Rwandan genocide in 1994, especially due to wars in Tigray and Ukraine.

The war in Ukraine may well beat these sorry records, with mounting military and civilian casualties and about 15 million Ukrainian refugees to date, not to mention the over 3 million who have fled Russia to escape political repression and the military draft. The Russian invasion of Ukraine is the first large-scale inter-state war in 20 years and the first inter-state armed conflict since the Second World War in which a major power in the international system seeks both territorial gains for itself and the subjugation of another state through regime change. We have witnessed an emerging trend of increased conflict between states in the last decade, including cases where major powers support opposite sides in internationalized intra-state conflict. UCDP recorded 55 active state-based armed conflicts in 2022, an increase of one compared to the previous year. Eight of these conflicts reached the level of war, and 22 of them were internationalized, meaning that one or both parties received troop support from an external state. While the fatalities caused by non-state conflict decreased somewhat when compared to 2021, the number of non-state conflicts, as well as both the number of civilians killed in one-sided violence and the number of actors carrying out such violence, increased in 2022 (Davies, Pettersson and Öberg 2023).

War in the decades before 1990 was between industrial powers fighting traditional wars on an industrial scale, a 'settlement' of conflict through war, though even here the evidence is of civil wars always having been more numerous than inter-state wars (Mack 2006: 6–8). During the last 20 years or so of the Cold War, the Powers (and especially the United States and the Soviet Union) slowly realized that the use of force had its limits as a way of creating peaceful outcomes to conflict. At the end of it, some even predicted that war would become 'obsolete', in an ironic echo of those who made the same prediction before 1914 (Angell 1910; Kaysen 1990). Mandelbaum was more accurate in saying that *major inter-continental* war may now be obsolete (Mandelbaum 1998), though again, this is a dangerous prediction given historical precedent. In the early post–Cold War era, with the exception of the intervention in Iraq in 1990–1, the levers of 'hard' power were less willingly used by democratically cautious Western states that saw no evident national interest at play for them in many 'obscure' conflicts. Powers then aimed to try and bring about internally 'sustainable' ends to wars, ones that did not need continual revisiting. It became more fashionable and responded well to a post–Cold War desire for a 'peace dividend' to discuss the use of 'soft' power (Nye 2005), and in particular a rise in the idea that conflicts could be talked down in workshops and off the battlefield. Joseph Nye defined this as follows:

> [Soft power] is the ability to get what you want through attraction rather than coercion or payments. It arises from the attractiveness of a country's culture, political ideals, and policies. When our policies are seen as legitimate in the eyes of others, our soft power is enhanced.
>
> (Nye 2005: Preface)

The peaceful transitions in South Africa, Mozambique, most of Latin America, even Israel/Palestine (with the Oslo Accords of 1993), and, most of all, in Eastern Europe in the 1990s led to a swathe of books touting the benefits of 'Track One' and 'Track Two' conflict resolution, mediation and the like (see p. 149 for more discussion on these distinctions). There was, in the words of Samuel Huntington, a 'third wave of democratization' (Huntington 1993), though Huntington soon reversed his own optimism by declaring the 'Clash of Civilizations' (Huntington 1996). A real change began to happen at the end of the 1990s as a result of the terrible events in the former Yugoslavia and Rwanda in 1994–5. In the late 1990s, and even more so after that, there was a resurgence in the use of hard power in 'pre-emptive' mode in Afghanistan and Iraq and a refusal to negotiate with 'insurgents' in the 'war on terror'.

The truth of it is that the 'logic of violence' had not changed all that much, even if the response to it had. As a number of commentators have noted, violence has an internal logic of its own; it is 'rational' in the sense that it is often a response to local or more widely felt senses of injustice against governments, even resentment about global inequality or perceived cultural humiliation (Kalyvas 2006; Balcells 2010, 2011). The 'Boko Haram' insurgency by Islamist militants in Northern Nigeria, Cameroon and Chad that specializes in massacres and kidnappings (most infamously of 274 schoolgirls in Chibok in April 2014) garners huge publicity and provokes terror well beyond the borders of its area of activity (International Crisis Group Africa 2014). The organization known as 'Islamic State' (ISIS or *Daesh*) used both social and conventional media to publicize its aims through slick videos of beheadings and other executions, as well as the encouragement of such acts in public places in Western cities like London, Madrid, New York and Paris. These 'spectaculars' also gave it a platform in the Western media which aided its recruitment of disaffected Muslim youth in the West, many of whom have died in Syria and Iraq.

As a result of the rising tide of violence since 2010, Western public opinion has played an important role in reducing the desire and the appetite for intervention by Western and other Powers, in spite of the often-violent aftermath of the 'Arab Spring'. The civil wars (or 'revolutions') that erupted in, especially but not exclusively, Syria (2011–present) and Libya (2011), and the civil unrest and violent changes of government in Egypt and some Gulf states, added to already unstable and violent situations in Afghanistan and Iraq. The aftermath of American and other troops being withdrawn from Afghanistan and Iraq after the election of President Barack Obama in 2009 and a Conservative government in Britain in 2010 led to a reduction of the interventions by American and British administrations wedded to 'liberal internationalism' or 'neo-conservatism'. But it did not noticeably lead to a reduction of violent conflict or an improvement of levels of economic, social and political development in those countries, as 'insurgent' forces took advantage of the power vacuums thus created.

Some writers believed that the UN and some major Western states had embarked after 2001 on a policy of 'international pacification' and 'pre-emptive regime change' (Duffield 2007 is one such example). It might be more accurate to say that a belief in the use of hard power, as opposed to softer versions, as a way of bringing about an end to conflicts in developing (and more 'developed') countries fluctuates in effect in line with optimism or pessimism about the prospects for different tools to end conflicts and the self-confidence of the underlying belief in prospects for a 'liberal peace'.

In trying to fashion responses to conflict both in the 1990s and since, much ink has been spilled trying to show the differences between various possible approaches. Conflict 'settlement', 'management' and 'resolution' have often been referred to as the 'three generations' trying to bring about peace. However, it must be understood that different approaches are often tried simultaneously or in sequence depending on the conflict being dealt with. So any attempt to be prescriptive about the use of such approaches has to be cautious and hedged around with caveats. History has an unfortunate habit of proving wrong any linear explanation of 'what is happening'.

So why attempt to define such terms? The reason is because they are so described in much of the literature on conflict and because they tend to coincide with particular approaches to international politics more generally. But first, we need to put these concepts within the broader emerging paradigm of 'peacebuilding'.

Conflict 'settlement', 'management' and 'resolution'

There has been a long-standing tradition in international relations and conflict analysis of saying that different practical approaches to dealing with conflict can be seen as existing in parallel with certain 'paradigms' of international relations theory. This was particularly striking in the 1980s and early 1990s in the so-called inter-paradigm debate in international relations. This rather simplistic thinking has been superseded by a much more subtle conversation about 'critical', even 'postmodern' theorizing about the nature of international activity. These latter thinkers have in particular attacked the ethos of thinking that 'we' can intervene to resolve 'their' conflicts, a way of thinking that has accelerated with the recent actions in Afghanistan, Iraq and the Middle East more generally. We will touch on these issues in this chapter. They are nonetheless key ideas, for as Jeong points out, all of these techniques and theories aim for a 'just outcome that is not only fair to participants but also meets the broader ethical concerns of a given society, and those of humanity' (Jeong 2008: 243). That is a tall order, and, unsurprisingly, there are many who scoff at its practicality.

However, the terms that were developed then are still currently and widely used, so we need to understand how they might apply to particular ways of dealing with conflict by theorists and practitioners alike. It might also be mentioned that this is not a *linear* debate. Conflict analysis has its fashions and obsessions like every other field of human endeavour, and certain theorists and practitioners use the terms we will now explore in ways that defy exact categorization. There is no clearly and widely accepted *one* approach that all in this field agree on, just an ongoing, circular debate (Ramsbotham et al. 2011 give a good overview of this, as they do in the 2016 edition). Indeed we now have collections that point to myriad definitions about what is meant by even the expression conflict 'resolution' (Sandole et al. 2009; Allen et al. 2012; Wallensteen 2012). An innovative approach to concepts of peace, justice and coexistence is the focus of study in the research NGO Everyday Peace Indicators (EPI, see p. 106).

We can break this debate down in many ways, but the authors think it most useful to look at it first in the simple terms of 'top-down' or 'bottom-up' and then look at how that might translate into more precise terminology.

Top-down/bottom-up?

By *'top-down'* is meant all those strategies that have been deployed by third-party representatives of a state or IGO. This is often referred to as 'Track One' diplomacy. These kinds of action are in most cases undertaken before the conflict/war has come to the stage of a ceasefire. Sometimes this can be supplemented by 'Track Two' action, which is usually conducted by unofficial players, such as NGOs or officials acting in an unofficial capacity and behind the scenes, or even academics acting entirely on their own without any form of support.

Track One has been given huge prominence by the activities of American presidents and their advisers (like Henry Kissinger) in the Arab–Israeli context since the 1970s, starting with the Camp David discussions of 1977 (Touval 1982) and by President Trump in the 2020 Doha Agreement (Box 4.1). It is now being used again in the many and varied offers to 'solve' the Ukraine crisis, notably by Chinese president Xi-Jinping (see Box 4.2). The main advantages of Track One are the legitimacy such discussions provide: the resources that can be marshalled in support of any agreement (military and economic aid, for example, as has been the case of Israel and Egypt since the late 1970s), and the guarantee of media interest in, and usually support for, such efforts. The disadvantages are that the Powers will use such occasions for their own interests and will cajole, bully or otherwise intimidate the parties and not in any way transform their basic unhappy relationship.

Box 4.1 The Doha agreement of 2020

This might even be said to have become a 'norm' to echo Jabri's term. Smith was discussing the drawbacks of the 'Doha Peace Agreement', signed in 2020, and like Oslo, attempting to bring a hitherto intractable conflict to an end. It was signed by the United States and the Taliban insurgency group and intended to bring an end to the war in Afghanistan that started with the invasion of 2001 by the United States and the UN, but which had far deeper roots going back at least to the Soviet invasion of 1982.

The aim of the Agreement for the US government, which in 2018–19 was led by President Donald Trump, was to ensure the survival of the official Afghan government. Afghanistan, however, was to fall to the 'Islamic Emirate of Afghanistan' (the 'Taliban'), always mentioned in the Agreement as 'not recognized by the United States as a state and is known as the Taliban.' The Agreement was based on the fiction that President Ashraf Ghani could resist Taliban military pressure without American airpower or troops, reduced to 8,600 and left with no airpower to rely on, thus negating all the training in 'combined arms warfare' given to the Afghan Army by the United States. The resulting speed of the collapse of Kabul and the flight of Ghani surprised even the Taliban. The fall of Kabul happened in August 2021 (see Doha Agreement 2020; CFR/Maizland 2020).

Box 4.2 Conflict management and settlement in Ukraine

The conflict in Ukraine since 2014 has led to a number of important initiatives by powerful states. The 1994 'Budapest Memorandum' guaranteed the borders of all signatories, including Russia and Ukraine. After the Russian annexation of Crimea and part of the Donbas in 2014, a framework was established, the 'Minsk agreements', which included Russia, Ukraine and the OSCE, with Germany and France acting as mediators. This failed to stop fighting over the 'border'. On 24 August 2022, Ukrainian president Volodymyr Zelensky said, 'We will not play these games, we have lost part of our territories this way … it is a trap'. Since then, the most significant initiative (to date – October 2023) has been China's 'Position on the Political Settlement of the Ukraine Crisis', which purports to offer a 'balanced position'. But, in the words of Stefan Wolff, it 'is an important statement of China's vision for global, Eurasian and European Security'. The emphasis by Beijing on the 'sacrosanct nature of sovereignty and territorial integrity' is an attempt to accept both the Russian allegations that their rights in this domain have been violated by NATO's expansion toward Russia's borders, as well as a condemnation of Russia's violation of Ukrainian rights. Seemingly all parties are at loggerheads about who should mediate and about what. Is the lesson that such attempts tend to be the manoeuvres of powerful states, not those who have to express those manoeuvres? There must be mutual trust.

Sources: Budjeryn and Bunn 2020; Wolff 2023; OSCE 2022

By *'bottom-up'* is meant the introduction of a methodology for the local participants and victims of war or conflict to resolve their own problems using their own methodologies and systems. In many cases, but not all, these have indeed been elaborated locally with minimal outside help. Such actions usually occur after the fighting has ended, though hostilities may still simmer and flare up. This approach was particularly popularized by practitioner-theorists like John Burton and John Paul Lederach (Burton 1979, 1997; Lederach 1997). Outsiders in this way of framing conflict help to 'facilitate' the reduction of violent conflict by the locals themselves and do not impose outside ideas or practices on them. Both the aforementioned thinkers assert that conflict is caused by frustrated 'human needs', not by any innately violent tendencies. Before trying to deal with it, we need to 'capture the complexity of conflict' (Sandole 1999).

Conflict settlement

'Settlement' is usually seen within a power-based realist paradigm where force is the arbiter. This can include 'mediation with muscle' as in the context of the United States using its power to broker agreements over the Middle East. The conflict between Russia and Ukraine (2014–present) has led Kissinger to make new suggestions in a type of conflict that was seen as 'ended'. Interest in mediation, even 'with muscle', has thus assumed a new acuity. There was an attempt organized within the Organization of Security and Cooperation in Europe (OSCE), known as the 'Minsk Process' to reduce levels of violence in Eastern Ukraine after the annexation of parts of the Donbas region and that of Crimea in 2014 (OSCE 2022). This led nowhere, and it has been suggested that it was mere window-dressing for the next round of annexation by Russia, attempted in 2022.

Aim

Conflict settlement's aim is order and often does not shrink from a 'realist' use of military force, sometimes on a huge scale, as in the First and Second World Wars, or from zero-sum outcomes. Realists take the view that all peace is a lull between wars and that trust between states and nations is a plausible but unlikely scenario (Mearsheimer 2001; Waltz 2001). They also deny that there is any such thing as an 'international society' with clear rules and processes but that if such a thing exists, it is due to a temporary balance of power and the dominance of a hegemon for a given historical period. Indeed, it has been pointed out that 'War appears to be as old as mankind, but peace is a modern invention' (Sir Henry Maine, quoted by Howard 2000: 1). High-level diplomacy, even including heads of state, has played a very large role in past, and maybe in future, conflicts. The difficulty of finding a head of state who is 'neutral' in the conflict and thus agreeable to the warring parties is difficult, with a prominent role of this kind assumed by the president of Türkiye, Tayyip Erdogan, early on in the Ukrainian war of 2022. His ability to control the Straits of the Bosphorus (which by international treaty Turkey has the right to do) makes him a mediator with muscle, as does his provision of highly effective 'drones' to Ukraine which have killed many Russian troops (Reuters 2023).

Problems

Conflict settlement often does not work, in the sense that only the maintenance of force keeps the 'lid' on the problem but does not address the underlying problems, indeed cannot, as states are always getting ready for the next war. We might also say that hard power is difficult to use when the areas where it must be applied are ones in a state of civil war ('realism' assumes a functioning state system on the whole) and the conflict has a regional nature. As the United States has discovered in Afghanistan (2001–2021) and Iraq (2003–2011), the most formidable military machine can get very bogged down in an area wracked by an insurgency with no identifiable 'head'. As Iraq also shows, the idea of 'ending' a war by force can often create a bitter illusion. Wars that are finally 'ended' by force alone are in a distinct minority and arguably always have been (Coker 1997). Even the hardest of realists understand that, as Clausewitz et al. (2012) commented, in war, all is '*friktion*', which might be translated as 'it's easy to get into, but unpredictable in the extreme once engaged upon' (see, for example, Gray 2005). The ethical issues for such thinkers and practitioners are clear, and they are largely ignored. So the main trouble with this approach has been highlighted by the recent wars in Afghanistan, Iraq and Ukraine. The 'locals' do not particularly like being forced to make peace or to desist from war. One true believer in the use of force, Donald Rumsfeld, US deputy secretary of defense under President G. W. Bush, admitted in November 2001 that

> one of the lessons of Afghanistan's history, which we have tried to apply in this campaign, is if you are a foreigner, try not to go in. If you go in, don't stay too long, because they don't tend to like any foreigners who stay too long.
>
> (Roberts 2009: 29)

Perhaps the 'lesson' of American intervention in the Middle East more broadly can be traced back to the first real attempt to interfere in a war in the area: the First World War. The young William Yale started his association as an employee of Standard Oil ('Socony'), which led to the discovery of oil in the Sinai Desert. By 1917, he was acting as the key American diplomatic and military official in the area, as there was no one else available: in the words of Scott Anderson,

> William Yale was establishing a tradition of fundamentally misreading the situation in the Middle East that his successors in the American intelligence community would rigorously maintain for the next ninety-five years.
>
> (Scott Anderson 2013: 463)

Yale later became a distinguished academic specialist in the Middle East and the State Department's chief expert on Palestine, 1941–45. He admitted to his earlier ignorance in later life:

> I lacked a historical knowledge of the background of the problems I was studying.... I had no philosophy of history, no method of interpretation, and very little understanding of the fundamental nature and function of the [local] economic and social system.
>
> (Daily Beast 2013)

Plate 7 President Volodymyr Zelensky No trust exists between Ukraine and Russia, so mediation is very difficult credit: Wikimedia Commons.

Conflict management

A still essentially realist view is taken by conflict 'managers' who believe that it is usually impossible to do anything but try and grease the wheels of a conflict to make it less violent through the judicious use of a combination of hard power, soft power (usually economic sticks and carrots like sanctions) and peacekeeping forces. Roland Paris argues that such 'management' approaches can encompass 'liberalization', which he explicitly equates with 'democratization', a dual concept which we have linked to the 'liberal peace' theory and will be explored further in the next chapter (Paris 2004: 5). The ethical problem with this is of course that the 'targets' may not want to be so dealt with.

A feature of such approaches is also the use of 'Track One' mediators, usually in the form of 'good offices' or 'mediation' between leaders of insurgent groups, the established 'state' and any other regional players, a time-honoured use of a third party to help negotiation between states. Scholars have calculated that there have been visible mediation efforts of this kind in over 70 per cent of conflicts since the Second World War, and that does not take into account more discreet efforts (Bercovitch and Fretter 2004). In the post–Cold War period, mediation has come to mean 'an extension and continuation of peaceful conflict management' (for a robust defence of this, see Bercovitch, in Zartman and Rasmussen 1997: 127 and *passim* and Bercovitch and Rubin 1992). It should be noted that the expression 'mediation' can also be used without 'muscle' in the conventional sense of the latent or actual use of power. Religious leaders often have huge influence that they wield in a public or private capacity (Little 2007; Tenenbaum 2011). The degree to which this influence can be brought to bear can depend on imponderables like personality and credibility, the role of Archbishop Desmond Tutu in South Africa being one notable example (Sandal 2011).

Aim

The aim is the minimization, but not necessarily elimination of, violence; stability insofar as it can be achieved; and 'realist' expectations of the difficulty of getting people to agree on underlying causes, symptoms and solutions to their problems. The recent resurgence in the United States in the study of conflict management has much to do with the feeling that the optimism of the 1990s was misplaced and 'recognizes that the global environs are both conflict-ridden and often bloody, and it harbors no illusions that murderous enmity will suddenly give way to universal comity' (Solomon in Crocker et al. 2007: xi).

Problems

If conflict management is too centred on domestic US obsessions (the 'war on terror', etc.), we might question whether it therefore sings to a tune not appreciated by others. It does nonetheless ask some very important questions about whether democracy is the 'answer' or stability the main aim and also proposes interesting ideas about new forms of sovereignty and other ways of organizing societies to make them less war prone (Ottaway and Marten in Crocker et al. 2007: Part VI).

A more fundamental problem is raised by scholars of peace making like Oliver Richmond. He reinforces the problem, cited earlier, that the mediators themselves have often unstated but evident aims that have more to do with their own understanding of the world than the parties to the conflict. He sees peace as an 'inter-subjective concept' that is 'then open to ideological and functional manipulation.... It may well be that dominant forms of peace represent the methods and objectives of hegemonic actors' (Richmond 2007: 197).

Conflict resolution

So how can peace be achieved? Can we suppose that by listening to the 'local', rather than any external, actors, we could get to an approximation of a more acceptable agreement? This was an idea that was widely explored in the years immediately following the end of the Cold War. Often delivered by 'Track Two' intermediaries, who could be academic groups, NGOs or local community activists, conflict 'resolution' was usually held to mean an attempt to fully (or mainly) resolve the underlying root causes of a conflict by hopefully resolving (or indeed transforming) the relationship of the parties so that they develop an entirely new and peaceful relationship. The theoretical and quite a lot of the practical impetus for this idea can be traced back to the early 'conflict' and 'peace' researchers and practitioners of the 1960s and since. These included Elise and Kenneth Boulding and John Burton, as well as the scholars who worked with him and used his ideas, like A.J.R. Groom and Christopher Mitchell (Mitchell 1981, 2014). This heterogeneous group developed 'problem-solving' workshop techniques that were, and indeed are, claimed as the best tools for examining 'intractable' conflicts, of which the Israel–Palestine conflict is the best single example. Institutionally, and also one linked to this particular conflict, one of the best examples was the 'Oslo Process' of the early 1990s, which seemed at the time to be capable of addressing the underlying problems between Israel and Palestine (Corbin 1994), but by 2015, it was seen as a false dawn as has been discussed earlier in the book (**p. 119**). Other initiatives have included processes to address the issues of local climate change and land ownership. These come from newer 'emergent' conflicts which we will increasingly need to address (Miall 2006).

Aim

Full airing of differences, usually in a closed and confidential environment, with no preconceived agenda and a belief that problems and parties to conflict can be 'transformed' so that a future relationship will be better, less conflictual and ultimately stable (for a wider summary see Miall et al. 1999: 58). It purports to see *all* parties to a conflict in an equal light and does not exclude any potential solution to a conflict situation. So, for example, the activities of the Centre for Conflict Analysis (established by John Burton among others) did not seek to impose a pre-ordained solution but led the parties to better communication of their needs (Burton 1997 *passim*; Williams 2005). Ideally, it will also 'highlight the wider social and political sources of a conflict in seeking to break the perpetuating cycle of oppression and resistance' (Jeong 2008: 244). One of the best summaries of the characteristics and methodologies used can be found on the Berghof Foundation's website, a resource handbook that is updated periodically (Berghof Foundation 2023 (continuous) http://www.berghof-foundation.org).

Problems

The idea of such conflict transformation is as old as the ideal of human rationality. Christopher Mitchell quotes Mary Parker Follett as suggesting in 1940 that we can arrive at an 'integrative solution ... [where] both desires have found a place, that neither side has had to sacrifice anything'. But Mitchell suggests that '"conflict resolution" really makes very strong demands' in terms of the necessary ending of violence and a further routinization of peaceful behaviour (Mitchell 2014: 220, 290). There is also an often-naïve belief in human perfectibility, a belief that 'basic human needs' can be identified and rectified (Mitchell 1981; Burton 1990). But many conflicts are intractable in the sense that (for example) both the majority of Palestinians and Israelis would claim the West Bank of the Jordan as 'their' territory. The idea of a 'two-state' solution has been mooted for several decades, but one look at the map attached to such formulas illustrates the key problem – where to draw the borders that will satisfy everyone? Even Mitchell seems to have given up on this likelihood ever existing: 'Conflicts consist of behaviour, issues and attitudes, and unless all three of these aspects are resolved there seems little chance of any deep-rooted and intractable conflict finally being terminated' (Mitchell 2014: 290).

Other problems with analysing this approach have been the secrecy that is inherent in these processes, which has led to few of the case studies being written up for public analysis. This has been somewhat addressed by, for example, examining 'transfer effects' that look at how Track Two experiences can be of use in Track One discussions and vice versa (Fisher 1997, 2005; Fisher, Tadevosyan and Cuhadar 2023). Another more intractable problem is that it is seen by some as having been demonstrated to be just another, if more subtle, use of power by the West to coerce the 'rest'. The idea that several parties to a conflict can *ever* be given equal attention is also naïve, especially as the international system as now constituted gives a clear priority to *established* nation states, not peoples seeking secession or self-determination. It has also been suggested that the insurmountable barrier to successful conflict resolution lies in the impossibility of understanding another culture, which is an 'ontological' barrier to any real possibility of transformation, a difficult charge to refute as conflict resolution teams rarely even understand the language of those they are dealing with, in developing countries in particular (Väyrynen 2001; Avruch 2003; Jabri 2007; Ron 2009).

Another criticism that could be directed against such 'naïvety' is that in some cases, it could be argued that success in transformation is only possible because some form of 'settlement' was already going on. The successful transition in Ethiopia/Eritrea in 1993 (states that subsequently resumed their conflict, 1998–2000; though they are now once again allies) was maybe because the military victory of the Eritrean and Northern Ethiopian insurgencies had brought about a radically new situation. It is now evident after the Ethiopian civil war of 2021 in Tigray (ongoing in 2023) that underlying attitudes have hardly changed; this remains an 'intractable' conflict.

All of these categories of third-party intervention have as a central aim the improvement of communication, whether it be to signal war or threat, or to enhance understanding of the parties' views, needs and wishes. In the European case, that was provided initially by 'hegemonic powers' and then reflected in local civil society. Law has played a role as well: Much of the theory and practice in third-party intervention has been influenced by what could be described as the Western, legal, rational perspective. It has also been influenced by theories of bargaining and rational choice – some of which have limited application in the real world. And, as several analysts (Keashley and Fisher 1990; Webb et al. 1996) have stressed, multiple conflict transformation approaches can be pursued together, or one at a time or in different orders, depending on the conflict in question. This has also been described as a 'multi-track' approach (with 'Track Three' being local community groups), justified when, as is usually the case, the conflict is particularly complex. What we tend to see is a hybridity in peacemaking whereby a variety of initiatives – both internationally and locally inspired – are simultaneously deployed.

We must seriously doubt that the halcyon days of second-track approaches will return any time soon. Geopolitics is back in ways that seem to preclude the possibility of non-state actors being able to mediate in any conflict, past, present or future. That is not to say that they won't ever return. After all, during the Cold War, the CCA demonstrated that such approaches did help, if to a limited extent. But those years, especially the 1990s, were exceptional ones of 'unipolarity'. The United States is now under attack for its supposed arrogance and 'hegemonic' attitudes in Ukraine, by the Russian government and in Taiwan and on Chinese territory proper in Xinjiang and Hong Kong.

For the purposes of this book, there is a strong whiff of liberal hypocrisy in the air, fully exploited, cynically no doubt, by those in China and Russia who see no difference between US claims that it was 'liberating' the Iraqis and their doing likewise in Hong Kong, Taiwan and Ukraine. China's peace plan (discussed in Box 4.2) gives another important cover to show that the two leaders are genuinely interested in a peaceful final solution to the Ukraine crisis. The two leaders meeting in Moscow in late March 2023 was to show that they both stand up for peace and those who reject US bullying and support for illegitimate regimes, a line that gets a great deal of acceptance from those former colonies of Western states who were equally 'hegemonic' for many decades (Rachman 2023a).

These accusations are also normatively backed by accusations that the USA was able to drop any real pretence of liberal restraint and a good deal of hubris in the attack on Iraq of 2003, which was not approved by the UN Security Council. The invasion was a clear breach of the UN Charter, which alone can authorize the declaration of war. Russia's invasion of Ukraine is a similar breach. The resurrection of the geopolitical language of 'containment', one coined by George Kennan at the beginning of the Cold War in 1947, by the Chinese and Russian governments is intended to show that Washington is up to its old tricks, hemming in the rest of the world while claiming its own right to dominate the economic, military and other parts of the global system of states. These are difficult charges to entirely refute in light of the bungled operations in Iraq and Afghanistan and the proclamations of 'America First' domestically. Liberals have to work much harder to convince global populations that the liberal values that underpinned the actions of the West in Iraq and Afghanistan were truly intended to benefit the local population or world peace.

Conflict transformation?

So is the ending of conflict doomed to failure? Not necessarily, as the much-underrated example of peace in Western (and most of Eastern Europe from 1945 until 2014 can be said to show). Germany and France had an 'intractable' conflict for hundreds of years. Can we explain why they now cooperate at all levels of politics and society? It cannot solely be explained by the help given to Germany after comprehensively defeating it in 1945. Real friendship between Germany and France now exists. The authors, and many others, believe that the more overarching category of 'peacebuilding', leading to conflict 'transformation', is ultimately a more useful description of the process for creating a lasting peace, above settlement, management or resolution. The inclusion of 'peace' to the aid and development nexus indicates its recognition as an agent for conflict transformation.

Conflict settlement and management can be likened to holding a lid on a boiling pot, not attempting to turn the heat down, and conflict resolution has its own problems, as itemized earlier. So 'peacebuilding' is also often referred to as a 'third generation' approach to dealing with conflict (Richmond 2006) as we grope toward a successor to both conflict 'settlement' and conflict 'management', again because the underlying attitudes, causes and structures of the conflict need to be properly addressed.

It is on the ideal approach to dealing with deep-rooted violent conflict that we will now concentrate, that of 'transformation', as we believe that should be the central aim of all approaches to conflict, even if not to the exclusion of the others, which are still the most widely used. It is therefore our preferred term, but we are aware that no term can be satisfactory, given the extreme complexity of many conflicts in developing countries and the limited resources available for dealing with them (see Chapter 1 for more discussion of these distinctions). Mitchell also argues that 'transformation' would require that enemies become 'partners, colleagues or, finally, friends so that their relationship can be said to be genuinely "transformed"' (Mitchell 2014: 190).

The issues faced: the need for 'mapping' and 'deconstructing' conflicts

So what kinds of conflicts can be potentially 'transformed', and are they amenable to the kinds of theory and practical tools that we have at our disposal? It is now, for example, widely believed that we are dealing with conflicts that are not between *unitary* societies (usually states), as was believed to be largely the case between 1939 and until about 1990, but now within ethnically or ideologically *divided* societies. A number of theorists have put this down to the resurgence of ethnic violence (Gurr and Harff 1994, 2003; Kaufman 2001). Certainly, compared to the period before 1990, one of the areas most identified as being truly 'new' in the post–Cold War conflict mix is the huge rise in what is termed 'ethnic violence'. Ted Gurr was one of the first to recognize this phenomenon in the context of conflict studies, but of course there was a huge discussion of nationalism and self-determination well before this (for an excellent introduction to this and the links to the ethnicity question, see Oberschall 2007: Chapter 1).

We might nonetheless question this now practically embedded wisdom about the differences between pre- and post-1990 (Kaldor 2012). First there were many civil wars *before* 1990 (Kalyvas 2001; Newman 2004). And, very unfortunately, ethnicity has often been assumed since 1990 to be a key component of a conflict situation both for the participants and for outside supposed 'conflict resolvers'. This 'framing' can have disastrous effects, when it arguably is not the determinant factor by any means.

'Names', 'identity' and 'ethnicity'

Outsiders often misunderstand the conflict they are faced with and often have very different understandings of even 'who' they are dealing with. They will tend to come up with 'names' that may or may not correspond with the names used by local people. As Anthony Ware and Costas Laoutides write in their analysis of the 'Rohingya' conflict with the 'Burman' state of Myanmar: 'Conflict routinely involves attempts to control perceptions of identity, of one- self and 'the Other'. Names are often hotly contested'. They may even be used to deny the very existence of a group, as in the Myanmar government stating 'there are no Rohingya in Myanmar', which is not so much to deny there are no people who call themselves 'Rohingya' but that they 'do not compromise an "ethnicity" who are "indigenous" to Myanmar, with the group rights that would imply' (Ware and Laoutides 2018: xv–xvii).

To take a more banal example: the tennis champion Novak Djokovic saw no problem saying at the opening of the French Open Championship (and writing on a TV camera) on 1 June 2023 that Kosovo is the 'heart of Serbia', a notion that is ingrained in the belief system of many Serbs. 'Kosovars' would clearly take exception to that statement. But both would agree with Djokovic's next statement, 'Stop the violence' (Camut 2023). The conflict between Serb and Kosovar identities are far from being resolved, so any such statement is bound to be inflammatory.

Another even more serious example is of journalists who routinely used to parrot the idea that 'ethnic Russians' were being persecuted by 'Ukrainians', also often referred to as 'Nazis', a favourite claim by Vladimir Putin and his cheer leaders in the Russian press and used to justify the invasion of Ukraine in 2022 with a detailed 'analysis' broadcast over state television in the summer that preceded the invasion (Putin 2021; Weir 2022). So how can Volodymyr Zelensky, the president of Ukraine, happen to be both a Russian speaker and a Jew, as well as a 'Nazi'? That use of names by Putin morphed quite rapidly into claims that there was no such thing as a 'Ukrainian' identity, that anyone that believed it was justifying 'fascism'. Kaufman (Kaufman 2001) was not the only commentator who wrongly sees ethnicity as the root of most recent conflict in Eastern Europe. Moldova, which the author knows well, was *not* an ethnic conflict, nearly everyone is multilinguistic, as they are in Ukraine.

A classic example of this misnaming was in Bosnia as pointed out by David Campbell, who blamed Western politicians for in effect fanning the flames of inter-ethnic violence (Campbell 1998a, 1998b), but such misjudged preconceptions have been widely and unfortunately repeated. As the cover explanation for 'National Deconstruction', Campbell's book, states, 'How did Bosnia, once a polity of intersecting and overlapping identities, come to be understood as an intractable ethnic problem?' Partly because of the ascribing of 'names'. 'Ethnicity' is rather like 'gender', it can be self-described or attributed. And like 'gender', it can indeed generate problems, for the press and governments need to have a hook on which to hang their prejudices.

The problem is arguably worse in developing countries where the obsession with 'identity' has tended to obscure the continuities of local (for example African) history and traditions that pre-date the colonial era. Our tendency to forget about historical precedent and memory has led us, in Richard Werbner's words, to 'mark the end of an epoch falsely by placing a break where none exists' (Werbner in Werbner and Ranger 1996: 5). The result has been to both marginalize developing countries as mere adjuncts of Western historical experience, 'post-colonial' child-like entities without local agency capable of creating their own destinies. But, most significantly, the creation of states themselves has *always* been a violent process. Massacres, civil wars and the like have attended the birth of all the 'classic' states of the past (Tilly 1985; de Tocqueville 1856). Edward Newman and others ask whether the big difference now is that the 'statebuilding' project (see Chapter 5, pp. **205–208**) is now mainly different from that of the past in that these processes are supposedly being shepherded by IGOs like the UN with the aim of liberal statebuilding (Newman 2013: 142–3).

So, although the analysis of all conflict as 'ethnic', or even 'civil', is a neat and workable shorthand in many conflicts, it is not sufficient (Brubaker and Laitin 1998). As Oberschall puts it, 'These theories lack specificity and context'. His preference is for what he terms 'conflict and conciliation dynamics', which essentially seek to 'map' the conflict by looking at issues, players and their strategies and implementation (Oberschall 2007: 29). This idea of mapping a conflict has also been used quite widely in conflict resolution and transformation approaches (including those of Mitchell and Webb 1988; Burton 1990; Fisher 1997; Kelman 2005, etc.). For all these theorists we might suggest a widespread belief that a first step *has* to be to 'deconstruct' a conflict before it can be properly understood and analysed, as a precursor to identifying possible solutions to it. (See also Chapter 1 for more discussion of this vital theme.)

If this is true, then unravelling the motives for perpetrators, victims and interveners alike is far more daunting than dealing with mere ethnic violence, bad as that obviously is. It is obvious that the use of violence has multiple motivations, with correspondingly difficult decisions to be made about how to 'map' these motivations. It is equally the case that those who try to intervene to solve a conflict can have the motivation to help themselves or those they care about who are affected by the conflict. A good example is the (first ever) seven-nation delegation of African leaders to both Moscow and Kiev to try and find a way to resolve the Ukraine conflict in June 2023. Africa has been greatly affected by the food shortages that have resulted worldwide from the Russian invasion, the leaders also want to show that they have agency to influence events beyond Africa (Jones 2023). It is important to note that, as in this case, the actions do not necessarily advance the cause of peace, but they do enlarge the field of those who wish for it. That in itself is potentially transformational.

Gender

Examining the issue through the lens of gender is another essential approach that is far more fruitful than ethnicity. First, because the use of sexual violence and rape as a weapon of warfare has long been noted (see Enloe 1993; Thornhill and Palmer 2000). Second, as our colleague in this series, Janet Henshall Momsen, writes the vital point is that women are usually seen as 'victims of war' and not as agents of potential conflict transformation. Third, rape is now recognized as a 'crime against humanity', an advance of sorts. In a study of the subject Margot Wallström is cited as saying 'this is a security problem.... It's not an inevitable consequence of war.... It's not inevitable' (Baaz and Stern 2013: 1).

But women are as often enough not credited with being able to 'participate fully ... in conflict prevention and resolution, as well as in peace initiatives', in the words of Ethiopia's minster of state in charge of women's affairs. These words were written in 2002. We doubt the reality is much different in the current Tigray conflict (Momsen 2004: 102). Claire Duncanson and Jane Parpart both make the point that such attitudes are due to the same power imbalances that afflict all political and social systems. Whether the failure of vision is due to entrenched attitudes within wider society or have been particularly exacerbated by the kinds of 'neo-liberal' policies we have noted are most widely favoured by Western donor states and IGOs, women's role in peacebuilding has usually been sidelined and continues to be so in the current century (Duncanson 2016; Parpart 2017). We might, however, also point out that sexual violence in war was as much a feature of the (dominantly male) repertoire of war before 1990 as after it. Historians such as Anthony Beevor have catalogued the widespread use of rape by Russian soldiers in Berlin in 1945 (Beevor 2002) and Edward Newman also made a similar point, again in criticizing the 'new war' thesis (Newman 2004). Have neo-liberal policies made matters worse? Britain's first woman prime minister Margaret Thatcher, an active initiator of war on occasion but also a person with huge agency for change in East-West relations, would have argued that neo-liberal policies were the solution, not the problem.

Protracted social conflict?

As we stressed in the Introduction, an important input into thinking about conflict transformation has been the work of Edward Azar, who believed that 'protracted social conflict' (Azar 1990) needed 'problem-solving', not military solutions. Azar's approach was one that accepted much of what has been written by theorists of 'basic human needs' like John Burton. Their aim was to get a wider acceptance that most conflicts were due to 'the prolonged and often violent struggle by communal groups for such basic needs as security, recognition and acceptance, fair access to political institutions and economic participation'. For Azar and others, it is far more likely that economic or ideological factors underpin much 'ethnic' conflict (Nordstrom 2004; Pugh et al. 2004).

Azar's approach allowed for the analysis to be on a variety of levels – 'identity' at the 'communal level' with an analysis of the 'deprivation of human needs', the crucial factor of the 'governance and the state's role' and the need to look at 'international linkages' (Azar after Ramsbotham et al. 2005: 84–8). All of these need a long-term process that is often beyond the means of 'Track Two' teams and beyond the mandate of 'Track One' mediators. We cannot yet be said to be anywhere near resolving this conundrum (but for some understanding of it, see Fisher 2005).

Strategies of conflict resolution and transformation

So although purists will immediately see flaws in such a narrative, we can therefore, cautiously, describe most attempts by (usually but not always) outside parties at mitigating or 'solving' conflicts through two major series of approaches: 'top-down' (conflict settlement and management) and 'bottom-up' (conflict resolution or transformation). We can also suggest there is now in place an overall idea of 'peacebuilding', defined by Miall, Ramsbotham and Woodhouse as 'the attempt to overcome the structural, relational and cultural contradictions which lie at the root of conflict in order to underpin the processes of peacemaking and peacekeeping' (Miall et al. 1999: 56–7). This can be seen in terms of actions by states or IGOs, and as such, it is usually used in the 'reconstruction' attempts we will describe in the next chapter. But the emphasis in the rest of this chapter is on 'peacebuilding from below', which is what the term often means in the context of conflict resolution, transformation or attempts at reconciliation, all of which often, though not exclusively, are used by changing the attitudes and practices of the base.

To put that into simple language: how would you, our reader, think it best to deal with (a) two warring factions within a government or (b) the reintegration of people who had been killing your neighbours or family? Would you advocate a policy of justice (or revenge) for paramilitary forces that may have destroyed your village or killed your relatives, or their 'reintegration' into society (as suggested in the next chapter) or getting them to undergo a process of much more personal reconciliation?

Reconciliation: memory, truth-telling and forgiveness

One way to bring about a lasting peace is now seen as being through 'reconciliation', either through the mechanism of 'truth and reconciliation commissions' (TRCs), often referred to as 'restorative' or 'transitional justice' (probably the more used of these terms nowadays), or through the use of an older institution, the 'war crimes tribunal' (WCT), often referred to as a form of 'retributive justice'. The first usually assumes a non-retributional and non-jural outcome; the latter assumes the use of some form of sanction, including imprisonment, or even the death sentence, for 'guilty' parties (Rigby 2001; Bell et al. 2007). We will discuss them in turn, but first look at what they can be said to have in common in addressing conflicts in developing countries.

Both of these institutional frameworks attempt to work based on the proposition that we need to establish the 'facts' of any conflict, including the hardest of all: those to do with killing, torture and other massive abuses of human rights. This, it is assumed, will somehow help both victims and perpetrators to come to terms with their experiences and actions and help them to understand, possibly forgive, and hopefully 'move on', to a better and more functional relationship. It also assumes that the bringing of 'justice' will indeed reduce the levels of violence and contribute to peacebuilding (Lambourne 2009). These propositions are drawn from essentially three sources:

- First, from psychological insights that a 'talking cure' (Freud's term) will enable individuals and societies to make a transition from hostility to peace through the establishment of clearly identified memories and contending 'truths' and that this will then lead to a better ability to deal with collective and individual trauma (Jeong 2005: 155; Fierke, in Bell 2006; Fierke 2013; Fierke and Mackay 2020).

- Second, it bears more than a passing resemblance to ideas in conflict resolution/transformation theory and practice that a 'ventilation' of grievances and an establishment of the history(ies) of a conflict will help establish a new point of departure for the players and thus a transformation of their relationship discussed earlier.
- Third, its proponents often link it explicitly to 'liberal peace' arguments that 'transitions to democracy' necessitate an adoption of market reforms, constitution building and the establishment of a rule of law and a consideration of 'what to do about the past'. Justice and accountability will go hand in hand with prospects for a stable peace (Hayner 2002: xi, Chapter 7).

Transitional justice/TRCs

The first attempts at setting up TRCs to deliver transitional justice were in Latin America (Chile, El Salvador and Argentina in particular), to attempt to draw a line under the 'dirty wars' of the Cold War era (Whittaker 1999: Chapter 2 on El Salvador; Hayner 2002). After 1990, the most prominent examples have been in South Africa after the end of apartheid (see Box 4.3 for more detail), the Rwandan genocide of 1994 (Prunier 1995; Corey and Joireman 2004; Moghalu 2005, see also Box 4.4) and in Sierra Leone after the Lomé Peace Accords of 1999 (http://www.trcsierraleone.org/drwebsite/publish/index.shtml). Suggestions have been made for similar commissions in the former Yugoslavia, in Northern Ireland and elsewhere.

Box 4.3 The South African Truth and Reconciliation Commission (TRC)

The end of apartheid in South Africa came after an all-white referendum in 1992 and the freeing of Nelson Mandela by the National Party, led by F. W. de Klerk. De Klerk voluntarily handed over power to a new majority administration of the African National Congress (ANC). The ANC has followed a very 'liberal' path since then, holding a number of elections, encouraging foreign multinational companies to set up and develop in South Africa and avoiding all of the worst excesses of neighbouring Zimbabwe by encouraging a process of reconciliation. This has been concentrated in the TRC, set up in 1995, which has heard hundreds of testimonies from former perpetrators from the South African defence forces and many more victims and their relatives. The idea has been that suggested by Archbishop Desmond Tutu, that the truth will heal the trauma of the past and that it will help in the 'nation-building' of South Africa. Some have argued that the process, which has never really addressed the problems of reparation for the crimes committed (although there is provision for this in the TRC's statutes) lets off the perpetrators too lightly. The Commission finished its work in 2002. The government is pursuing a policy of gradual and cautious reform of land, for example, which it is hoped will help the reconciliation process along.

Sources: Rotberg and Thompson (2000); Hayner (2002); Moon (2009); Gready (2011)

Box 4.4 Rwanda

By many calculations around 800,000 mainly 'Tutsi' men, women and children were massacred by their 'Hutu' neighbours in 1994, and many women gang-raped, resulting in a huge subsequent dispersal of the population that committed these massacres as the new government, which was also ethnically mainly Tutsi, came to power and demanded retribution. The international community turned a largely blind eye to the massacres, in spite of anguished pleas to the UN by the UN military commander in the country, General Roméo Antonius Dallaire (Dallaire 2005; Melvern 2006). The French government (through 'Operation Turquoise') has been accused of harbouring the 'génocidaires'. The massacres were of an ethnic nature but might be argued to have been about other issues, as the two groups were practically indistinguishable. They were certainly about elite competition and demonstrated the power of a centrally organized bureaucratic state in coordinating such events, as Nazi Germany had previously shown. The post-massacre 'Gacaca' courts, set up in 2002, have attempted to get lesser (that is to say not leading) perpetrators to confront the reality of what they have done, in exchange for a lesser or even no punishment. The post-genocide era has seen the imposition of another authoritarian state and massive intervention by the Rwandan army in neighbouring Congo, though with a lot less violence being committed within Rwanda. About 110,000 were incarcerated awaiting trial in 2001. An International Criminal Tribunal for Rwanda has also been set up by the UN, but its effectiveness has been less than impressive in terms of the numbers of those brought before it or punished, though important legal precedents have been established, against sexual violence for example.

Sources: Straus in Pouligny et al. (2007), Kayumba and Kimonyo in de Zeeuw and Kumar (2006): 211–36, Booth – on sexual violence in Rwanda – in Sands (2003): Chapter 5, Moghalu (2005)

Scholars who have assessed the successes and failures of TRCs have come to mixed conclusions (Olsen et al. 2010; Gready 2011). Nonetheless, the aftermath of the civil wars in Sri Lanka, Myanmar and other parts of the Asia-Pacific region has led to a renewed interest in mechanisms of transitional justice, especially given the intercommunal nature of the violence and the often-brutal attempts to 'manage' or 'settle' them (on Sri Lanka, see Sriram in Jeffery and Kim 2014; on Myanmar see International Center for Transitional Justice (2023) and Annan (2017)). Transitional justice is now a standard part of the tool kit of peacebuilding.

One way of assessing the usefulness of such attempts is to analyse the underlying essential elements of such thinking, and especially the ideas of 'memory' and 'truth'.

Memory, history and 'truth'

An essential element of reconciliation lies in having common, or at least mutually comprehended, memories of what happened in a conflict or war that needs to be reconciled. History is a powerful weapon in the hands of those who would establish their version of the 'truth' but also a potentially even more useful tool for shedding light on the wrongs experienced by all sides in a conflict, thus allowing them to (potentially at least) 'move on' in their relationship. The lack of a historical dimension to much of the study and practice of international relations can be identified as a grave flaw in policymaking and actions in a large number of recent and older conflicts and wars (Williams et al. 2012). The role of historical factors in decision-making by the US government about how to 'reconstruct' Iraq after 2003 will be dealt with in Chapter 5.

Transformation by historians?

The aim of much conflict transformation and reconciliation is thus to establish the 'individual truth' of what happened but also to interrogate what might be called 'collective' truths. This, it is hoped, will then lead to mutually acceptable changes in the versions of the 'truth' that have long dogged the peoples involved in a long-running feud. If, for example, the Irish and the English or the English and the Scottish people persist in 'remembering' their past histories as exclusively ones of mutual persecution, with 'signposts' along the way that encourage a prolonged feeling of mutual grievance, or even hatred, there is little chance of them being able to make lasting peace with one another. It might be reasonably asserted that the popular film *Braveheart*, which portrayed the heroic and allegedly largely blameless William Wallace as fighting a guerrilla war against the heartless and brutal English monarchy, did little to foster Anglo–Scottish understanding, and even contributed to the much remarked upon resurgence of Scottish nationalism since the 1960s (Webb 1978). Equally, folk memories of Cromwell's massacres at Wexford and Drogheda in the 1650s or the Battle of the Boyne, won by William of Orange against the rump of James III's army in 1690, have been constantly evoked in nationalist propaganda by Republicans and Loyalists alike in Ireland to justify horrific atrocities. Even more recently the memories of the Great War of 1914–18 have been used in Ireland to justify Protestant violence against Catholics in the North (Mac Ginty and Williams 2007).

In recent times, historians and public opinion has started to shift towards a more even-handed reading of this history. This has now been corrected politically by the 'Good Friday' Agreement of 1998 and other agreements, and an acceptance that 'unionism' is a legitimate political position. It has been rectified historically by a new acceptance that Irish soldiers from the Republican and Unionist traditions fought in the Great War, often together (Ferriter 2015; Walsh 2015), that there might well have been a 'constitutionalist' alternative to civil war and even exit from the British Empire (now the Commonwealth) by Ireland. Historians have played a large role in this (Bew 2009; Laffan 1999), acknowledged by Irish politicians like John Bruton (Taoiseach, 1994–7), Irish presidents Mary Robinson (1990–97) and Mary McAleese (1997–2011), as well as Queen Elizabeth II and key politicians in the United Kingdom.

Many, probably most, other conflicts figure a contested understanding of a shared history. Since 1990 the war in the former Yugoslavia the Croat and Serb leaderships have both evoked historical grievances dating back to the fourteenth century to justify ethnic cleansing and other brutality. We can point to evidence of what is often called 'inter-generational violence', where a new conflict between recently entirely peaceful neighbours can erupt if there is a historical memory of similar violence between neighbours in the recent past (Argenti and Schramm 2010). Outsiders can fuel the fires of resentment. For example, the way the rest of the world has written about the 'Balkans' has informed a lot of the decision-making about how outsiders should try and intervene to bring about peace there (Goldsworthy 2013), a process called 'the imperialism of the imagination'. Locally, many former Yugoslavs listened as children to their grandparents telling stories about the brutality of their neighbours' during the Second World War. All of these internal and external influences make a real difference in the construction of 'enemy' and 'friend' narratives, which in turn fuel the conflict. Thus the very act of observation changes the conflict, what might be called a 'quantum' analysis of conflict (Fierke and Mackay 2020).

The same phenomenon has been noted in many conflict areas. Many of these studies are anthropological, and many address difficult

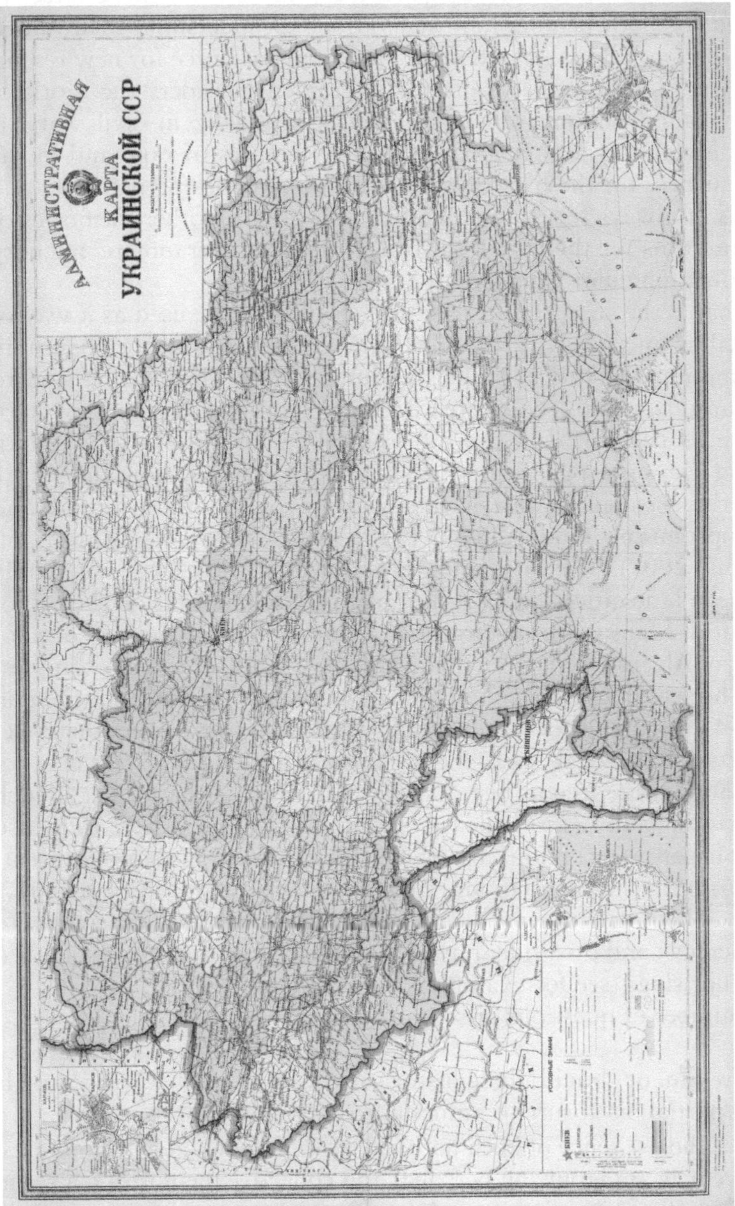

Plate 8 Ukraine and Russia share a common history: is that a key part of the reason why they now disagree on so much?

Source: Wikicommons (Soviet era map).

issues such as memories of sexual violence in war, as in one study of the attitudes of the children of women raped in earlier wars in Peru and Columbia (Theidon 2022). Some look at the way that such trauma have been transmitted through literature, often a far more effective method for both helping new generations to understand their ancestors experience but also potentially providing a source for new resentment (see, for example, Kalisa 2009, which considers the work of Frantz Fanon, among others). Violence, especially in civil wars, is nearly always an intimate act, especially in cases of rape and in the consequences for the lives of those who survive it as perpetrators and victims. It always forms the main backdrop for all other memories of war, and thus for the hoped for transformation of traumatic memory into something more hopeful.

If that is the case, can history and storytelling be used as a way of reconciling and resolving conflicts? There has been 'memory boom' in contemporary historical studies (Winter 2000). This can be said to have originated in a widespread rememorization of war in Western Europe as part of a process of reconciling the peoples of Western Europe who only a generation back (and for several before that) were regularly engaged in internecine warfare of the most brutal and uncompromising kind (Winter 1995; Evans 2006; Mac Ginty and Williams 2007). The First and Second World Wars were not purely European innovations, even if they had profound effects on the rest of the world. Many of the insurgents/freedom fighters in wars of decolonization (Algeria, Kenya, Vietnam, arguably Israel, are good examples) learnt how to fight their colonial masters by fighting alongside them in 1914–45. Sadly, they often used similar tactics after their decolonization to deal with their enemies, as in FLN treatment of Algerian *Harkis*, seen as being 'collaborators' with the former French authorities. Many *Harkis* fled to France after 1962, where they also met a negative reception, and their children have harboured resentment to both the Algerian and French states ever since (Evans 2017, 2014). Their resentment at the pronouncement of adherence to 'democracy', and the evident hypocrisy inherent in such claims, fuelled many uprisings (see for example Lewis 2007 on Kenya).

A number of questions arise immediately:

- How can, or should, these historical ghosts be put back to sleep? Is full/partial disclosure or amnesia better (Rigby 2001)?
- Does the existence of both national mythologies and, in some cases, extremely antagonizing personal and collective memories matter in potentially fostering future violence?
- What if there are 'contested memories' (no agreement on what happened) (Beneduce in Pouligny et al. 2007)?

Plate 9 Napoleonic era commemorative arch, in this case hoping to see Britain's eventual defeat, Tuileries Gardens, Paris credit: author's own picture.

- Who has 'ownership' of a historical narrative and could therefore be trusted to write an account of it that would not (inevitably perhaps) be merely a version to suit the 'victors', or at least a dominant view of events or put their spin on commemorative activities (Mac Ginty and Williams 2007)?

The main issue is the existence of two opposing historical narratives. For example, if we accept the Israeli narrative, which essentially sees Jews having fled the Holocaust in Europe and settling an 'empty' land and developing it where before there had been desert and poverty, to the benefit not just of the Jewish population but all those within and indeed without Israel's borders, then a picture of a blameless Israel being attacked by fanatical hoodlums makes absolute sense. If we accept the Palestinian mainstream view that they were expelled in large numbers from their homeland by an alien invader backed by the Great Powers in 1947–8, Powers which were themselves motivated more by their guilt about the Holocaust than any genuine sympathy for either Jews or Arabs, then obviously we can sympathize with the Palestinians. As historians have pointed out (notably Edward Said, see Ruthven 2003; Bunzl 2008), these 'mirror images' or incompatible mythologies are the basis of many protracted conflicts.

'Ruptured histories'

The idea that history is 'dead' or that we have an 'end' (Fukuyama 1992) to it could not be less true than it is today. History informs most, if not all, of the enduring deep-rooted violent conflicts that afflict the developing world. Some of them, like those that pit Japan as a former imperial power against South Korea and China, are vital for understanding the future stability of the whole of East Asia (Miyoshi Jager and Mitter 2007; Kim in Jeffery and Kim 2014).

So what can be done? Should those identified as having responsibility for crimes of war be brought to book and put before a court of law, national or international, as mentioned earlier and often referred to as 'retributive justice'? Or should the miscreants be seen as victims as much as perpetrators and allowed or encouraged to reinsert themselves into societies that they once sought to destroy? We will now go through the arguments for both of these courses with the use of historical examples. It should be noted that there are significant differences that exist in the theory and practice of how a conflict can be de-escalated and a peacebuilding process given a chance to begin.

Some general issues need to be addressed first.

A conflict might be seen on one level as an inability to trust. Ramsbotham et al. put it thus,

> Legitimacy, acceptance and trust ... are integral to the functioning of any reasonably stable socio-political system, invisible and often taken for granted when differences are being settled relatively peacefully, but palpably lacking when they are not.
>
> (Ramsbotham et al. 2011: 206)

They also point out that 'one of the main obstacles to social and psychological healing is the cumulative hurt' (Ramsbotham et al. 2011: 207). So a spiral of distrust develops in an uncanny echo of the 'security dilemma' in international politics (for an explanation of this, see Booth and Wheeler 2007). Because you believe that the other side has more weapons than you (or is less trustworthy), you arm yourself more (and trust less). The classic example of a global conflict is the arms races of the Cold War; a more current and local one is the ever-increasing use of more and more sophisticated small arms in pastoralist conflicts in Africa (Mkutu 2008; Wepundi et al. 2012).

Truth and reconciliation commissions

The question we have asked can be summed up as can/do TRCs help in dealing with the aftermath of deep-rooted violent conflicts? The South African case has required the unravelling of nearly a century of apartheid policies, and that may well take generations.

What can we learn from such episodes for the future of reconciliation attempts in developing countries? It is evident that in Rwanda, 'even by strict definitions the episode qualifies as "genocide"' (Straus in Pouligny et al. 2007: 123). The aftermath of such events is that much more difficult to deal with than, but also different from, the long-drawn-out agony of South Africa. In both cases we could argue that these were 'unforgivable' crimes. In South Africa, the fight against apartheid came to be seen as the sole prerogative of the ANC. Mahmood Mamdani has asked the pertinent question as to whether 'if truth has replaced justice in South Africa, has reconciliation turned into an embrace of evil?' He also points out that 'in Rwanda there are a lot of perpetrators and a few people who benefited; in South Africa there are a few perpetrators, but lots of beneficiaries' (Mamdani, quoted by Krog 1998: 146–7).

This conundrum has led many to ask if the reconciliation in either country is more than skin-deep. The post-apartheid crime figures in South Africa (where hundreds of white farmers have been killed and the country has one of the highest murder rates of any country on the planet) suggest a society ill at ease with itself. So is reconciliation the 'lesser of two evils?' Mothers who lost their children have a different view of the idea that reconciliation can help us 'turn the page'. One such mother who testified to the TRC, a Mrs Kondile, commented that she could not forgive a South African policeman, Dirk Coetzee, for killing and burning her son: 'It is easy for Mandela and Tutu to forgive ... they lead vindicated lives. In my life, nothing, not a single thing, has changed since my son was burnt by barbarians ... nothing. Therefore I cannot forgive' (quoted in Krog 1998: 142). Others have claimed that, in the case of the TRC, even if the whole truth is not found out, a great deal is learnt and the process is therefore 'therapeutic' (Christie 2000: 173–5).

Retributive justice: war crimes tribunals

Some have suggested that a better solution would be to opt for the forensic truth-seeking of a legal retributive tribunal, as was the case with the Nuremberg (1946–9) and Tokyo (1946–9) trials after the Second World War (Sands 2003; Browning 2005; Friedlander 2008). There are many detailed discussions about the nature of both German and Japanese war crimes (on Japan, see Rees 2012), and the subject has led to sharp exchanges between historians and political figures in Japan, with a notable contribution to truth-seeking from the Japanese historian Chihiro Hosoya (Hosoya et al. 2001). As we discovered in discussion with the excellent Japanese translator of the first edition of this book (Mac Ginty and Williams 2012), there are still deep divisions about how these issues should be interpreted. There is also now some scholarly discussion in English as to whether we can in fact compare

Plate 10 British prosecutors at the Nuremberg Tribunal, 1946. NB. No women are present! © Imperial War Museums 2023.

the Nuremberg and Tokyo 'WCTs' trial processes with any ease (Futamura 2008).

A number of criticisms have been made of these trials. At the time they were held (1945) and since, the process has been criticized for trying defendants for crimes which had not yet been defined as such when they were allegedly committed. Radhabinod Pal, Indian justice at the Tokyo trials in 1946, found the Japanese accused all 'not guilty', and referred to the judgements, which led to many executions, as 'victor's justice', an expression still used today by critics of the process. One important element of it was that, as historian John Dower expressed it, the "trial was fundamentally a white man's tribunal" (National WW2 Museum 2021).

A corollary and serious problem is that evoked by Sands among other jurists that there is a real intellectual problem in the use of two perhaps contradictory charges levelled in such tribunals. The crime of 'genocide', enunciated initially by Rafael Lemkin in 1945 in *Le Monde*, refers to one 'intended to destroy a national or racial minority within the population'. The (then) recently revealed Nazi 'Holocaust' of the Jewish population of Europe clearly qualified for this description. But the Nuremberg and Tokyo trials largely prosecuted defendants for 'crimes against humanity', a term elaborated by British professor of international law Herch Lauterpacht and American chief prosecutor Robert Jackson in a Cambridgeshire pub garden. Its remit would cover

Plate 11 'Victor's justice': Saddam Hussein of Iraq was hanged after a widely criticized trial.

Source: SSGT D. MYLES CULLEN, USAF / WikiCommons.

'atrocities against individual citizens'. This included a stress on individual responsibility 'before as well as after commencement of the war' and was included in the final text used in the trials as Article 6I. After the Russian prosecuting team agreed to this clause, the expression became international law.

It is worth noting that the expression 'crimes against humanity' was not entirely new. An excellent Wikipedia page on the topic cites several cases where it was evoked before 1945. Of particular interest to this book was its use in 1889 by US president Benjamin Harrison who used the same term about slavery (Wikipedia 2023). It was also used by American George Washington Williams about the appalling atrocities committed in the name of King Leopold II of Belgium in the Congo Free State in the 1890s. This cause was notably taken up by the journalist E.D. Morel, who founded the 'Congo Reform Association' (CRA). He was later a prominent opponent of the First World War in the Union of Democratic Control (UDC) and Labour M.P. The CRA was able to recruit a long list of what we would now call 'celebrities' to condemn the atrocities including novelists Joseph Conrad, Arthur Conan Doyle, Ford Madox Ford, John Galsworthy and Mark Twain, as well as Irish nationalist Roger Casement. Morel's activities can be seen as a precursor of the leaders of many other campaigning groups against slavery and other practices that 'shock the moral conscience of mankind', a term often used by Michael Walzer, as in Hothschild (2019) and Waltzer (2015).

On 11 December 1946, the UN General Assembly consecrated the expression 'crimes against humanity' in Resolution 95. In the next Resolution 96, the Assembly accepted that 'genocide' was also a crime in international law. This was followed by a 'Convention and Punishment of the Crime of Genocide' on 9 December 1948 (Sands 2016:55, 111–3, 377). The problem that was to arise was not that 'crimes against humanity' were difficult to detect, all wars have multiple examples of them, but that 'genocide' is notoriously difficult to define. The Genocide Convention of 1948 has also been identified as flawed in many ways, most notably because 'only national, ethnical, racial and religious groups [are] protected' (Moses 2021).

Even in the case of the Rwanda 'genocide' of 1994 the then American secretary of state Madeleine Albright and various other members of the Clinton administration were unclear about what was happening in Rwanda and what to do about it. The massacres were carried out by (largely) 'Hutu' inhabitants of Rwanda against their 'Tutsi' neighbours, encouraged by the 'Hutu' government's *Radio Mille Collines*. But it has been pointed out that these 'names' were given to sections of the population by the former Belgian colonial authorities as part of a policy of divide and rule and very suspect racial theories (Prunier 1995; Dallaire 2005). On the face of it, the targeted groups *did* therefore fit the definitions of the Genocide Convention. The Security Council nonetheless decided to withdraw the small UN force, UNAMIR, when an increase was undoubtedly needed. Albright explained later that there was such chaos that no clear consensus emerged that the killing that is now called a 'genocide' was as bad as some were alleging, including the commander of UNAMIR, General Dallaire. She had had to talk to the relatives of those 'who went down in the Black Hawk helicopter', an event during the earlier Somalia intervention of 1991. That was widely denounced as a disaster giving birth to the expression 'Vietmalia syndrome'. In short, no one in a position to act could agree if this was indeed a 'genocide', though it was the clearest example of one since 1945 (Dallaire 2005; Albright 2004).

The end of the Cold War saw a number of retributive tribunals set up. It is noticeable that there were no international tribunals during the Cold War but a large number run by national governments, often, as in Germany, against their citizens accused of crimes against humanity. Arguably, these were most effective in making a new generation of Germans aware of crimes committed by previous generations and generating a real horror of ever repeating them. Although many of them in the Western section of a divided Germany resulted in mild sentences they instilled a greater sense of the need for a democratic population to take responsibility for its members' actions and to restrain any repeat of its behaviour. It also led, as a consequence, to Germany reaching out to neighbouring countries, former enemies in many cases like France, thus contributing to the peace of Europe through integration that has marked the post-1945 years.

Unfortunately, in other parts of Europe the nationalisms of the past re-appeared and often led to violent dis-integration. This was most notable in parts (but not all) of the former Soviet Union and former Yugoslavia. The latter led to atrocious crimes against humanity by all sides and led to the establishment of an International Criminal Tribunal for the Former Yugoslavia (ICTY). This has generated a huge amount of academic and media attention (Gow 1997; Gow et al. 2014) and; has led to the prosecution of many alleged perpetrators of war crimes in the Balkans, most famously of former Yugoslav president Slobodan Milosevic and, in 2008, Radovan Karadzic, the Bosnian Serb leader (Gow and Zverzhanovski 2004; Donia 2014; Baker 2015). They, and many others, were all prosecuted using the precedent of Article I (c) of the ICTY: 'ordering, soliciting, inducing, aiding, abetting or otherwise assisting in or contributing to the commission or attempted commission of such a crime'.

This development further enshrined the idea that states (as in Germany after 1949) were responsible for the acts of their own citizens in war and the citizens of any other state, so opening the national courts to the prosecution of individuals committing crimes in any other state. One of the first major prosecutions under this new principle became former president Augusto Pinochet of Chile, for crimes committed in Chile during the brutal repression after the overthrow of President Salvador Allende in 1973, as well as of Milosevic. Many more heads of state, former and recent, were to be affected in the same way, as we shall see next.

We can thus see a logic that performs a cross over with debates of our time with the demands for 'reparations' for slavery. Someone has to assume 'responsibility' and acknowledge 'guilt', the self-same logic of all restorative and retributive justice (Williams 2024 forthcoming).

The ICC

This emerging logic led in turn to the establishment of a permanent International Criminal Court (ICC), one that serves the international community as a whole. It was formally founded in 1998 under a treaty known as the Rome Statute and directed against individual misdemeanours, potentially also by peacekeepers (which is why a number of states have not ratified the treaty setting it up). Individual criminal responsibility before the ICC nonetheless builds on the aforementioned logic of WCT in general and the *ad hoc* tribunals such as the ICTY (Goy 2012). The ICC website as of June 2023 tells us that 123 (out of a possible 193 full members of the UN) countries are full signatories and that

> [t]he International Criminal Court (ICC) investigates and, where warranted, tries individuals charged with the gravest crimes of concern to the international community: genocide, war crimes, crimes against humanity and the crime of aggression.
>
> https://www.icc-cpi.int/about/the-court (accessed 14 June 2023)

The scale of the ICC is not great at all. It has a modest budget: for 2023, Euros 169,649,200 and about 900 staff. To date, it has had 31 cases before it and issued 40 arrest warrants, which has led to 21 detainees and 14 still at large. There have been ten convictions and four acquittals. So its activities have been modest, indeed compared to the Nuremberg Trials very modest indeed. From November 1945 to 1 October 1946 these judged 21 defendants and 19 convicted (of whom 12 were executed and 3 acquitted). But Germany saw in the later 1940s 12 subsequent trials. One hundred seventy-seven more defendants were prosecuted with 24 death sentences and 20 more imprisoned for life. More domestically run trials followed into the 2000s (National WW2 Museum, 2023).

From its inception, there has been a problem of the jurisdiction of the ICC and which cases it prosecutes and for what crimes. We have already noted that 'genocide' is a particularly difficult crime to define since it is so all-encompassing. In addition, the ICC does not have universal jurisdiction, as it is only states that have ratified the Rome Statute. This excludes some of the major potential offenders and has led to ambiguous decisions that over-ruled ICC jurisdiction. In most cases, the state whose citizen is implicated in an alleged war crime has to 'self-refer' to the court and then only when domestic procedures have been exhausted. So, for example, quite a few British military personnel whose alleged crime was committed in British uniform were not referred to the ICC but rather tried and, in some cases, convicted by British courts. This does not of course mean we can say that the ICC has 'failed' in its mission of shining a light on offences within its remit.

One important pre-ICC case, though the Court was not up and running until 2002, involved former Chilean dictator Augusto Pinochet, who was responsible for the deaths in Chile between 1973 and 1990 of 2,279 people and torture of 27,255 more (according to the Chilean TRC) was of worldwide interest. His old friend former prime minister Margaret Thatcher protested vigorously when Pinochet was arrested on his way to see her in London under an international warrant from a Spanish magistrate, Baltazar Garzón, at Heathrow Airport. Pinochet was allowed to stay in London under house arrest and visited by

Thatcher before finally being sent back to Chile on health grounds in 2000. Thatcher is said to have taken Pinochet some whisky, stating later that 'Scotch is one British institution that will never let you down' (Youkee 2019). The important legal point that led to the arrest was that it was under a new principle of 'universal jurisdiction', a principle that the ICC tries to uphold. Pinochet's arrest means that any international warrant, including by the ICC, can be activated whenever a state party to the Rome Statute or similar is involved.

Now that Putin has been indicted by the ICC he can be open to arrest in 127 countries at least. Since several other members of the BRICS (made up of Brazil, Russia, India, China and South Africa) are in an economic alliance Putin could therefore be arrested in Brazil or South Africa. None of the other states are signatories, as indeed is also the case of the United States. We might ask if Afghanistan (ICC member since 2003) would now extradite a new Osama bin Laden?

The use of the WCT has of course been explored in the context of developing countries and their conflicts. The International Criminal Tribunal for Rwanda (https://unictr.irmct.org/) has been in existence since 1994, just after the massacres, and has as its official brief 'to contribute to the process of national reconciliation in Rwanda and to the maintenance of peace in the region'. The criticisms of its breathtakingly slow pace and ineffectiveness have led to some commentators thinking it brings the rule of law into disrepute (Moghalu 2005). There is a similar one for Sierra Leone, the Special Court for Sierra Leone, which used the ICC's facilities in The Hague to prosecute Charles Taylor, widely accepted as being a prime mover in the horrors that afflicted much of West Africa (for more on this see Wierda 2006: 183–207).

One potential problem for such tribunals' effectiveness in developing country contexts is that their roots clearly lie in the Nuremberg and Tokyo Tribunals after the Second World War, when the circumstances of global politics were very different to now and when the justice being meted out was in the context of inter state war, not in our predominantly 'new' war era. Another problem is that which can also be levelled against WCTs in the industrialized world – on occasion, the needs of 'justice' and those of 'peace' are in distinct opposition. How can you expect a warlord to lay down his weapons if he knows he will be subsequently accused of war crimes and probably executed or imprisoned for life? Equally, there is some sense in the US criticism of the ICC that it potentially criminalizes soldiers doing their best to do the bidding of the UN. Against this, it could be argued that all 'proper' armies prosecute their troops for criminal behaviour, as have the Americans in Iraq and Vietnam, for example, and that the ICC stricture about it being a court of last resort answers that criticism. Maybe more seriously, the ICC was not used in the prosecution of those deemed to be war criminals in Iraq – the executions and other punishments inflicted there were by domestic judicial instances and might be accused of being essentially 'victors' justice' (Bass 2000; Williams 2006 and 2024 *forthcoming*). We feel there is not enough evidence, for or against, to yet have a judgement on the effectiveness of such tribunals in the context of conflict in the developing world.

To summarize, the aims of TRCs and WCTs are the following:

- Accommodate the vital role of historical memory. Each player in the conflict believes that history is on their 'side' and evokes it to justify ever-increasing cycles of violence and revenge.
- To deflate this spiral, the inverse process of historical truth-telling has to take place, whether through a formal system of justice ('retributive' justice – WCTs, for example) or a non-judicial process ('reformative' justice, such as TRCs).

It is important to note that the main aim of both TRCs and WCTs is to make an *attempt* to address the aforementioned aims. Hayner quotes Michael Ignatieff's caveat: 'The past is an argument and the function of truth commissions, like the function of honest historians, is simply to purify the argument, to narrow the range of permissible lies' (Ignatieff in Hayner 2002: 25). This has given rise to frustrations and potential future conflicts.

Conclusions

Anthony Oberschall sums up the substance of the issues dealt with in this chapter in two sentences: 'Collective Myths are the enemy of truth and justice ... conflict management and peace building are hampered by crisis framing, the victim syndrome and denial' (Oberschall 2007: 2). While we might disagree about the exact wording, this seems to us a fair summary of dealing with the kinds of problems faced in trying to manage or resolve most conflicts in developing countries, as it has also been in some in the 'West' like Northern Ireland, other parts of the European Union (the Basque Country most notably) and the former Yugoslavia and the former Soviet Union. So a first conclusion has to be that the problems of developing countries are not *sui generis*. They can happen anywhere, wherever there are people struggling to find economic, political or cultural identity and meaning. One of the worst obstacles that needs to be overcome if we are to improve what is a dismal track record of dealing with complex, deep-rooted conflicts is to assume that they are easy to resolve but also that we should not try.

Second, we might also ask whether there are lessons to be learnt from looking at *non*-conflict situations, as any thesis must be falsifiable. Of course, this is the basis of liberal peace arguments. Such theorists would say that the effective deployment of democratic institutions is what makes a state succeed in avoiding destructive conflict and enhancing successful development (Paris 2004; Kaldor 2012). Although such considerations are very interesting, we feel we need in this book to concentrate on attempts to address what does *not* work. What seems clear to us is that the official discourse of participation and empowerment through democracy often falls far short of the reality. In most cases, the evidence seems to be that such discourse and practice can actually depoliticize and demobilize the poor, who are the subjects and objects of much conflict behaviour (Moore 2001; Williams 2007).

Third, we would argue that the previous discussion, and what we have discussed in all of our chapters, reveals what can be called a 'crisis of imagination' in international relations and world politics, where the default position is a military intervention for the settlement of problems (Nyabola 2023a). Nanjala Nyabola, a Kenyan scholar who asked herself that question, was not the first to observe that despite the billions of dollars that have now been spent trying to 'fix' the problems of Haiti, not only does no 'progress' seem to be made, but the problems the people face there seem more and more intractable. Why is this? Part of her answer is that the interventions made from the outside have actually lessened, not increased, the population's sense of having any agency over their lives. She suggests using the digital revolution not to solely increase the divide between rich and poor both in developing countries and between the South and the North (Nyabola 2023b, in Roberts and Bosch 2023).

How might this need for more local agency manifest itself? One obvious way is by asking why the short-term effects of peacebuilding processes can be so perverse. When we in the Western 'conflict community' talk about 'progress', what do we mean? Usually, the answer is some form of 'metric'. Can it be measured? Progress itself is a term that liberals assume can be achieved. The authority on liberal ideology Richard Bellamy points out that 'progress … has theological foundations' (Bellamy 1992. 2–3). It assumes that societies have our kind of civil society, our kinds of economic relationships, our moral codes, not something we could even claim about near neighbours in Europe. The Hungarian prime minister Victor Orban likes to rail against 'progressive liberals, neo-Marxists, intoxicated by the dream of wokeness.… They want to abolish the western way of life'. And of course, it depends what *kind* of liberal we are talking about. One other extreme version might be Ayn Rand, who is described as a liberal by one commentator due to her belief that 'progress requires freedom', which many might say suggests we should all always act selfishly (Butler 2019; Rachman 2023b).

Another answer to the question of agency comes out very strongly in the earlier discussion of justice. Whose definition of justice are we taking? TRCs and WCTs assume identifying and dealing with the accused is the metric of choice. But what about those for whom the crimes committed will never be over? Many have accused the South African and Rwandan TRCs of asking the impossible – forgiveness by victims for unforgivable crimes. So an area that needs much more research is the long-term effects of such attempts and what might be done to improve the prospects for success. The phenomenon of 'intergenerational violence' has been noted. Although the evidence is often anecdotal, for example, in areas of the former Yugoslavia where there were widespread atrocities during the Second World War, there were higher levels of 'revenge' violence in the 1990s (Dragojevic 2013). One area that is now being explored is that of the effects of conflict on (even) unborn children (Devakumar et al. 2014) and the possibility of education for succeeding generations. The German example of education changing wider attitudes is often cited (Dierkes in Cole 2007). Likewise, the negative evidence of such educational attempts not being made in the teaching of history in Japan, Korea and China has also elicited widespread comment (Bleiker 2007; Miyoshi Jager and Mitter 2007; Yoshida 2007). Not many analyses have yet been done on developing country cases, however.

Another is to continue to ask the really difficult questions. For example, we must ask whether an outside party *should* interfere in the warfare/conflict raging within a developing country, of which there are many examples. And will their efforts stand any chance of succeeding, and why or why not? As John Darby (2006) has put it, 'The signing of a multiparty peace accord is unlikely to end violence'. Should we now just let a bit of 'chaos' run its course, naturally a suggestion that makes all humane liberals (like we are) very queasy? But when those like Darby had so invested their lives in the whole study and practice of peacebuilding seemed to despair over it, we have to take this despair seriously. It can easily turn into an often-incoherent demand for 'resistance' against what comes to be seen as a foreign invader, not a saviour.

We need to look beyond such 'institutional' concepts as conflict settlement and management, discussed in this chapter, and democratization, disarmament and the reintegration of former combatants, reconstruction and institutional change, which we will discuss at length in Chapter 5, to see how deeper mechanisms can be used to try and coax whole populations and especially elites towards a more gentle and less confrontational way of solving their inevitable societal conflicts.

Finally, and with a germ of hope, we can quote the words of Pierre Hazan, a remarkable French journalist who witnessed war crimes in (among other places) Rwanda and Bosnia in the 1990s. In many interviews with victims and survivors of conflict and war, he makes the following statement (my translation): '[T]he immense majority of victims agree on one point: they all demand to know the truth and find justice,

Plate 12 'Resistance' in Iraq: A search by US troops in Iraq has revealed hidden weapons.

as a prerequisite to reconstruction' (Hazan 2000: 12). Peacebuilding, conflict resolution and the transformation of societies and individuals from the state of violent conflict or war is never going to be an easy task. It is difficult in developed countries; it is all the more difficult in conditions of underdevelopment, poverty and state collapse. Chapters 5 and 6 will address these vital corollary issues. Can development 'from without' through reconstruction and humanitarian aid help societies become more economically and politically stable so that peace can have a chance of succeeding?

Summary

- Theories of how to deal with conflict (settlement, management and 'transformation') are often seen as dependent on the observer's underlying theory of international relations.
- Most, but not all, conflicts in developing countries are civil wars.
- Peacebuilding and conflict 'transformation' approaches can be 'top-down' or 'bottom-up'.
- Protracted social conflict cannot simply be explained by reference to 'ethnic' differences.
- Reconciliation within divided societies can be attempted by internally organized jural or non-jural tribunals or by international WCTs.
- We need to understand the historical roots of conflicts in order to have any chance of resolving them.

Discussion questions

- Is it true to say that 'new wars' have now largely replaced the 'old wars' of the era before 1990?
- Should peacebuilding be approached in a 'top-down' or 'bottom-up' way?
- Can the ICC provide impartial justice, or does it rather prosecute individuals from states that are vulnerable to pressure?
- Are commissions, such as the South African TRC, a better way to bring about peace than WCTs?
- What role do you think the 'history' of a place plays in helping us understand the causes of and potential solutions to conflict and war?

Note

1 General Mark Milley is chairman of the [US] Joint Chiefs of Staff.

Further reading

(Possibly) the best introduction to the entire field of conflict resolution and management can be found in: O. Ramsbotham, T. Woodhouse and H. Miall (2016) *Contemporary Conflict Resolution: The prevention, management and transformation of deadly conflicts*, 4th edition, Cambridge: Polity. Also still valuable is: J. Bercovitch, V. Kremyenuk and I. W. Zartman (eds) (2008) *The Sage Handbook of Conflict Resolution*, Thousand Oaks, CA: Sage Publications; Still good, but a bit dated, is: C. A. Crocker, F. O. Hampson and P. Aall (eds) (2007) *Leashing the Dogs of War: Conflict management in a divided world*, Washington, DC: USIP Press. On mediation the best survey is: A. Lempereur, J. Salzer, A. Colson, M. Pekar and E. B. Kogan (2021) *Mediation: Negotiation by Other Moves*, Chichester: Wiley. The best introduction to the logic of conflict resolution is still E. Azar (1990) *The Management of Protracted Social Conflict: Theory and cases*, Aldershot: Dartmouth; or C. Mitchell (2014) *The Nature of Intractable Conflict: Resolution in the twenty-first century*, London: Palgrave Macmillan. An excellent (if rather traditional) summary of the aims of peacebuilding can be found in H.-W. Jeong (2005) *Peacebuilding in Post-Conflict Societies: Strategy and process*, Boulder, CO and London: Lynne Rienner; and H.-W. Jeong (2008) *Understanding Conflict and Conflict Analysis*, London: Sage. A good discussion of the causes and potential resolution of conflicts can be found in A. Oberschall (2007) *Conflict and Peace Building in Divided Societies: Responses to ethnic violence*, London: Routledge. Sung Yong Lee presents a study on alternative approaches to conventional conflict transformation, examining national peace negotiations in Aceh and Bougainville (Lee, S. 2023). Transformative peace negotiation. In J. Senehi, I. M. Scott, S. Byrne & T. G. Matyók (Eds.), *Routledge handbook of peacebuilding and ethnic conflict* (pp. 292–303). Abingdon, UK: Routledge). One critical review of conflict resolution approaches can be found in V. Jabri (2007) *War and the Transformation of Global Politics (Rethinking Peace and Conflict Studies)*, London: Palgrave Macmillan. A good link between conflict resolution and peacebuilding is: Kevin Avruch (2012), *Context and Pretext in Conflict Resolution: Culture, Identity, Power, and Practice*. London: Routledge. The best critique of 'liberal' peacebuilding is by E. Newman, R. Paris and O. Richmond (2009) *New Perspectives on Liberal Peacebuilding*, Tokyo: United Nations Press. Two good overviews of reconciliation after wars are: A. Rigby (2001) *Justice and Reconciliation: After the violence*, Boulder, CO: Lynne Rienner; and D. J. Whittaker (1999) *Conflict and Reconciliation in the Contemporary World*, London: Routledge. On the possibilities of locally organised conflict transformation see: R. Mac Ginty (2021), *Everyday Peace: How So-called Ordinary People Can Disrupt Violent Conflict*. Oxford: Oxford University Press.

Useful websites [all accessed for this edition in November 2023]

On conflicts themselves, see the unsurpassed series that has been produced by the International Crisis Group since 1995: http://www.crisisgroup.org. The best website on conflict resolution/transformation is *The Berghof Handbook for Conflict Transformation* (https://berghof-foundation.org/library/berghof-handbook-for-conflict-transformation). Various websites related to particular commissions and tribunals mentioned earlier include: South African TRC: http://www.doj.gov.za/trc/ [though this is now mainly a more general South African legal website]; for Rwanda, the official government website: The International Criminal Tribunal for Rwanda (https://unictr.irmct.org/ [alas also now non-operational]); http://www.ictj.org/our-work/regions-and-countries/sierra-leone (Sierra Leone, but also covers Kenya, Uganda, Sudan and a number of other countries worldwide). The International Commission on Transitional Justice has the most comprehensive remit of an INGO for such matters across the world: http://www.ictj.org/ [again has worldwide coverage]. The International Criminal Court can be found at http://www.icc-cpi.int. For the wars in the former Yugoslavia, see: Humanitarian Law Center: http://www.hlc-rdc.org/?lang=de and 'Balkan Transitional Justice' (Balkan Insight): http://www.balkaninsight.com/en/page/balkan-transitional-justice-home

Bibliography

Albright, M. (2004) Ghosts of Rwanda, Public Broadcasting System (PBS), 25 February. https://www.pbs.org/wgbh/pages/frontline/shows/ghosts/interviews/albright.html (accessed 3 November 2023).

Allen, Nan S., Mampilly, Z. and Bartoli, A. (eds) (2012) *Peacemaking: From Practice to Theory*. Santa Barbara, CA: Praeger.

Anderson, S. (2013) *Lawrence in Arabia: War, Deceit, Imperial Folly and the Making of the Modern Middle East*. London: Vintage.

Angell, N. (1910, 1912) *The Great Illusion: A Study of the Relation of Military Power in Nations to Their Economic and Social Advantage*. London: William Heinemann.

Annan, K. (2017) Advisory commission on Rakhine state. https://www.kofiannanfoundation.org/our-work/initiative-advisory-commission-rakhine-state/

Argenti, N. and Schramm, K (2010) *Remembering Violence: Anthropological Perspectives on Intergenerational Transmission*. Oxford: Berghahn.

Avruch, K. (2003) Context and pretext in conflict resolution. *Journal of Dispute Resolution* 2: 353–65.

Avruch, K. (2012) *Context and Pretext in Conflict Resolution: Culture, Identity, Power, and Practice*. London: Routledge.

Azar, E. (1990) *The Management of Protracted Social Conflict: Theory and Cases*. Aldershot: Dartmouth.

Baaz, M.E. and Stern, M. (2013) *Sexual Violence as a Weapon of War? Perceptions, problems in the Congo and Beyond*. London: Nordiska Afrikainstitutet/Zed Books. https://www.diva-portal.org/smash/get/diva2:617148/FULLTEXT01.pdf (accessed 4 November 2023).

Baker, C. (2015) *The Yugoslav Wars of the 1990s (Studies in European History)*. London: Palgrave Macmillan.

Balcells, L. (2010) Rivalry and revenge: Violence against civilians in conventional civil wars. *International Studies Quarterly* 54(2): 291–313.

Balcells, L. (2011) Continuation of politics by two means: Direct and indirect violence in civil war. *Journal of Conflict Resolution* 55(3): 397–422.

Barr, J. (2012) *A Line in the Sand: Britain, France and the Struggle that Shaped the Middle East*. London: Simon & Schuster.

Bass, G.J. (2000) *Stay the Hand of Vengeance: The Politics of War Crimes Tribunals*. Princeton, NJ: Princeton University Press.

Beevor, A. (2002) *Berlin: The Downfall, 1945*. London: Penguin.

Bell, C., Campbell, C., and Ní Aoláin, F. (2007). Transitional Justice: (Re)conceptualizing the field. *International Journal of Law in Context* 3(2): 81–8.

Bell, D. (ed) (2006) *Memory, Trauma and World Politics: Reflections on the Relationship between Past and Present*. Basingstoke: Palgrave Macmillan, pp. 116–34.

Bellamy, A. (2012). *Massacres and Morality: Mass Atrocities in An Age of Civilian Immunity*. Oxford: Oxford University Press.

Bellamy, R. (1992) *Liberalism and Modern Society: A Historical Argument*. Pennsylvania: The Pennsylvania State University Press.

Bercovitch, J. and Fretter, J. (2004) *Regional Guide to International Conflict and Management from 1945 to 2003*. Washington, DC: Congressional Quarterly.

Bercovitch, J. and Rubin, J.Z. (1992) *Mediation in International Relations: Multiple Approaches to Conflict Management*. London: Macmillan.

Bew, P. (2009) *Ireland: The Politics of Enmity 1789–2006*. Oxford: Oxford University Press.

Bleiker, R. (2007) On the Use and Abuse of Korea's Past: An Inquiry into History Teaching and Reconciliation. In Cole E.A. (ed.) *Teaching the Violent Past: History, Education and Reconciliation*. Lanham, MD: Rowman and Littlefield.

Booth, K. and Wheeler, N. (2007) *The Security Dilemma. Fear, Cooperation, and Trust in World Politics*. London: Palgrave Macmillan.

Brooke-Holland (2023) *UN ends Peacekeeping Force in Mali*. London: House of Commons Library https://researchbriefings.files.parliament.uk/documents/CBP-9827/CBP-9827.pdf (accessed 3 August 2023).

Browning, C. (2005) *The Origins of the Final Solution: The evolution of Nazi Jewish policy September 1939–March 1942*. London: Arrow.

Brubaker, R. and Laitin, D.D. (1998) Ethnic and nationalist violence. *Annual Review of Sociology 24*: 423–54.

Budjeryn, M., and Bunn, M. (2020) *Budapest Memorandum at 25: Between Past and Future*. Harvard: Belfer Center. https://www.belfercenter.org/sites/default/files/2020-03/budapest/BM25.pdf (accessed 3 March 2023).

Bunzl, J. (2008) Mirror Images; perception and interest in the Israel/Palestine conflict. *Palestine–Israel Journal 12*(2–3): 8–14.

Burton, J. (1979) *Deviance, Terrorism and War: The Process of Solving Unsolved Social and Political Problems*. Oxford: Martin Robertson.

Burton, J. (1990) *Conflict Resolution and Provention*. New York: St Martin's Press.

Burton, J. (1997) *Violence Explained: The Sources of Conflict, Violence and Crime and Their Prevention*. Manchester: Manchester University Press.

Butler, E. (2019), *Schools of Thought: 101 Great Liberal Thinkers*. London: Institute of Economic Affairs. https://iea.org.uk/wp-content/uploads/2019/08/Great-Liberal-Thinkers-Interactive.pdf (accessed 21 February 2023).

Campbell, D. (1998a) *National Deconstruction: Violence, Identity and Justice in Bosnia*. Minneapolis, MN: University of Minnesota Press.

Campbell, D. (1998b) *Writing Security: United States Foreign Policy and the Politics of Identity*. Manchester: University of Minnesota Press/Manchester University Press.

Camut, N. (2023) 'Djokovic doubles down on "Kosovo is the heart of Serbia" remark', *Politico*. 1 June 2023: https://www.politico.eu/article/would-say-again-djokovic-insists-kosovo-row-serbia/ (accessed 10 June 2023).

CFR/Maizland L [Council on Foreign Relations, Washington DC] (2020) U.S.-Taliban Peace Deal: What to know. https://www.cfr.org/backgrounder/us-taliban-peace-deal-agreement-afghanistan-war (accessed 2 November 2023).

Chandler, D. (2013) Peacebuilding and the politics of non-linearity: Rethinking 'hidden' agency and 'resistance'. *Peacebuilding 1*(1): 17–32.

Christie, K. (2000) *The South African Truth Commission*. London: Macmillan.

Clausewitz, C. Von, Heuser, B., Howard, M., Paret, P. (2012) *On War*. Create-Space independent publishing platform, Amazon.

von Clausewitz, K. et al. (2012) *On War*. Princeton: Princeton UP.

Coker, C. (1997) How wars end. *Millennium; Journal of International Studies 26*(3): 615–29.

Cole, E.A. (ed.) (2007) *Teaching the Violent Past: History, Education and Reconciliation*. Plymouth: Rowman and Littlefield.

Corbin, J. (1994) *Gaza First: The Secret Channel to Peace between Israel and the PLO*. London: Bloomsbury.

Corey, A. and Joireman, S.F. (2004) Retributive justice: The Gacaca courts in Rwanda. *African Affairs 103*: 73–89.

Crisis Group (2023) *What Future for UN* 10 July 2023: https://www.crisisgroup.org/global-mali/what-future-un-peacekeeping-africa-after-mali-shutters-its-mission (accessed 3 August 2023).

Crocker, C.A. Hampson, F.O., and Aall, P. (eds) (2007) *Leashing the Dogs of War: Conflict Management in a Divided World*. Washington, DC: United States Institute of Peace Press.

Daily Beast (2013) America, once again dazed and confused in the Middle East. *The Daily Beast*, 22 September 2013. http://www.thedailybeast.com/articles/2013/09/22/america-once-again-dazed-and-confused-in-the-middle-east.html (accessed 15 January 2015).

Dallaire, R. (2005) *Shake Hands With The Devil: The Failure of Humanity in Rwanda*. London: Arrow.

Darby J. (ed) (2006) *Violence and Reconstruction*. Notre Dame, IN: University of Notre Dame Press.

Davies, S., Pettersson, T., and Öberg, M. (2023) Organized violence 1989–2022 and the return of conflicts between states? *Journal of Peace Research* forthcoming? https://ucdp.uu.se/year/2022 (accessed 4 August 2023)

De Tocqueville, A. (1856) *The Old Regime and the Revolution*. Chicago, IL: University of Chicago Press. [*L'ancien evolu et la evolution*, 1856]

De Zeeuw, J. and Kumar, K. (eds) (2006) *Promoting Democracy in Postconflict Societies*. Boulder, CO: Lynne Rienner.

Devakumar, D., Birch, M., Osrin, D., Sondorp, E., and Wells, J. (2014) The intergenerational effects of war on the health of children. *BMC Medicine* *57*(12): 15.

'Doha Agreement' (2020): *Agreement for Bringing Peace to Afghanistan*. The full text can be found at: https://www.state.gov/wp-content/uploads/2020/02/Agreement-For-Bringing-Peace-to-Afghanistan-02.29.20.pdf (accessed 3 August 2023).

Donia, R.J. (2014) *Radovan Karadzic: Architect of the Bosnian Genocide*. Cambridge: Cambridge University Press.

Dragojevic, M. (2013) Memory and identity: Inter-generational narratives of violence among refugees in Serbia. *Nationalities Papers: The Journal of Nationalism and Ethnicity 41*(6): 1065–82.

Duffield, M. (2007) *Development, Security and Unending War: Governing the World of Peoples*. Cambridge: Polity.

Duncanson, C. (2016) *Gender and Peacebuilding*. Cambridge: Polity.

Enloe, C. (1993) *The Morning After: Sexual Politics at the End of the Cold War*. Berkeley, CA: University of California Press.

Evans, M. (2006) Memories, monuments, histories: The re-thinking of the second world war since 1989. *National Identities 8*(4): 317–48.

Evans, M. Stopping the cycle of violence: Political transition in Algeria since 1988 in a comparative perspective. In Johnson, R. and Clack, T. (eds) (2014) *At the End of Military Intervention: Historical, Theoretical and Applied Approaches to Transition, Handover and Withdrawal*. Oxford: Oxford University Press.

Evans, M. (2017) Reprisal violence and the Harkis in French Algeria, 1962. *International History Review 39*(1): 89–106.

Ferriter, D. (2015) *A Nation and not a Rabble: The Irish Revolution 1913–23*. London: Profile.

Fierke, K.M. (2013) *Political Self Sacrifice: Agency, Body and Emotion in International Relations*. Cambridge: Cambridge University Press.

Fierke, K.M. and Mackay, N. (2020) 'To "see" is to break an entanglement: Quantum measurement, trauma and security. *Security Dialogue 51*(5): 405–66.

Fisher, R. (1997) *Interactive Conflict Resolution*. Syracuse, NY: Syracuse University Press.

Fisher, R. (ed.) (2005) *Paving the Way: Contributions of Interactive Conflict Resolution to Peacemaking*. Lanham, MD: Lexington.

Fisher, R., Tadevosyan, M., and Cuhadar, E. (2023) *The USIP Learning Agenda: An Evidence Review: Track 2 Dialogues*. Washington, DC: U.S.I.P.

Fox, J. (2013) Religious armed conflict and discrimination in the Middle East and North Africa: An introduction. *Civil Wars* 4(15): 407–10.

Friedlander, S. (2008) *Nazi Germany and the Jews: The Years of Extermination: 1939–1945*. London: Weidenfeld and Nicolson.

Fromkin, D. (2000) *A Peace to End All Peace: The Fall of the Ottoman Empire and the Creation of the Modern Middle East*. London: Phoenix.

Fukuyama, F. (1992) *The End of History and the Last Man*. New York: Free Press.

Futamura, M. (2008) *War Crimes Tribunals and Transitional Justice: The Tokyo Trial and the Nuremberg legacy*. London: Routledge.

Gawerc, M. (2006) Peace-building: Theoretical and concrete perspectives. *Peace and Change 31*(4): 435–78.

Goldsworthy, V. (2013) *Inventing Ruritania: The Imperialism of the Imagination*. London: Hurst.

Gow, J. (1997) *Triumph of the Lack of Will: International Diplomacy and the Yugoslav War*. London: Hurst.

Gow, J., Kerr, R., and Pajic, Z. (2014) *Prosecuting War Crimes: Lessons and Legacies of the International Criminal Tribunal for the Former Yugoslavia*. London: Routledge.

Gow, J. and Zverzhanovski, I. (2004) The Milosevic trial: Purpose and performance. *Nationalities Papers 32*(4): 897–919.

Goy, B. (2012) Individual criminal responsibility before the international criminal court: A comparison with the *Ad Hoc* Tribunals. *International Criminal Law Review*, January.

Gray, C. (2005) *Another Bloody Century: Future Warfare*. London: Weidenfeld and Nicholson.

Gready, P. (2011) *The Era of Transitional Justice: The aftermath of the Truth and Reconciliation Commission in South Africa and Beyond*. New York: Routledge.

Gurr, T. and Harff, B. (1994) *Ethnic Conflict in World Politics*. Boulder, CO: Westview Press.

Gurr, T. and Harff, B. (2003) *Ethnic Conflict in World Politics*. London: Routledge (later ed).

Hayner, P. (2002) *Unspeakable Truths: Facing the Challenge of Truth Commissions*. New York: Routledge.

Hazan, P. (2000) *La justice face à la guerre: De Nuremberg à la Haye*. Paris: Stock.

Holt, S. (2011) *Aid, Peacebuilding and the Resurgence of War: Buying Time in Sri Lanka*. London: Palgrave Macmillan.

Hosoya, C. A50 Editorial Committee (eds) (2001) *Japan and the United States – Fifty Years of Partnership*. Tokyo: The Japan Times.

Hothschild, A. (2019) *King Leopold's Ghost: A Story of Greed, Terror and Heroism in Colonial Africa*. London: Picador.

Howard, M. (2000) *The Invention of Peace: Reflections on War and International Order*. New Haven, CT: Yale University Press.

Huntington, S. (1993) *The Third Wave: Democratization in the Late Twentieth Century*. Oklahoma, OK: University of Oklahoma Press.

Huntington, S. (1996) *The Clash of Civilizations and the Re-making of World Order*. New York: Simon & Schuster.

ICTJ (2023) International center for transitional justice. http://www.ictj.org/our-work/regions-and-countries/burmamyanmar [Myanmar]

International Crisis Group Africa (2014) Curbing violence in Nigeria (II): The Boko haram insurgency. Report 216, April.

International Institute for Strategic Studies (2015a) *Armed Conflict Database*. London: IISS.
International Institute for Strategic Studies (2015b) *The Military Balance*. London: IISS.
International Institute for Strategic Studies (2022) *The Armed Conflict Survey*. London: IISS: https://www.iiss.org/online-analysis/online-analysis/2022/11/acs-2022-introduction/ (accessed 3 August 2023).
Jabri, V. (2007) *War and the Transformation of Global Politics*. London: Palgrave Macmillan.
Jabri, V. (2013) Peacebuilding, the local and the international: A colonial or a postcolonial rationality? *Peacebuilding 1*(1): 3–16.
Jeffery, R. and Kim, H.J. (2014) *Transitional Justice in the Asia-Pacific*. Cambridge: Cambridge University Press.
Jeong, H.-W. (2005) *Peacebuilding in Post-Conflict Societies: Strategy and Process*. Boulder, CO and London: Lynne Rienner.
Jeong, H.-W. (2008) *Understanding Conflict and Conflict Analysis*. London: Sage.
Jones, D. (1999) *Cosmopolitan Mediation? Conflict resolution and the Oslo Accords*. Manchester: Manchester University Press.
Jones, M. (2023) Africa's Ukraine-Russia peace mission: What was achieved? https://www.bbc.com/news/world-africa-65951350 (accessed 25 June 2023).
Kaldor, M. (2012) *New and Old Wars: Organized Violence in a Global Era*, 2nd and 3rd edn. Cambridge: Polity.
Kalisa, M.-C. (2009) *Violence in Francophone African and Caribbean Women's Literature*. University of Nebraska Press.
Kalyvas, S. (2001) 'New' and 'old' civil wars – a valid distinction? *World Politics 54*(1): 99–118.
Kalyvas, S. (2006) *The Logic of Violence in Civil War*. Cambridge: Cambridge University Press.
Kaufman, S.J. (2001) *Modern Hatreds: The Symbolic Politics of Ethnic War*. Ithaca, NY: Cornell University Press.
Kaysen, C. (1990) Is war obsolete? A review essay. *International Security 4*(14): 42–64.
Keashley, L. and Fisher, R. (1990) Towards a contingency approach to third-party intervention in regional conflict: A Cyprus illustration. *International Journal 45*(2): 425–53.
Kelman, H.C. (2005) Interactive Problem Solving in the Israeli–Palestinian Case: Past Contributions and Present Challenges. In Fisher, R. (ed) *Paving the Way: Contributions of Interactive Conflict Resolution to Peacemaking*. Lanham, MD: Lexington, pp. 41–63.
Krog, A. (1998) *Country of My Skull: Guilt, Sorrow and the Limits of Forgiveness in the New South Africa*. Johannesburg: Random House.
Laffan, M. (1999) *The Resurrection of Ireland: The Sinn Féin Party, 1916–1923*. Cambridge: Cambridge University Press.
Lambourne, W. (2009). Transitional justice and peacebuilding after mass violence. *The International Journal of Transitional Justice 3*: 28–48.
Lederach, J.P. (1997) *Building Peace: Sustainable Reconciliation in Divided Societies*. Washington, DC: United States Institute of Peace Press.
Lee, S. Y. (2023) Transformative Peace Negotiation. In Senehi J., Scott I. M., Byrne S., & Matyók T. G. (eds.) *Routledge Handbook of Peacebuilding and Ethnic Conflict* (pp. 292–303). Abingdon, UK: Routledge.
Lewis, J. (2007) Nasty, brutish and in shorts? British colonial rule, violence and the historians of Mau Mau. *Round Table: The Commonwealth Journal of International Affairs 96*(389): 201–23.

Little, D. (ed.) (2007) *Peacemakers in Action: Profiles of Religion in Conflict Resolution*. New York: Cambridge University Press.
Long, W.J. and Brecke, P. (2003) *War and Reconciliation: Reason and Emotion in Conflict Resolution*. Cambridge, MA: MIT Press.
Mac Ginty, R. (2011) *International Peacebuilding and Local Resistance: Hybrid Forms of Peace*. London: Palgrave Macmillan.
Mac Ginty, R. (2021), *Everyday Peace: How So-called Ordinary People Can Disrupt Violent Conflict*. Oxford: Oxford University Press.
Mac Ginty, R. and Peterson, J.H. (2015) *The Routledge Companion to Humanitarian Action*. London: Routledge.
Mac Ginty, R. and Richmond, O. (2013) 'The Local Turn in Peace Building: A critical agenda for peace. *Third World Quarterly 34*(5): 763–83.
Mac Ginty, R. and Williams, A. (eds) (2007) Commemoration and remembrance in the Commonwealth. *Round Table: The Commonwealth Journal of International Affairs 96*(393).
Mac Ginty, R. and Williams, A. (2012) *Conflict and Development* [in Japanese]. Tokyo: Kuniaki Asomura.
Mack, A. (2006) *Human Security Brief 2006*. Vancouver: Human Security Center, University of British Columbia. www.humansecuritybrief.info (accessed on 13 March 2008 but no longer available).
Mandelbaum, M. (1998) Is major war obsolete? *Survival 40*(4): 20–38.
Mearsheimer, J.J. (2001) *The Tragedy of Great Power Politics*. New York: Norton.
Melvern, L. (2006) *Conspiracy to Murder: The Rwandan Genocide*. London: Verso.
Miall, H. (2006) *Emergent Conflict and Peaceful Change*. London: Palgrave.
Miall, H., Ramsbotham, O., and Woodhouse, T. (1999) *Contemporary Conflict Resolution*, 1st edn. London: Polity.
Milley, M. (2022) Miley provides highest US estimate yet of Ukraine war casualties. *The Hill* 11 October 2022: https://thehill.com/policy/international/3729609-milley-provides-highest-us-estimate-yet-of-ukraine-war-casualties/ (accessed 20 February 2023).
Mitchell, C. (1981) *The Structure of International Conflict*. London: Macmillan.
Mitchell, C. (1996) *Structure of International Conflict*. New York: Macmillan.
Mitchell, C. (2014) *The Nature of Intractable Conflict: Resolution in the Twenty-First Century*. London: Palgrave Macmillan.
Mitchell, C. and Banks, M. (1996) *Handbook of Conflict Resolution*. London: Frances Pinter.
Mitchell, C., and Webb, K. (eds) (1988) *New Approaches to International Mediation*. New York: Greenwood Press.
Miyoshi Jager, S. and Mitter, R. (2007) *Ruptured Histories: War, Memory and the post-Cold War in Asia*. Cambridge, MA: Harvard University Press.
Mkutu, K. (2008) Disarmament in Karamoja, Northern Uganda: Is this a solution for localised violent inter and intra-communal conflict? *Round Table: The Commonwealth Journal of International Affairs* 97(394): 99–120.
Moghalu, K. (2005) *Rwanda's Genocide: The Politics of Global Justice*. London: Palgrave.
Momsen, J. H. (2004) *Gender and Development*. London: Routledge. [A third edition of this work was published in 2019]
Moon, C. (2009) Healing past violence: Traumatic assumptions and therapeutic interventions in war and reconciliation. *Journal of Human Rights 8*(1): 71–91.
Moore, M. (2001) Empowerment at last? *Journal of International Development 13*(3): 321–9.

Moses, D. (2021) *The Problems of Genocide: Permanent Security and the Language of Transgression*. Cambridge: Cambridge University Press.
National WW2 Museum (2021) *Justice Radhabinod Pal and the Tokyo Tribunal*. https://www.nationalww2museum.org/war/articles/justice-radhabinod-pal-tokyo-war-crimes-trial (accessed 4 August 2023).
National WW2 Museum (2023) *The Nuremberg Trials*. https://www.nationalww2museum.org/war/topics/nuremberg-trials (accessed 14 June 2023).
Newman, E. (2004) The 'new wars' debate: A historical perspective is needed. *Security Dialogue 35*(2): 173–89.
Newman, E. (2013) The violence of statebuilding in historical perspective: Implications for peacebuilding. *Peacebuilding 1*(1): 141–58.
Nordstrom, C. (2004) *Shadows of War: Violence, Power, and International Profiteering in the Twenty-First Century*. Berkeley, CA: University of California Press.
Nyabola, N. (2023a) The Crisis of Imagination in International Relations. *The Nation*, 11 April 2023: https://www.thenation.com/article/world/international-relations-haiti-militarism/ (accessed 7 August 2023).
Nyabola, N. (2023b) 'Citizenship, African languages and digital rights: The role of language in defining the limits and opportunities for digital citizenship in Kiswahili-language communities', *in* Roberts, T. and Bosch, T.
Nye, J. (2005) *Soft Power: The Means to Success in World Politics*. New York: PublicAffairs.
Oberschall, A. (2007) *Conflict and Peace Building in Divided Societies: Responses to Ethnic Violence*. London: Routledge.
OECD (2007) *Handbook on 'Security Sector Reform'*. Paris: OECD.
Office of the Historian (US State Department) (1993) *The Oslo Accords and the Arab – Israeli Peace Process*. https://history.state.gov/milestones/1993-2000/oslo#:~:text=On%20September%2013%2C%201993%2C%20Israeli,%2C"%20at%20the%20White%20House (accessed 3 August 2023).
Olsen, T.D., Payne, L.A., and Reiter, A.G. (2010) Transitional justice in the world, 1970–2007: Insights from a new dataset. *Journal of Peace Research 47*(6): 803–9.
OSCE (2022) *OSCE Minsk Group*. https://www.osce.org/mg
Oslo Accords (1993) UN General assembly/Security Council. *Declaration of Principles on Interim Self-Government Arrangements* 11 October 1993: https://history.state.gov/milestones/1993-2000/oslo#:~:text=On%20September%2013%2C%201993%2C%20Israeli,%2C"%20at%20the%20White%20House. (accessed 3 August 2023).
Paris, R. (2004) *At War's End: Building Peace after Civil Conflict*. Cambridge: Cambridge University Press.
Parpart, J. (2017) *Gender and Development 25*(1).
Plaut, M. and Vaughan, S. (2023) *Understanding Ethiopia's Tigray War*. London: Hurst.
Pouligny, B., Chesterman S., and Schnabel, A. (2007) *After Mass Crime: Rebuilding States and Communities*. Tokyo: United Nations Press.
Prunier, G. (1995) *The Rwanda Crisis: History of a Genocide*. New York: Columbia University Press.
Pugh, M. and Cooper, N. with Goodhand, J. (2004) *War Economies in a Regional Context: Challenges of Transformation*. Boulder, CO: Lynne Rienner.
Putin, V. (2021) Article by Vladimir Putin "On the historical unity of Russians and Ukrainians". 12 July 2022, President of Russia website: http://en.kremlin.ru/events/president/news/66181 (accessed 10 June 2023).
Rachman, G. (2023a) The real meaning of Xi's visit to Putin: Talk of a pace plan for Ukraine will obscure the tightening of ties between China and

Russia. *Financial Times*. 20 March 2023. https://www.ft.com/content/e4cac940-ada7-454f-ae02-20054be84aeb

Rachman, G (2023b), How the war in Ukraine met the war on woke. *Financial Times*, 30 January 2023.

Ramsbotham, O., Woodhouse, T., and Miall, H. (2005) *Contemporary Conflict Resolution*, 2nd edn. London: Polity.

Ramsbotham, O., Woodhouse, T., Miall, H. (2011) *Contemporary Conflict Resolution: The Prevention, Management and Transformation of Deadly Conflicts*, 3rd edn. Cambridge: Polity.

Rees, S. (2012) *Passion for Peace*. Sydney: UNSW Press.

Reuters News Agency (2023) *Putin tells Erdogan: Ukraine must accept loss of territories for there to be dialogue*. https://www.reuters.com/world/europe/putin-tells-erdogan-ukraine-must-accept-loss-territories-there-be-dialogue-2023-01-05/ (accessed 30 January 2023).

Richmond, O. (2006) The problem of peace: Understanding the 'liberal peace'. *Conflict, Security and Development* 6(3): 291–314.

Richmond, O. (2007) *The Transformation of Peace*. London: Palgrave.

Rigby, A. (2001) *Justice and Reconciliation: After the violence*. Boulder, CO: Lynne Rienner.

Roberts, A. (2009) Doctrine and reality in Afghanistan. *Survival 51*(1): 29–40.

Roberts, T. and Bosch, T. (2023) *Digital Citizenship in Africa: Technologies of Agency and Repression*. London: Bloomsbury.

Ron, A. (2009) Peace negotiations and peace talks: The peace process in the public sphere. *International Journal of Peace Studies 14*(1): 1–16.

Rotberg, R.I. and Thompson, D. (eds) (2000) *Truth v. Justice: The Morality of Truth Commissions*. Princeton, NJ: Princeton University Press.

Ruthven, M. (2003) Edward Said: Controversial literary critic and bold advocate of the Palestinian cause in America. *The Guardian*. 26 September.

Sandal, N.A. (2011) Religious actors as epistemic communities in conflict transformation: the cases of South Africa and Northern Ireland. *Review of International Studies* 37(3): 929–49.

Sandole, D. (1999) *Capturing the Complexity of Conflict: Dealing with Violent Ethnic Conflicts of the post-Cold War Era*. New York: Frances Pinter.

Sandole, D., Byrne, S., Sandole-Saroste, I., and Seheni, J. (2009) *Handbook of Conflict Analysis and Resolution*. Abingdon: Routledge.

Sands, P. (2003) *From Nuremberg to The Hague: The Future of International Criminal Justice*. Cambridge: Cambridge University Press.

Sands, P. (2008) *Torture Team: Uncovering War Crimes in the Land of the Free*. London: Penguin.

Sands, P. (2016) *East-West Street: On the Origins of Genocide and Crimes Against Humanity*. London: Weidenfield and Nicolson.

Sarotte, M.E. (2023a) *Not one inch: America, Russia and the Making of the Cold War Stalemate*. London: Yale University Press.

Sarotte, M.E. (2023b) Not one inch': Unpicking Putin's deadly obsession with the details of history, *Financial Times* 18 February 2023 https://www.ft.com/content/24f81b4d-420e-4217-b498-cf13c6e254f2 (accessed 20 February 2023).

Sebastiàn-Aparicio, S. (2014) *Post-War Statebuilding and Consitutional Reform: Beyond Dayton in Bosnia*. London: Palgrave Macmillan.

Smith, S. (2021) *Afghan Peace Talks: Could a Third-Party Mediator Help*. Washington, DC: United States Institute for Peace.

Tenenbaum, C. (2011) Mediation by Inter-governmental Organisations. In Devin, G. (ed.) *Making Peace: The Contribution of International Institutions*. New York: Palgrave Macmillan, pp. 67–92.

Theidon, K. (2022) *Legacies of War: Violence, Ecologies and Kin.* New York: Duke University Press.
Thornhill, R. and Palmer, C.T. (2000) *A Natural History of Rape: Biological Bases of Sexual Coercion.* Cambridge, MA: MIT Press.
Tilly, C. (1985) War Making and State Making as Organized Crime. In Evans P., Rueschemeyer, O. and Skocpal, T. (eds) *Bringing the State Back In.* Cambridge: Cambridge University Press, pp. 169–91.
Touval, S. (1982) *The Peace Brokers: Mediators in the Arab–Israeli conflict, 1948–1979.* Princeton, NJ: Princeton University Press.
UCDP (2022a) https://ucdp.uu.se/ (accessed 4 August 2023).
UCDP (2022b) https://ucdp.uu.se/country/ 652 (Syria); 678 (Yemen); 530 (Ethiopia) 645 (Iraq).
Ulrichsen, K.C. (2014) *The First World War in the Middle East.* London: Hurst.
UNHCR (United Nations High Commission for Human Rights) (2023) https://www.statista.com/statistics/1293492/ukraine-war-casualties/ (accessed 20 February 2023).
United Nations (2023) Security Council Briefing, 12 April: 'Situation in Mali Remains Volatile amid Escalating Clashes between Armed Groups, Official Tells Security Council, Calling for Sustained International Engagement', https://press.un.org/en/2023/sc15253.doc.htm (accessed 3 August 2023).
Väyrynen T. (2001) *Culture and International Conflict Resolution.* Manchester: Manchester University Press.
Wallensteen, P. (2012) *Understanding Conflict Resolution*, 3rd edn. London: Sage (NB the 2nd edition, 2007 is also quoted here).
Walsh, M. (2015) *Bitter Freedom: Ireland in a Revolutionary World 1918–1923.* London: Faber and Faber.
Waltz, K. (2001) *Man, the State and War.* Columbia, NY: Columbia University Press.
Waltzer, M. (2015) *Just and Unjust Wars: A Moral Argument with Historical Illustrations.* 5th edn. New York: Basic Books.
Ware, A., and Laoutides, C. (2018) *Myanmar's 'Rohingya Conflict.* London: Hurst.
Webb, K. (1978) *The Growth of Nationalism in Scotland.* London: Harmondsworth.
Webb, K., Walters, M., and Koutrakou, V. (1996) The Yugoslavian Conflict, European Mediation and the Contingency Model. In Bercovitch J. (ed.) *Resolving International Conflicts – The Theory and Practice of Mediation.* Boulder, CO: Lynne Rienner, pp. 171–89.
Weir, F. (2022) (in *Christian Science Monitor*) April 20, filed from Moscow, so subject to Russian laws of the time: https://www.csmonitor.com/World/Europe/2022/0420/Russia-says-it-s-fighting-Nazis-in-Ukraine.-It-doesn-t-mean-what-you-think (accessed 2 November 2023).
Wepundi, M., Nthiga, E., Kabuu E., Murray, R., and Alvazzi del Frate, A. (2012) *Availability of Small Arms and Perceptions of Security in Kenya: An Assessment.* Geneva: Small Arms Survey.
Werbner, R. and Ranger, T. (eds) (1996) *Postcolonial Identities in Africa.* London: Zed.
Whittaker, D.J. (1999) *Conflict and Reconciliation in the Contemporary World.* London: Routledge.
Wierda, M. (2006) Transitional Justice in Sierra Leone. In de Zeeuw, J. and Kumar, K. (eds.) *Promoting Democracy in Postconflict Societies.* Boulder, CO: Lynne Rienner, pp. 183–207.
Wikipedia (2023) Crimes against humanity https://en.wikipedia.org/wiki/Crimes_against_humanity (accessed 3 November 2023).

Williams, A. (2005) 'Reconstruction' before the Marshall Plan. *Review of International Studies 31*: 541–58.
Williams, A. (2006/2024 *forthcoming*) *Liberalism and War: The Victors and the Vanquished*. London: Routledge.
Williams, A. (2007) Reconstruction: The bringing of peace and plenty or occult imperialism? *Global Society: Journal of Interdisciplinary International Relations 21*(4): 539–51.
Williams, A., Hadfield, A., and Rofe, S. (2012) *International History and International Relations*. London: Routledge.
Winter, J. (1995) *Sites of Memory, Sites of Mourning: The Great War in European Cultural History*. Cambridge: Cambridge University Press.
Winter, J. (2000) The generation of memory: Reflections on the 'Memory Book'. Contemporary historical studies. *German Historical Institute Bulletin* Fall: 69–92.
Wolff, S. (2023) Ukraine: Beijing's peace initiative offers glimpse at how China plans to win the war. *The Conversation*, 24 February 2023: https://theconversation.com/ukraine-beijings-peace-initiative-offers-glimpse-at-how-china-plans-to-win-the-war-200300 (accessed 3 March 2023).
Yoshida, T. (2007) Advancing or Obstructing Reconciliation: Changes in History Education and Disputes over History Textbooks in Japan. In Cole, E.A. (ed.) *Teaching the Violent Past: History, Education and Reconciliation*. Lanham, MD: Rowman and Littlefield, pp. 51–79.
Youkee, M. (2019) Thatcher sent Pinochet finest scotch during former dictator's UK house arrest. *The Guardian* 4 October https://www.theguardian.com/world/2019/oct/04/margaret-thatcher-pinochet-chile-scotch-malt-whisky
Zartman, I.W. and Rasmussen, J.L. (eds) (1997) *Peacemaking in International Conflict: Methods and Techniques*. Washington, DC: United States Institute of Peace.

5 Post-conflict reconstruction, democratization and development

Introduction

This chapter will analyse the often-made assumptions about reconstruction and its allied concepts, democratization, state and nation-building. It will also, though in less detail, examine the allied doctrines of disarmament, demobilization and reintegration (DDR) as they apply to post-conflict situations in developing countries. It will first look at a number of what might be termed 'generic' reconstruction attempts, starting with those of Germany and Japan (Japan in 1945 was still a 'developing country', and both operations are even now seen as ideal types), as well as the largely failed interventions in Afghanistan and Iraq. It will also examine Ukraine, which we will treat as an outlier to a more pronounced trend of Western attempts to change developing countries' patterns of governance and economic organization by force. In Ukraine, it has been more a case of Western states and international organizations helping existing patterns of democratic governance and capitalism to succeed in the face of a different kind of intervention, one motivated more by an imperial imposition more reminiscent of the British and French Empires than of the kind of liberal interventionism of the post-1990 era.

The period after the end of the Cold War was not characterized by a linear rise in intra-state wars. There was an important rise in 1991–2, but a decline in such conflicts until 2007 (Peace and Conflict website, University of Uppsala, www.pcr.uu.se). From the perspective of intervening powers and IGOs, these wars are often termed 'small wars' in that they required, after an initial large investment of troops and material (as in Afghanistan in 2001 and Iraq in 2003), relatively small numbers of troops, often in counter-insurgency (COIN) mode (Etzioni, 2015). However, the forceful peacemaking efforts of the 'coalitions' in Afghanistan and Iraq are widely perceived to have been ineffective, even counter-productive. And yet they were often conducted by troops with a great deal of experience in these types of operations, especially in Northern Ireland and in post-imperial operations, like the British

DOI: 10.4324/9781003353485-6

Plate 13 The Japanese surrender document, 1945 credit: Wikimedia Commons (National Archives, USA).

army (Ledwidge 2011, but see for a contrary view Sky 2015). They did not leave behind them democratic state structures.

The period since 2016 has led to major reappraisals of the various elements of the reconstruction debate, as it has in other areas covered by this volume. The withdrawal of the United States and its allies in Afghanistan in August 2021 and the invasion of Ukraine by Russia in 2014 and 2022 are causing a great deal of reflection on the topic. So while there have been some post–Cold War cases that are seen as 'successes', notably in Sierra Leone and Kosovo (both 1999), these have tended to be eclipsed by what has happened since, especially in the Middle East, where there has been a conspicuous lack of international intervention in the Libyan and Syrian civil wars, reinforced under the US presidencies of Barack Obama (2008–2016), Donald Trump (2016–2020) and Joe Biden (2020 to the present). The big question has to be whether reconstruction in Ukraine can relaunch the idea in a positive sense.

So what has gone wrong? How can we square the enormous resources deployed in wars in developing countries with their relative failures? One way is to look at what might be termed the changing 'semantics' of reconstruction, for it is in terms of a phenomenon's

Plate 14 What does 'victory' now mean? American tank destroyed by 'insurgents' 2003.

description that its core values can usually be determined. In turn, a presentation of a term can lead to its being seen as a use of power, positive and negative. To say as much is not to be 'anti-American' or to accuse the Western Powers of 'imperialism', but to identify how these actions can be seen as such, no matter how pure the intentions of the intervening Powers or even how effective their actions in helping the cause of long-term peace (Williams 2007). For a casual reader of the literature on how we should go about bringing peace to war-torn societies after, usually, civil wars are presented with a plethora of seemingly contradictory terms. Our aim is to show how the ideal of 'peacebuilding' described in Chapter 4 relates to 'reconstruction' in the 1990s and since, as well as to 'state-' or 'nation-building'.

'Nation-' and 'state'building are often conflated with 'reconstruction', and both usually figure in defences of 'peacebuilding' efforts. But, of course, they do have distinct implications. Many 'nations' do not have 'states'. We could see a nation as a group of people who identify themselves as a coherent current, as well as historical and usually linguistic grouping like the Kurds or the Kikuyu in Africa. But not all such groups have, or indeed wish to have, a 'state'. Hence studies of nationalism often differentiate between 'civic' and 'ethnic' national identity. Many states are multinational, like the United Kingdom. In the pages that follow, we will attempt to show how the two concepts of 'nation' and 'state' can and have been used or abused in the context of policies of intervention and reconstruction.

Changing definitions, changing historical contexts

There are many definitions of what we would now loosely call 'reconstruction'. We use a broader definition than that usually employed. Hence, we believe that:

> reconstruction encompasses short-term relief and longer-term development. It extends far beyond physical reconstruction to include the provision of livelihoods, the introduction of new or reformed types of governance, and the repairing of fractured societal relationships. Thus reconstruction is not merely a technocratic exercise of rebuilding shattered infrastructure. Instead, it is an acutely political activity with the potential to effect profound social and cultural change. Post-war reconstruction holds the capacity to remodel the nature of interaction between the citizen and the state, the citizen and public goods, and the citizen and the market.
>
> (Mac Ginty 2007: 458)

Even this catch-all definition has its problems. It leaves hanging what kinds of 'interactions' are envisaged and between whom. We hope to show that the usual default *model* that is most often used in statebuilding and reconstruction attempts is that of the 1940s and the reconstruction of Germany and Japan using the twin levers of military domination and economic mobilization. So what if this model is fundamentally flawed when we attempt to deal with civil war situations of the post-war period that do not lend themselves to comparison to those of the 1940s?

Colonial and civil war antecedents

The term 'reconstruction' has two main semantic roots, in the logic of colonial/imperial administration and in the American Civil War (Foner 1989; Williams 2005; Cramer 2006).

The first of these has maybe had the longest impact on thinking about the practice in developing countries and the erstwhile colonial, and especially the European Powers. In the nineteenth century, the idea was used a great deal by great liberal thinkers like John Stuart Mill to convey the need to 'civilize' the 'natives', as Mill put it, under the guidance of 'philosophical legislators' to create a 'Greater Britain' through colonial emigration to what are now developing countries in the Commonwealth (Bell 2007: 50). Unfortunately, the implantation of white farmers, which was often linked into 'punitive expeditions' to clear 'natives' from large areas of Africa, has had the long-term result of causing massive resentment that came out in such conflicts as the Mau Mau in Kenya in the 1950s and in inter-tribal violence as a sequel to colonially imposed 'solutions' to land and other problems (Lewis 2007). As the distinguished British political theorist John Plamenatz explained at the end of the colonial period, in 1960, 'No western people have cared more for freedom than the British and the French ... and to elaborate the rules and institutions needed to establish it. They have also created the largest empires in Asia and Africa. ... True, they did not create them to bring freedom to others so much as to increase their own wealth and power' (Plamenatz 1960: 17). Many of the issues now facing countries like Kenya, Zimbabwe and South Africa were in part caused by the unintended consequences of liberal attempts to create modern political and economic state structures, which are usually seen by indigenous peoples as being part of a system of institutionalized inequity.

The reality of reconstruction and nation-building in many post-colonial states was often in opposition to well-meaning colonial regimes. Their first leaders were invariably leaders of revolts against the occupier. In Kenya's case, it was Jomo Kenyatta; in Zimbabwe, Robert Mugabe; and in South Africa, Nelson Mandela. In Vietnam the opposition to the colonial French was led by Ho Chih-Min, still a revered figure; the examples are many. This phenomenon is not unique to those countries. The first leader of the new United States was the general who defeated the occupying British forces, George Washington.

We might even say that the historical record of reconstruction attempts since the Second World War reinforced suspicion of reconstruction efforts. When the Americans attempted to 'reconstruct' or 'nation-build' South Vietnam in the 1960s and 1970s while simultaneously fighting a war against the north (and much of the population of the south) of the country, those efforts led to a huge distrust of American military and political motives. That this was also tied into a new science of 'development', a new idea in the 1960s, meant that the suspicion was extended to economists as well (Gawthorpe 2018; Seitz 2018). There are even authors who believe that all development attempts are in essence 'occult imperialism'. George Kennan, national security advisor to President Truman, once complained that the 'emphasis on economic development was "irresponsible", "reckless[ly] plundering," and "vulgar"' (Williams 2007; Kennan 1962).

It might therefore be asked if there is any point in seeing reconstruction as anything more than a newer form of colonialism?

The Marshall Plan, mandates and 'precarious statehood'

The major partial exception to this rule was in the aftermath of (and during) the Second World War. The term 'reconstruction' became synonymous with the emergence of the first UN agency, the United Nations

Relief and Rehabilitation Agency (UNRRA) in 1943–47, and the Marshall Plan in 1948–52, its official title being the Economic Recovery Program. This programme did much to help resettle refugee and former concentration camp populations and rebuild much of the Western European and Japanese economies in the largest such enterprise to date (Hogan 1987; Williams 2005, 2006; Mayers 2018). The creation of the international financial architecture (the International Monetary Fund and the World Bank in particular) discussed in Chapter 2 also did much to establish the context in which German and Japanese reconstruction could take place. Hence macro-level reconstruction (of the international system) enables micro-level reconstruction at the country level (for more on this, see Box 5.1). This reconstruction by the United States after 1945 in the Marshall Plan became both a model for later attempts but also held within it a core of mistrust that reconstruction efforts are always tainted with imperial designs underlying them.

Box 5.1 German and Japanese reconstruction, 1945–55

At the end of the Second World War, the victorious Allies (the 'UN,' led by Britain, China, France, the USA and the USSR) decided to dismember Germany into 'zones,' dismantle the political, economic and social organizational structure of the National Socialist State and in effect rebuild the German nation. The same fundamental logic was used for Japan and, to a much lesser extent, Italy. The process was interrupted after 1947 by the onset of the Cold War between the former partners, but the Western zones of Germany, Japan and Italy all benefited from a huge ($50 billion) injection of soft loans and direct aid to rebuild the shattered infrastructure of the countries and more widely of Western Europe as a whole. This 'Marshall Plan' was largely successful in creating or at least enabling the 'economic miracles' in Germany and Japan in the 1960s. There was little or no obstruction from the previous power hierarchies that had been comprehensively defeated in battle and were subject to awful privation and violence on a continent brutalized by total war. Reconstruction took place on the following levels: physical infrastructure, institutions, political/cultural – 'de-Nazification' or 'democratization,' legal, military and civil and economic. The legal efforts aimed at 'reconciling' the defeated countries to their neighbours both internally and by a form of retributive justice (see pp. 149–51) in war crimes tribunals, some of which were directed by international legal teams.

Sources: Lowe (2013); Beschloss (2002); Dobbins et al. (2003); Killick (1997)

International governance of territories was also given an IGO imprimatur, as with League of Nations 'mandates', which transferred German and Ottoman colonial territories to (mainly) Britain and France to be held in 'trusteeship' until such as time as they were seen as being ready for self-rule. This policy was extended after the Second World War when Japanese and Italian former colonies were transferred to Allied control. So such a process was applied to Palestine and Transjordan from Ottoman rule to Britain and the Free City of Danzig from Germany to Poland under the provisions of the 1919 Treaty of Versailles from 1919 to 1939, and the Saar from 1922 to 1935. In addition, France was mandated to govern Lebanon and Syria; South Africa was given German South-West Africa (present-day Namibia), and so on.

In several ways, we could see this form of activity as having had the same effect as UNRRA in helping improve the normative image of reconstruction efforts in general. In theory, at least a 'mandate' was not the same thing as a 'colony', one reason why the nominally anti-imperialist Woodrow Wilson was able to square his conscience with his stated desire for full 'self-determination'. As Erez Manela has shown, many would-be nationalists in the Middle East and elsewhere were greatly disappointed that Wilson's formal stance did not in fact lead to independence but rather to a modified form of colonial administration (Manela 2009).

On a positive note, the League's administrators undoubtedly did much to protect minority populations, to act to deal with disputes between the territories and other nearby states and to develop constitutional guarantees for the population. This has continued to be seen as a successful strategy since the end of the Cold War by the EU, OSCE (Organization for Security and Cooperation in Europe) and UN among other IGOs in Cambodia (UNTAC, 1992–3), Bosnia-Herzegovina (UNMBIH, 1994–2002), Kosovo (UNMIK, 1999–2008) and East Timor (UNTAET, 1999–2002). These have tended to be called 'protectorates', but they are 'mandates' by any other name, existing in what Richard Caplan calls a 'precarious statehood' (Caplan 2005: 29 and Chapter 1; 2007).

These cases have become paradigmatic in thinking about nation-building and reconstruction ever since. To summarize, the reconstruction exemplars of Germany and Japan have to be seen in the context of a 'reconstructed' macro-international system (especially its public and private financial architecture) and a clear and enforced change in not only the political but also the cultural (including economic) mindsets of the countries involved. It was made clear to their populations that only certain beliefs and activities were acceptable and others deviant.

It might be argued that this logic has been applied to 'developed' as well as 'developing' countries since 1945. The tools are those of 'conditionality' for grants and soft loans, often imposed by the IMF, as was done to Britain in the 1970s and many least developed countries (LDCs). It might even be said to have been demonstrated by the way the EU Commission and the European Central Bank have insisted on 'conditionality' for loans given to Greece, Ireland and Portugal since 2009. But one evident example is that of Argentina. Argentina was declared bankrupt in 2002, leaving about eight million of the 35 million population 'destitute' (United Nations Economic Commission for Latin America and the Caribbean) (ECLAC 2002). It is possible to avoid the logic of such IMF conditionality, but it will result in economic misery, and this is naturally greatly resented by subject populations. Social unrest and a distrust of IGOs and national governments is an almost inevitable result.

Reconstruction after the end of the Cold War

From consensus to pre-emptive war

There are important differences to the normative background to discussions of efforts at 'reconstruction' in the post–Cold War period, in short to what we have previously called the 'software' of reconstruction. Primary among these is a changed environment for claims to, and discussions about, sovereignty, a subject we broached briefly at the start of the last chapter. Sovereignty has been termed the 'master noun' of international relations and derives much of its underlying meaning from classical Western political thought, and especially from thinkers like Thomas Hobbes and Jean Bodin. This has consequently greatly coloured the global view of what a state is and what rights and duties states owe their citizens and vice versa (Slomp 2008). It might be argued that, as with the debate on democracy, the Western Powers that dominate the global system have come to be seen in developing countries as moving the goalposts whenever developing countries interpret sovereignty in ways deemed unacceptable in Western capitals. So, as many writers (e.g., Caplan 2005; Jackson 2005) have pointed out, the 'new interventionism' and its allied concept of a 'responsibility to protect' (R2P) represent both a realist, national-interest-based justification for intervention to address UN Charter Chapter VII 'threats to international peace and security' and liberal impulses to protect vulnerable populations in civil wars. Governments and populations in developing countries may be wrong to see R2P as 'imperialism' on occasions, as in Afghanistan, Iraq or Sudan currently, but they nonetheless do often consider that to be the case (Bellamy 2014; Hehir 2012).

The 'new interventionism' has often used the doctrine of R2P to justify regime change. It coincided in the period of the late 1990s and until around 2008 with a new American emphasis on 'pre-emptive war'. The key, indeed obvious, factor to have changed since 1990 in the field of grand strategy is the absence of the USSR. This left the United States as a 'unipolar' power in the international system. This in turn encouraged the hubristic belief, most notably deployed in the invasions of Afghanistan in 2001 and Iraq in 2003, that the United States could 'go it alone' in solving world problems, although there was an attempt to present these actions as being embedded in coalitions. This was truer in 2001 when a Security Council Resolution backed the assault by mainly US forces on the grounds of self-defence after the attacks on New York, even backed by the Iranian state, no friend of Washington. In Iraq, this was not the case, and the invasion was dubbed a 'coalition of the willing' in which several major world powers refused to participate, notably France and Germany. We could add the actions of NATO in the former Yugoslavia in March 1999 – 'Operation Allied Force' – that led to the secession of Kosovo from Serbia, backed by NATO in spite of extensive opposition from Security Council members China and Russia.

Another notable change was in American attitudes to war in the period after 1990. After the relatively uncontested liberation of Kuwait in 1991, 'Operation Desert Storm', also agreed by the UN Security Council and led by the United States on the grounds that Iraq had flagrantly abused Kuwaiti sovereignty. After this campaign, the United States also intervened in the Somalian Civil War and withdrew quite quickly when US troops were attacked (as shown in the film *Black Hawk Down*, 2001). This cautious approach continued for a while, often referred to as the 'Vietmalia' syndrome, signifying a US desire not to be seen as a bully in developing countries. It changed in 1994 when President Bill Clinton, infuriated by the EU's pusillanimous inability to intervene in the bloody Yugoslav civil war, authorized the sending of US planes and troops, culminating in the 'Dayton Accords' of December 1995. Also notable was the US non-intervention in Rwanda in 1994, discussed in Chapter 4.

After this, a taboo was seemingly broken; for over the next ten years, the United States developed a much more muscular approach to dealing with totalitarian regimes, known as 'pre-emptive' war, and demonstrated by actions in Afghanistan and Iraq. This process has been meticulously tracked in, among other publications *The Iraq Papers* (Ehrenberg et al. 2010). Maybe its best expression was in a memorandum by Mathew Rycroft to the British Cabinet on the eve of the Iraq war of 2003. To avoid any more moral and military humiliations like Somalia and Rwanda and to show real leadership of the democratic West in the future, it had been decided by the Bush administration that there should be no more hesitation if the intelligence reports were ambiguous about such matters as whether the Afghan Taliban regime of 2001 or that of Saddam in 2003 were *in fact* a threat to regional or world peace. Rycroft reported that 'the intelligence and facts were being fixed round the policy. The NSC [US National Security Council of the President] had no patience with the UN route.... There was little discussion of the aftermath of military action' (Ehrenberg et al. 2010: 68–69).

Nation-building and statebuilding

As with the 1960s idea of 'development' discussed earlier, various new epithets emerged to join with the existing ideas of reconstruction and international administration that pre-date 1990. Primary among these are 'nation-building' and 'statebuilding'. The first of these is usually taken to refer to the 'people' of a given area and to an ethos, the second to the building of institutions, though the two have often been semantically confused in recent times. Caplan distinguishes 'international administration ... from nation- or state-building (the two terms are frequently used interchangeably) though state-building is very often an integral part of an international administration' (Caplan 2005: 3). We might therefore suggest that the definition of what exactly 'reconstruction' means changes in line with the state of the international system and the obsessions of the actors that are prominent power-holders within it, rather than the objective needs of the populations affected by conflict and war.

Nation-building

The successes of the essentially US-funded initiatives in the 1940s and the early1950s gave rise to a belief after the Cold War that the 'nation-building' of Japan and Germany described earlier were paradigmatic examples of how it could and should be done in the new era. These examples from the 1940s and 1950s were much evoked in subsequent reconstruction and development efforts, notably, and probably most misleadingly, in the aftermath of the 2003 invasion of Iraq. Francis Fukuyama, an advocate of the intervention in Iraq, is surely right when he says that both conservative (for which read mostly Republican) and Democrat (or 'liberal') Americans 'have come to support nation-building efforts at different times – conservatives as part of the "war on terrorism" and liberals for the sake of humanitarian intervention'. It might also be said that many Americans would defend the parallels of the intervention in Afghanistan and Iraq with that of Germany and Japan and refer to this as 'nation-building' – as in both cases, they see it as 'constructing political institutions ... coupled with economic development'. In American minds, 'nation-building reflect[s] the specifically American experience of constructing a new political order in land of new settlement without deeply rooted peoples, cultures and traditions' (Fukuyama 2006: 1–3). However, this was certainly not the case in Germany, Japan, Afghanistan or Iraq, where there were existing political and other traditions, albeit illiberal ones. Its very existence as an aim was denied by President Bush in Afghanistan as too burdensome and expensive (Whitlock 2021: 30). His secretary of defense, Donald Rumsfeld, was also explicitly against the concept, being called in 2021 'the Anti-Nation-Builder' by the Rand Corporation's James Dobbins (Dobbins 2021).

'Nation-building' was defined by this same American think tank as 'the use of armed force as part of a broader effort to promote political and economic reforms with the objective of transforming a society emerging from conflict into one at peace with itself and its neighbors' (Dobbins et al. 2007: xvii). This explicit espousal of military force sits uneasily with those who wish reconstruction to come from within a society, the basic desire of the liberal, and not to be imposed from without, the basic instinct of the realist. The Rand authors call this distinction 'co-option', where 'the local population are actively involved in all stages of the planning and execution of a nation-building' exercise and 'deconstruction' as was the case in Germany and Japan in 1945 (Dobbins et al. 2007: xx).

More fundamentally, one of the key problems with the logic deployed since 2001 is the implicit assumption that we can indeed 'build' states. For the key to asking that question is the analogous one of asking, Can we build 'nations'? Put like that, the answer has to be either that we think that nation states are somehow 'natural' units of political organization or that there is a model we can follow that will guarantee the success of any such endeavour.

The main results of 'nation-' or indeed 'statebuilding' have not been happy in many instances. In Rwanda in 1994, we saw the 'genocidal state' (Werbner in Werbner and Ranger 1996: 12–13; Lemarchand 2009). In Zimbabwe, the state started in 1979 with a bout of 'ethnic cleansing' around Bulawayo and continued in like fashion. This is a far cry from the democratic dream of many of the nation's new citizens. Even Western observers of the process of reconstruction in Afghanistan recognized as early as 2002 that 'we – the foreign governments and their partners – invariably knew much less, and could do much less, than we pretended ... [we] did not grasp just how poor, fragile and traumatised Afghanistan was – just how conservative and resistant to foreigners villages could be' (Stewart 2014: 357). The author of these words, a former British government minister, walked on his own across Afghanistan from Herat in the West to Kabul in the East over 36 days in 2002. He spoke local languages and had to submit to being attacked by dogs and people, as well as experiencing astonishing generosity. His insights are practically unique in our experience. Nearly all Westerners travelled in Afghanistan and Iraq on roads in Toyota Land Cruisers, the vehicle of choice for foreign organizations in developing countries. Most officials are 'rotated' into such war zones for a maximum of three to six months, hardly meet any locals except government officials and hardly ever speak local languages. Later on, he was made a Provincial Governor in Iraq, in charge of over 800,000 people and with scant means to do so (Stewart 2006). It is worth remembering that the now controversial British poet and colonial administrator Rudyard Kipling was given over two million people to govern in the nineteenth century in India. The attempted parallels with British India are striking, for in the heyday of the *Raj*, only 1,000 civil servants ruled over 300 million Indians (Gilmour 2019).

Non-locals like Stewart help give us realistic analyses of what, if any, reconstruction efforts can deliver in non-Western societies. But much more negative and prejudiced accounts are more common. One example of the usual kinds of opinions held by Western Powers who hope to 'reconstruct' a country is given in Box 5.2. It illustrates the

Box 5.2 Reconstruction attitudes: the case of Iraq

- *Iraq was made into a British 'Mandate' of the League of Nations, 1919–43, an imperial power in reality, but prepared for 'nationhood' post Second World War under US pressure to decolonize, in the spirit of the Atlantic Charter of 1941.*
- *Racist British attitude: 'Arabs actually prefer to die young … their religion prevents them from accumulating the capital with which to make the improvements they need in their way of life' [as related to US Ambassador Wendel Wilkie, 'One World' Life, 26 April 1943 (Barr, p. 16).*
- *'Iraq's post-war reconstruction projects have been inspired by a desire to raise the standard of living of the population, and so satisfy the widely felt urge to ensure a fair share of the fruits of progress to all' Iraqi government (Committee of Officials 1946: 109).*
- *'[O]ur strategic goal in the months ahead is to set in motion policies which will have the effect of allocating people and resources from state enterprises to the more productive private firms. A fundamental component of this process will be to force state enterprises to face hard budget constraints by reducing subsidies and special deals.… Reconstruction spending will create jobs and raise incomes, but sustained economic growth will depend on whether Iraq's future leaders pursue the pro-market approaches that the Coalition has advocated' Paul Bremer, Director of Coalition Provisional Authority, Iraq, 2003.*

See also: Foote, C., W. Block, K. Crane and S. Gray (2004) 'Economic Policy and Prospects in Iraq', *The Journal of Economic Perspectives*, 18(3): 47–70.

Western obsession from 1946 on to make Iraqis 'improve themselves' by adopting Western attitudes and to shed local (for example) 'religious prejudices' and make themselves more 'liberal'. Not even the *Raj* attempted these aims in quite such a brazen fashion, although Christian missionaries were omnipresent, and we can point to the 'Indian Mutiny' of 1857, which more than illustrated the appalling attitudes of white supremacy that dominated British imperial attitudes.

Neo-liberal ideologies of reconstruction?

Imperialism in the classic sense had been largely consigned to the history books by the 1940s. A new ideology, or 'software', came along that was to dominate thinking about intervention and reconstruction, arguably to this day. Its prophet was London School of Economics professor Friedrich Hayek, the main influence on both British prime minister Margaret Thatcher and US president Ronald Reagan in the 1980s (Hayek 2001; Butler 2012). The idea that Hayek encouraged in his *The Road to Serfdom* (1944, 2001) was not meant to contrast with the welfare economics of his contemporaries John Maynard Keynes and William Beveridge but was rather aimed at the 'planning' of the Soviet Union (Williams 2020).

Notwithstanding that nuance, neo-liberalism is often associated with the idea that 'the market knows best' and should be applied across economies and to all sectors even when natural monopolies exist, like in railways, water distribution and the like. The British experience denationalizing firms that are essential utilities (like railways and water) is now realized to have been a mistake. But neo-liberalism has been widely applied in reconstruction (and indeed development) policies by Western liberal IGOs such as the IMF and, most disastrously, in Iraq.

Hayek would have agreed that states with too many public-sector state-owned enterprises (SOE) and state-owned banks (SOB) are a danger to the country that relies on them. These ideas were strongly supported by Thatcher and Reagan, as well as think tanks like the Heritage Foundation when they were in power. The initial targets for these comments were countries of the former Soviet Union (FSU) and the satellite states of the FSU after 1991, then collapsing, so the states of Central, South-Eastern and Eastern Europe (CESEE). If anyone Googles the Iraq CPA 2003 statement noted earlier, they will read that 'allocating people and resources … to the more productive private firms' had become the mantra after the defeat of Saddam Hussein. These were critiques of the FSU and its centralized economic and trading patterns. This is a critique that has persisted. Hence we learn from a widely quoted paper by the IMF and the European Bank for Reconstruction and Development (EBRD) from 2019 that countries that have widespread reliance on SOEs and SOBs will inevitably encounter economic problems (IMF/EBRD 2019).

Rebuilding 'states'

Western advocates of reconstruction start with the premise from their own political DNA that they are dealing with established states with relatively homogenous populations, as Germany and Japan were in 1945. A frequent issue facing countries emerging from post-colonial and/or civil wars is that the constituent parts of the state have no binding 'identity' in that they often in no way see themselves collectively as a 'nation'. The multi-lingual and multi-confessional Swiss do, but only because they have a long history and devolve much of state power to the 'cantons' and 'communes'. Creating a shared identity cannot merely be done by creating common institutions, as important as they are. Consider the case of an unnamed young American who came to Iraq in 2003 to rebuild the university sector, one which, by general consent, had been badly damaged by the regime of Saddam Hussein. The efforts of this one raw recruit (probably a graduate of a long-established and well-endowed 'Ivy League' university) were reduced to ridicule when 'the [Iraqi] universities he was trying to rebuild and the faculties he wanted to invigorate [and visit] became more and more dangerous – and infrequent'. His initial youthful enthusiasm was often reduced to lying beside the pool while the universities continued to rot (Chandrasekan 2008: 5).

It was not meant to be like that. The intention of the State Department's *Future of Iraq Project* was to involve many of the local population in the post-war reconstruction, not just outsiders imported for the operation. The result has proved, in Iraq, Bosnia and Timor-Leste, to be a variant of international governance. In Bosnia, this governance is still dispensed through the Office of the High Representative, who can overrule the decisions of local politicians if they do not fit the desired model. Currently, the high representative (HR) is German Christian Schmidt, who is chivvying local Bosnians as he emphasizes the urgency for reforms and EU integration. Many Bosnian Serbs want to be integrated into Serbia, but none of them are keen on the 'values of dialogue, responsibility, and citizen-focused governance [to] create a brighter future for all'. Helpfully, the HR's telephone number is on his web page (OHR 2023). The 'reforms' will of course require democratic commitments, the adoption of the euro and compliance with the Single Market, a feature of the EU rejected by the United Kingdom in 2016, for example.

Some authors have suggested that such top-down pressure is the only way to 'remake' successful communities (Toal and Dahlman 2011). In talking about the attempts by the international community to bring justice through the ICC, Lara J. Nettlefield is more positive. In a counterblast to the argument that the Bosnian Dayton Accords are merely 'faking democracy' (Chandler 2000), Nettlefield suggests that we should not be so harsh in our judgements about 'pioneering institutions' (Nettlefield 2010: 7). But Nettlefield also acknowledges that these institutions can only hope to be some solace to those who have lost their 'youth' and 'political community' during a long and vicious civil war (Nettlefield 2010: 7). With all the old ones lost, those of the former Yugoslavia, shared 'symbols' have to be created (Keranen 2014) to overcome 'different visions of community and belonging'. This takes a very long time indeed to become anchored in the popular consciousness.

Other writers question the 'purpose, legitimacy, goals and effectiveness of "nation-building"' and point to its 'mixed' record. The imposition of democracy, a prime aim in all such efforts since 1990, has seemingly brought about a semblance of order (Bosnia, Namibia, Sierra Leone and Timor-Leste are key examples), while in others (like Cambodia) peace, but not democracy, has resulted, and in yet others (Somalia, Afghanistan, Iraq, the Democratic Republic of Congo (DRC), Haiti) neither peace nor democracy has resulted (Hampson and Mendeloff, in Crocker et al. 2007: Chapter 37). Hence the idea that the creation of liberal democratic institutions and the implementation of market 'reforms' (which often means helping an economic elite at the expense of the mass of people) will lead to a more equitable and legitimate division of power is often an illusion. As Meike de Goede has put it in a particularly apt analogy, many people in the DRC see politicians as 'thugs in ties' or '*Kulana en cravate*' (de Goede 2015: 2).

Occult imperialism?

So, one clear explanation of the popularity of the term and practice of 'reconstruction' is that it is a term that is deeply embedded in the historical consciousness and belief systems of the great Western Powers. Historically, this is mainly due to the successful use of the idea in the pacification of Germany and Japan after the Second World War and its adoption in the Mandate and Trusteeship ideal of the League and the UN. But it must be underscored that the term has a *different* resonance in other parts of the world, in particular in the developing countries that are our main focus. For them, 'reconstruction' and its successor terms, 'nation-' and 'statebuilding', have the tinge of 'imperialism'. What seems like a good idea to a victorious power is not necessarily experienced in the same way by a vanquished one. It also has direct consequences for interveners, as happened in Somalia, for example. To use Keen's expression, 'One of the most important roots of violence is a sense of having been *humiliated*' (Keen 2007: 50, his italics). Many of those the West aims to help through 'reconstruction', and indeed through the institutions that they use to exercise this help, do feel so 'humiliated' (Badie 2014).

It is therefore inescapable to avoid describing the intentions and actions of IGOs and major states alike in places like Afghanistan and Iraq as post-imperial *hubris*. Such has been this *hubris* that since 2003 and the invasion of Iraq, we have seen books emerging with snappy titles like *A Beginner's Guide to Nation-Building*, a kind of do-it-yourself approach to dealing with failed states, produced by the Rand Corporation (Dobbins et al. 2007) with useful graphs and chapters on every conceivable aspect of the task. Such manuals assert that although '[e]ach nation to be rebuilt may be unique ... the nation-builder has only a limited range of instruments on which to rely'. The 'instruments' are itemized as various kinds of military and administrative personnel, as well as 'experts in political reform and economic development' (Dobbins et al. 2007: vii).

This process tends to exclude the 'locals' at the expense of outside 'experts', and books like that by Dobbins et al. do tend to put such an emphasis on the outside agents concerned. But can the locals be ignored? When they are, they tend to see the institutions created as illegitimate. This in turn has influenced the strength of the state, and this can become a problem not just for one 'nation' but for many, even creating the kinds of *regional* destabilization that we see in Central Africa. As a consequence, there are destructive effects across huge swathes of territory for civil and political norms that states are supposed to uphold – the rule of law, civil society, human rights and so forth – as well as for social and economic norms such as economic stability, poverty reduction, health and many more besides.

Statebuilding

The difficulties of creating a 'nation', or healing one after a civil war, are just as complicated as building or mending a 'state'. This has been put into even sharper evidence by the events in the Middle East since the beginning of the Arab Spring in 2011. The revolution that has broken up the 'states' that were Iraq and Syria has turned the populations into warring factions, has ended the rule of the capitals and governments and created or reinforced new national entities like the Kurds (who now have both an effective 'nation' and 'state') and transnational entities like the 'Islamic State' (also known as ISIS or *Daesh*) (Schneider 2021). Liberalism has met with massive opposition in the Middle East, often in the form of religiously inspired uprisings termed *jihad*, where reconstruction is largely seen as another form of Western domination.

Both the Kurds (who are seeking statehood in opposition to jihadists, as well as local states) and Daesh (who wish for a transnational 'Caliphate' as was the Ottoman Empire) have had more or less effective power over the territories they control at various times, though neither is formally recognized by the international community, nor by any of the states or IGOs that make it up. Even before or coterminous with the current crisis, there were other entities such as 'Transnistria' (a part of Moldova) that were not recognized but operated as 'states'. We have other such territories in what is officially 'Ukraine' that have declared themselves independent ('The Donetsk People's Republic') or been annexed by other states (such as Crimea and later four other *oblasts* of Ukraine by Russia). The secession by Kosovo in 2008 from Serbia has been described as breaking one of the fundamental norms of UN-sanctioned conduct, the interdict on 'secession' by a region from a sovereign state (Ker-Lindsay 2011; Ker-Lindsay and Fabry 2023). What, say the opponents of such action, is to prevent other secessions? Or even the outright annexation of territory, as carried out in Ukraine by Russia in 2014 and now after 2022? This was also said when Eritrea seceded from Ethiopia in 1993. Before that, only one example of secession since 1945 is ever quoted, that of Bangladesh in 1972.

So who or what is to blame for this sorry state of affairs? One culprit has to be the 'international community'. The lack of international official acceptance of the changed geometry of state borders or jurisdictions on the ground has in the past led, for example, to the state of Eritrea being denied official recognition in spite of the evident loss of control of that area by the previous official state, Ethiopia. In that case, the capture of Addis Ababa by two (then) allied liberation movements led to a temporary agreement over power-sharing and border delimitation that ended a civil war that had raged for over 30 years in the Horn of Africa. It is likely that more conflict between Eritrea and Ethiopia will erupt, until at least Ethiopia has its desired corridor through Eritrea to the Red Sea. But what if such agreement is not possible due to the refusal of surrounding states or the international community to accept a new *status quo*?

Another culprit (if that is the right word) is that in developing areas of the world like Central Africa, the Horn of Africa, as well as in the FSU and Yugoslavia, globalization and state failure have combined to create a caste of freelance economic military entrepreneurs. One example is the followers of Foday Sankoh in Sierra Leone in the 1990s or of Charles Taylor in Liberia, many of them disaffected and very violent young men (Hoffman 2011; Wiafe-Amoako and Mazrui 2014). These groups often claim ethnic or cultural motivations but also often have distinctly economic motivations, such as the extraction of commodities for their own enrichment (Pugh and Cooper 2004). This combination of appeals to 'greed' or 'grievance' (Collier 1999) has 'state-disintegrating' consequences, as in West Africa and the DRC, that are difficult to deal with locally given the enormous resources that wars put in the hands of these warlords.

However, it must also now be recognized that many civil wars, such as those in Africa or those now raging in the Middle East, are to do much more with identity issues, sometimes due to pre-modern allegiance to clan or tribe, and often allied to ones of confessional allegiance. These kinds of allegiance are not ones that sit easily with the models of conflict settlement, resolution or management that we discussed in the previous chapter. The idea that 'power-sharing' is going to work, as is often suggested (as by Sisk 2013), is not always evident. IGOs and states like to deal with their peers, not with revolutionary movements who want to change borders, defy international conventions, even on such issues as human rights, and who do not accept the legitimacy of IGOs like the UN or its Security Council. In short, these new entities (such as Hamas in Gaza) want to subvert the international system as a whole.

Of course, this does not mean to say that all intra- or inter-state wars are due to a total rejection of the notion of the 'state', just the state some of the population find themselves in. The section of the population that accepted the annexation of Crimea to Russia was undoubtedly keen to benefit from the enhanced security, economic and identity reinforcement that they perceived Russia to be able to provide over that of the 'official' state of Ukraine. There may even be those who welcome ISIS as a better guarantor of their basic needs than the official states of Syria and Iraq. Even Charles Taylor and Foday Sankoh promised those who followed them a better life, albeit with such slogans as Taylor's in 1997: 'He killed my ma, he killed my pa, but I will vote for him'. He won the election with 74.3 per cent of the vote (Meredith 2011).

As a further twist in a complicated tale of statebuilding, it could also be said that we need to temper the 'new war' emphasis on the negative economic results of globalization on many wars, as it could also be claimed that economic factors have persuaded some warriors to lay down their arms in return for the economic benefits of peace. This was partly the case in Northern Ireland and, maybe, in the former Yugoslavia. The salutary effects of economic interdependence on war and peace are much as they were when first described by Norman Angell in 1910 (Angell 1910) and now by scholars who follow a broadly similar logic. But we would still maintain that the logic of economic gain generally tends to reinforce Pugh and Cooper's thesis about economic drivers in civil wars and Carolyn Nordstrom's findings of the essentially economic nature of the motivations of many who fuel the 'new' wars in the context of globalization (Nordstrom 2004).

In many ways, the features of, and the solutions to, these processes are analogous to what happened in Europe in the late Middle Ages when state structures grew and eventually led to the emergence of systems of security and law. Other writers have referred to a 'neo-Medievalism' that characterizes the wars we are now seeing, defined by Hedley Bull as 'a system of overlapping authorities and criss-crossing loyalties' (Bull in Winn 2004). Indeed one of the differences between many developing states and those of established 'developed' ones is that they precisely are often not 'a sovereign entity dominated by a single predominant national culture. An enormously compelling mixture of legitimacy and efficiency'. Instead, much of the developing world more resembles the city states of Renaissance Italy or Europe before 1630 than the settled nations of modern Europe (Winn 2004: 2–3).

But, as Herfried Münkler has pointed out, the difference with the Europe of the Thirty Years' War, which devastated huge areas of Europe between 1618 and 1648, or the English Civil War (1642–51), is that 'the *state-building wars* in Europe or North America ... took place under almost clinical conditions, with no major influences "from outside", whereas this has not been the case with the *state-disintegrating wars* in the Third World or the periphery of the First and Second Worlds' (Münkler 2004: 8). These were unbelievably violent European wars, easily on a par with any we have seen in Africa and elsewhere. We might also add that although the Westphalian system of states was the ideal after 1648, in many parts of Europe, it was not the reality until very recently (Simms 2014). What we are seeing now in the Middle East, even in Ukraine, might also be said to be a return to the instability of states and the potential for transnational movements that wish to recreate not states but pre-modern entities like the Ottoman Empire or even a 'Caliphate', even a revitalized Russian Empire. This trend might get worse as the tensions and civil wars that we are now seeing in Nigeria and surrounding states, as well as in the Horn of Africa, lead to analogous demands for local 'caliphates'. Many states in Africa have not been characterized by economic or political stability in the past few decades, which might prove to have been a harbinger of their radical transformation or even collapse.

So, how can *statebuilding* be brought about? As we have seen, the end of the Cold War has led to a very different set of problems than that which prevailed until 1990. Caplan has summed it up as follows:

> State-building refers to efforts to reconstruct, or in some cases to establish for the first time, effective and autonomous structures of governance in a state or territory where no such capacity exists or where it has been seriously eroded.
>
> (Caplan 2005: 3)

The next section will explore how the underlying rationale(s) used in such efforts may be said both to hold out hopes of success but also to point to inherent dangers.

The 'stages' of peacebuilding and reconstruction: towards 'sustainable' peace?

Many writers on nation-building, statebuilding or peacebuilding see the process of achieving that in somewhat mechanistic terms. So, for example, Roeder and Rothchild (2005) lay down a series of 'stages' that they believe are essential if the international community is to bring 'sustainable' peace to an area. Jeong (2005) sees them as necessarily 'sequential'. The touted solutions usually have to do with the encouragement or imposition of 'democracy', which corresponds with the liberal mantra that 'democracies do not go to war with each other'. Empirically, that is true, but the concern is that countries in a state of *transition* to democracy, or even more so in a transition to *statehood*, do. Developing countries are usually in states of major transition of some kind and are consequently more violent than those that are not.

Paris (2004) has suggested that given the difficulties of transition to democracy, it would be better to concentrate on 'institutions' before changing political practices. The initial problem in the reconstruction of any developing country, he notes, is the lack of basic security. In successful states, the monopoly of power is held by the state; in states that have experienced a civil war, this is certainly not the case. So 'the first task of peacebuilding is to restore the monopoly as a foundation and precondition for all further institution-building efforts' (Paris 2004: 206–7).

We shall therefore now outline how that initial logic usually unfolds in terms of policy prescriptions and suggest how many writers have seen problems in the implementation of those prescriptions.

Actors

The primary actors in the processes of reconstruction/statebuilding since 1990 have been as follows.

States

In spite of George W. Bush's famous dictum, '[W]e don't do nation-building', during a presidential candidates' debate in 2000, that is precisely what the United States has been doing in many parts of the world. In fact the budget for reconstruction in Afghanistan was to top $143 billion, more than that expended during the entire Marshall Plan (Whitlock 2021:30). This is of course not including the other budgetary expenditure on military measures, for example.

The European Union has in many ways been an enthusiastic supporter of such efforts, taking a leading role in the former Yugoslavia through the European Agency for Reconstruction (http://ec.europa.eu/enlargement/archives/ear/home/default.htm), as well as in Cambodia. The UK played a determinate role in Sierra Leone with military, civil and other assistance on a large scale, latterly through the 'Conflict, Stability and Security Fund' (Ginifer 2004; CSSF 2022). So has France in its former colonies, such as Ivory Coast and Chad, often through the African Development Bank (ADB). The EU and other groupings of states, like the Commonwealth, have played an important role in election monitoring. Even Japan has played a rather more minor role (in Cambodia and Iraq: Japan MOFA 2007), one that may now expand with the changing of Japan's constitution by Prime Minister Shinzo Abe in September 2015.

IGOs and IFIs

The role of IGOs and IFIs, like the IMF in Russia in the immediate aftermath of the Cold War, has been followed by a subordinate role for that organization in other places. There has been the establishment of such specialist bodies as the EBRD and the OSCE.

The role of the World Bank has undoubtedly been the most important in reconstruction efforts, providing technical assistance in myriad ways. The way this has been applied has drawn criticism even from within the Bank itself (Stiglitz 2003) for advocating policies that tended to kill the patient rather than cure it. The World Bank has moved in recent years from the much-criticized structural adjustment programmes (SAPs), which often reduced the social safety nets for the poor in developing countries, in favour of 'Poverty Reduction Strategy Papers' (though neither of these is specifically for post-conflict countries). IFIs also act as important coordinators of effort through such mechanisms as 'donor conferences'. As we have noted, the UN's 'Millennium Development Goals' of the period 2000–15 have now been succeeded by 'Sustainable Development Goals', which have much the same aims of eliminating poverty, but with a new emphasis on 'resilience' and 'sustainability', hopefully, a more realistic approach than the first 'Goals' (United Nations Development Programme 2015).

Most such organizations are Western inspired and based, if nominally international. There are comparable economic organizations in the developing world, such as the Economic Community of West African States (ECOWAS) and regional development banks (such as the AfDB). There are also embryonic political organizations, like the African Union, which played an important role in Darfur in Sudan.

These are all organizations that have their roots in Western relationships, and changing that power relationship without damaging the source of the funding by giving the impression of *rejecting* the West is bound to be difficult. So attempts to expand the 'BRICS' grouping of nations after a summit in South Africa in August 2023 under what can be called Chinese leadership may lead to a massive increase of funding (AFP 2023). But the source of new BRICS funding would mainly come from the Chinese Belt and Road Initiative (described on p. **85-86**). The extension of membership from the initial grouping of Brazil, Russia, India, China and South Africa is to include Iran, Saudi Arabia, Ethiopia, Argentina and Egypt.

The risks are twofold. First, the context of a combination of countries that are currently supporting Russia's illegal invasion of Ukraine is giving at least the impression of the new grouping being anti-West, even anti-UN. Second, nearly all of the new BRICS members (and several of the original members) have been censured by UN agencies and Western donor countries alike for various infringements of liberal norms. Some old and new members have dubious records on democracy and human rights. So is this money 'tainted' by association? Or should poor countries accept cash and other assistance that does not require the usual conditionality agreements imposed by the World Bank and other IFIs?

These moral and economic choices will undoubtedly form part of future debates about development in general and such topics as reconstruction in particular. Should Ukraine, for example, accept money from an organization that includes the country that has invaded it? Should it take funds that are provided 'with strings' by benevolent donors like the EU or the United States? There is much discussion of a 'greener' rebuilding of Ukraine, for example (Malo 2022). That would certainly enlist support in Congress and the European Parliament from those who believe that every development should have the environment in mind. But in a country like Ukraine that has carbon fuel reserves (and the largest gas storage facilities in Europe), while it has high levels of poverty, is being 'green' really top of its priorities? Gaza, with even higher levels of poverty, has few places to physically install wind farms and obvious political constraints with its neighbours. Is there not a danger that reconstruction will merely gain itself a new layer of ontological notoriety to go with the ones we outlined earlier?

NGOs

On the ground, NGOs and the Red Cross and Red Crescent Movement have played vital roles both in primary assistance, such as the alleviation of hunger and with medical intervention, in war and post-war situations, and in monitoring roles (as with the International Committee of the Red Cross in its role as guardian of the Geneva Conventions). There has long been tension between the human rights and civilian focus of such groups and the military/security focus of peacekeepers and military personnel more generally in the final stages of a conflict (see Chapter 6). 'Soft' and 'hard' NGOs and IGOs have on occasion found it difficult to cooperate, as in Iraq before 2003 when there was some friction between human rights NGOs and IGOs engaged in monitoring the (alleged) weapons of mass destruction.

The press

The press presence is a phenomenon that has drawn much attention since the first Gulf War of 1990–1, and conflicts in Somalia, Rwanda and Bosnia in the early 1990s, with the 'CNN' effect, the effect on Western public opinion of the presence of media reporters in the midst of disaster areas (Hammond 2007: 12–13), being seen as a major pressurizer of governments to be 'seen to be doing something'. This is part of what might be called the 'framing' of international conflicts, affecting the types of intervention undertaken in significant ways (Hammond 2007). The media is also seen as a major need in the creation of civil societies by such bodies as the EU (Loewenberg and Bonde 2007), and modern forms of digital media are playing an increasingly prominent role in the construction of social movement and organization in contemporary conflict and post-conflict environments (Chernobrov 2018).

Demobilization

After the 'ceasefire', the job has only just begun, a realization that first really dawned at the end of the First and Second World Wars. How do you pick up the pieces left by violent conflict? The job of disarming the German armies in those wars' aftermaths was relatively simple, as they were disciplined organizations that obeyed orders and could be largely trusted to comply with agreements. The disorganized militias of the current period have a tendency to split into ever-smaller factions. There are also 'spoilers' in that they often hold on to their weapons, a rational choice in a way that an army demobilizing to return to a normal civilian life of work and family would not contemplate.

Disarmament, demobilization and reintegration

It is now accepted that 'development requires security and that without security, development cannot be achieved' (Muggah, 2009: xx). The resulting policy trio, 'disarmament, demobilization and reintegration', often referred to by the acronym 'DDR', is directed at reducing the most obvious proximate source of violence in a conflict (for a full discussion of DDR, see Özerdem 2008; Muggah 2009; McMullin 2013). As Durch et al. (2003: 26) put it, 'DDR programs are essential elements of post-conflict stability that reduce the likelihood of resumed conflict'.

Removing weapons through their collection and subsequent destruction ('disarmament') is important symbolically, as it signals the parties are willing to pursue political rather than violent means. This is especially so if the process is accompanied by registration and monitoring of all factions and their return to former home districts or barracks ('demobilization'). In civil wars in developing countries, this will quite often require the official state forces to accept a diminution of their role as sole arbiters of control.

However, the main system-level solution is always the 'reintegration' of ex-combatants, which Özerdem describes as 'becoming civilian' (Özerdem 2008). This often, even usually, requires the implementation of training programmes to enable former soldiers who may never have known anything but war to find different kinds of jobs, though that is often in the 'official' state army or police. It must be remembered that many wars in developing countries have been fought by children, notably in Central and West Africa (Beah 2008; Özerdem and Podder 2011). They therefore merit a place in the post-conflict discussions (Watson 2015; McMullin 2013). But to provide a satisfactory alternative to the profession of arms is notoriously difficult, and former child soldiers often have to be 'reintegrated' into populations that were previously their victims. Education also proves difficult for young people who are themselves traumatized by what they have seen and done (Betancourt et al. 2008).

The stated aim of reintegration is to help create a working democracy. This includes reviewing the civil/military relationship, a feature in many coup-prone countries (as with Fiji or Pakistan), or with countries where dictators have an unhealthy relationship with their armed forces, as for example in Zimbabwe, which is arguably worse in the long run for local and regional prospects for peace and prosperity alike. An important corollary to this is 'security sector reform' (SSR), where the army and police are reformed to make them more reflective of democratic norms and practices, not mere protectors of the elite and their own livelihoods. SSR has been defined as follows. 'Security sector reform aims to develop a secure environment based on development, rule of law, good governance and local ownership of security actors' (GFN-SSR http://www.ssrnetwork.net/about/what_is_ss.php, accessed 14 July 2015).

These efforts can collectively be summed up as attempting

> [t]o … devise transitional arrangements for the short term. These must provide the modicum of political stability necessary to conduct elections to a constitutional assembly and the security for delegates to assemble, conduct constitutional debates and craft political institutions to maintain stability and foster democracy for the longer term.
> (Roeder and Rothchild 2005: 2)

Second, these efforts also attempt to build arrangements for the longer term in the form of rewriting constitutions, even devising new parameters for the 'state'. Most prominent among these have been the 'Taif Accord' in Lebanon of 1989 and the Good Friday Agreement in Northern Ireland of 1998 (see Useful Websites).

Third, they attempt to do so in a very public way to try and enhance popular support for the democratization process, as well as to persuade those who have been demobilized that they have to change their behaviour. DDR mechanisms are thus required for the peacebuilding process to be seen to be working. Not to do so will lead to 'stigmatizing' ex-combatants (McMullin 2013).

The importance of, and potential for, DDR and SSR in using the tools of development to try and transform conflicts is therefore huge. Their aims are both on the individual and the collective level. On the individual level, the aim is the transformation of rebel movements into peacetime security forces or other economic groups and the demobilization of huge numbers of young people who have often known of nothing but war and killing in their short lives. This individual demobilization has to provide extensive educational, social and psychological help to young ex-combatants (Peters 2007). The group from which they come then has to be transformed into a political party that relies on the ballot box, not the gun, for its legitimacy. De Zeeuw is succinct in his assessment of the likely result of this: 'Experiences from people directly involved in the transformation of such movements show that the process is extremely complex and time-consuming and has a high risk of failure' (de Zeeuw 2008: 1). This seems to be particularly true of individuals who have spent a great deal of time in the field, often as children who then graduate to leadership roles in insurgent armies.

The issue is so important because the notion of criminal liability is said to be difficult to prove against a child, yet there are differing views about the age a 'child' becomes an 'adult'. The question is often posed as to how 'innocent' such young adults are, how much they should be held responsible for their actions and to what extent they are 'victims' as well as 'perpetrators' (Shepler 2005). In addition, IGOs tend to reduce help to 'adults' once they come of age, so consigning them to a nebulous hinterland of stigmatization at an arbitrary age. The risks of recidivism are very strong (McMullin 2011, 2011a).

The results have been mixed. In Sierra Leone, the difficulties of transforming the Revolutionary United Front (RUF), the rebel group led by Foday Sankoh, into a 'less violent political organization' failed until the UN peace mission effectively detained 'a large proportion of the movement's more politically articulate elements' (Richards and Vincent in de Zeeuw 2008: 91). RUF foot soldiers have not, on the whole, been retrained, except within the new regime's armed forces and police. Richards and Vincent conclude that '[u]nless the problems of an expanding and increasingly impoverished youth underclass are addressed, violent instability is likely to return to the forests and diamond districts of rural Sierra Leone' (Richards and Vincent in de Zeeuw 2008: 100). There is also a long-term legacy of mental instability among many of the former child soldiers in Sierra Leone, whose results will take many years to be fully appreciated (Betancourt et al. 2010).

On the collective level, DDR and SSR aim to provide the basis for a lasting political settlement through such processes as constitution writing (Caplan 2005: 29), which have occurred in many states in the past 35 years, nearly always internationally brokered by IGOs like the Commonwealth or think tanks like the United States Institute for Peace (Widner 2005). It is a complicated procedural process that often drags on for years during or, sometimes, after the conflict is mainly 'over', and aims 'to develop a sense of inclusion and trust (social capital)' in a process that will hopefully encourage peaceful dialogue, not conflict, through providing frameworks of cooperation where different sections of the population feel their interests are being acknowledged. This content of representation (after Pitkin 1967) has been found by some researchers to have 'no major effects on post-ratification levels of violence in some parts of the world, such as Europe, but do[es] make a difference in Africa, the Americas and the Pacific together' (Widner 2005: 516). This would seem to indicate that such efforts have some empirical evidence to back up their claims.

As Roeder and Rothchild admit, the underlying rationale for these actions is a belief that democracy is the essential tool that can calm the fervid spirits of war. But what if the erstwhile combatants have been divided by religious, ethnic or other problems that might lead them democratically to want to *divide from, not unite with*, each other? 'Majoritarian democracy can be a potential source of heightened inter-ethnic conflict' (Roeder and Rothchild 2005: 5).

One much more negative view has been expressed by Rory Stewart about Afghanistan, where he saw what happened to former militias who agreed to lay down their arms in return for their protection. That promise evaporated very quickly as new militias formed and the Taliban regrouped. These new militias were then approached to disarm again. For him DDR, 'this madness'... 'had become a policy of imperial "divide and rule", which was as foolish as it was morally disturbing. I saw how it had been directed against a group previously armed and supported by the international community'. Others have criticized the effect DDR had in the attempted neutralization of the Afghan Taliban, some analysing the 'contradictions and dilemmas' that emerged (Stewart 2014: 353–4; Bhatia and Muggah *in* Muggah, 2009).

What (else) is missing from the democracy, DDR and reconstruction analyses?

Having a peace to keep?

Democratization, DDR and reconstruction processes assume that the war is 'over' or at least in its final throes. But is there not (again) an 'artificial distinction between armed conflict and post-conflict'? And are the international organizations putting an excessive emphasis on DDR policies as 'magic bullets' (Muggah 2005)? There are therefore a number of grounds on which the aforementioned democracy and DDR orthodoxy can be challenged. Some of them are generic, and some (even many) more are based on the empirical observation of case studies.

We can also point to a more complex problem. As we have stressed, the causes and results of wars are intimately linked (Blainey 1988). But what if the DDR process itself actually makes people who are already very upset even more so? Grievances from before a conflict are carried into it and beyond it. Can the bringing of democratic institutions persuade combatants to feel that they can now lay down their weapons and all these hatreds will just evaporate? Keen's analysis that 'greed' and 'grievance' are not sufficient to explain why conflict breaks out, with which the authors broadly agree, means that a very deep-rooted analysis has to be made about all actors' responsibility for the fighting that has taken/is taking place, not just the militias (Keen 2007). Maybe, as with Sierra Leone, the elders provoked violence by their insensitivity over a long period to the demands of youth. Are they to be the new 'Members of Parliament'? How does a democratic process address a perceived lack of 'respect' agenda? How does the DDR process address a cycle of abuse, maybe over many generations? How do they know they will not be further 'betrayed' (Keen 2007: Chapter 3)?

'Spoiler violence' in post-conflict reconstruction

In the aftermath of a number of civil wars and conflicts in developing countries, it has been found that the main occupation of many of the youthful population of the country has ceased to be 'normal' but rather revolves around membership of various militias, groups that often indulge in 'freelance' work on the side. As Podder writes (2012: 199), 'Reintegration is accepted as the weakest link in DDR efforts'. Partly, this is due to the mental scars that the former combatants carry with them (as discussed earlier), partly to the difficulty of getting communities to re-engage with those who have previously done them harm. And partly it is due to former combatants not being so 'former': in Afghanistan, for example, former Taliban and other militia, or 'non-state' armed forces, members took a long time to demobilize after the Bonn Accords of 2001 and often resumed their former activities under cover of nightfall, for example (Giustozzi, in de Zeeuw 2008). This was in spite of there being a new constitution in 2004 and elections. On a more positive note, although youthful, badly 'demobilized' soldiers can become the thugs of the new 'parties' who intimidate their opponents, they can also form the nucleus at pro-democratic party rallies.

One related obvious area (dealt with elsewhere in this volume) is that of 'post-conflict crime' and what is referred to as 'spoiler violence' by disaffected paramilitary groups or even by 'official groups'. Sometimes it is driven by a deliberate attempt to undermine peace accords, sometimes by more venal motivations, but the results are identical. The point is that such violence seems attendant on most, if not all, post-conflict situations. It has been defined as 'violence that deliberately attempts to undermine peacemaking processes and peace accords' (Mac Ginty 2006), and it clearly will have a major impact on attempts at post-conflict reconstruction. However, as John Darby has put it, 'although substantial research attention has been paid to the origin and dynamics of ethnic violence, to the first moves towards negotiations, and to spoiler violence, the threat to post-accord reconstruction is under-researched' (Darby 2006: 6).

Spoiler violence can be generated by unofficial actors (militias, criminal groups, etc.) who feel that the DDR process has failed them, as in Liberia in the 1990s; by the state itself provoking violence, as in Zimbabwe with the taking of land from white farmers in clear violation of the 1979 Lancaster House Agreement; or by the (then) government of Rwanda in 1994 by provoking the genocide. Often, this is because a conflict in its final or initial post-conflict stage is still one where groups contesting established governmental power are jockeying for position for both a negotiated outcome and on the 'battlefield' while the state is trying to do the same (Hőglund and Zartman in Darby 2006). The Taliban was not 'defeated' in 2001 and successfully bombed its way back into Afghan political life; it certainly did not consider the war 'over'. Neither do the various parties who have been destabilizing Iraq since 2003, whether they be disaffected ex-soldiers and officials of the Saddam Hussein regime or other Islamic insurgents like ISIS *(Daesh)*. In other words, the parties and those attempting to 'manage' or 'resolve' the conflict do not operate on one sole path but rather on parallel paths. Conflicts are not resolved in nice, neat, linear ways but by a process of trial and error, more like a spiral than anything else (Lederach 1997).

It must also be said that other forms of 'spoiler violence' are also deeply ideologically satisfying for groups that feel they have not been given adequate recognition in the process of reintroducing structures of government. In the chaos brought about by the 2003 invasion of Iraq, for example, very little thought was given to what would be done about demobilized members of the Iraqi armed forces (see Box 5.3). It was certainly also not considered that foreign insurgents would join in the aftermath of the 'end' of the war proclaimed in May 2003 by

President Bush, again for ideological reasons. Insurgents in Iraq, like *Daesh*, have taken the concept of spoiler violence to a terrible new level and made a mockery of the idea that the 'liberal peace' can easily bring democracy to such divided societies. Indeed, it could be argued that the only place where democracy is now flourishing in the area is Kurdistan, where a group that had been marginalized and oppressed since at least 1919 by three states (Iran, Iraq and Türkiye) has finally found its voice, but through the barrel of a gun not through patient democratic peacebuilding (King 2013; Voller 2014).

Box 5.3 The problems of reconstruction and DDR in Iraq, 2003–15

The official 'end' of the war in Iraq in May 2003 was followed by a demobilization of nearly all the Iraqi armed forces and a purging of practically all of the former regime's officials, with an interim Coalition Provisional Authority that became an Iraqi government after elections in January 2005. The country subsequently collapsed into civil war with an estimated 200,000 civilians killed in intra-communal violence, an Al-Qaeda-inspired insurgency, the actions of the international interventionary forces and general criminality. Very little of the many billions of dollars disbursed in reconstruction funds has been used for that purpose but rather has been lost to corruption or to the sheer cost of security. The CPA only spent a fraction of the $18.6 billion allocated for reconstruction by the US Congress for that purpose, for example, in 2003–4 alone (Diamond in Fukuyama 2006: 176). It has since emerged that all the preparation for the transition by the Future of Iraq Project within the State Department has been essentially ignored, and the Pentagon has been allowed to dictate a failed post-war policy, the rule of law flouted by the occupying forces (epitomized by atrocities by US troops in Abu Ghraib prison) and the Shia and Sunni militias alike. The only relatively peaceful area of Iraq after 2003 is that of Kurdistan, which was effectively independently governed from the mid-1990s anyway. A 'surge' in US forces in 2007 led to a reduction in violence, but the arrival of Islamic State (Daesh) *and the spillover of the Syrian civil war has led to overt civil war between different factions, especially between Shia and Sunni hardliners. Tensions that already existed with Iraq's neighbours Iran (accused of aiding Shia violence by the United States), Türkiye (which has been attacking Kurdish irredentists in the north) and Saudi Arabia (accused of encouraging extreme Sunni ideology, if not of actually supporting Sunni insurgents) have grown into widespread terror and civil war, displacing millions of people and killing and maiming many more. Iraq has become a battleground for regional and global differences, thus complicating the reconstruction efforts immeasurably.*

Sources: Dobbins et al. (2007); Allawi (2007); Dodge (2013); Duffield (2007); Dodge (2006); Ismael and Ismael (2005); Rogers (2013); Sky (2015)

We shall return in the conclusion to this chapter to what this might say about future prospects for such massive 'nation-building' projects backed by military force and with a clearly defined economic and political agenda.

The challenge of small arms proliferation

Many writers (Duffield 2001; Muggah 2005; Greene and Taylor 2012; Kaldor 2012; Batchelor and Kenkel 2013 are just a few examples) have noted that recent conflicts leave countries awash with small arms. In Northern Ireland, where great efforts have been made to 'put weapons beyond use', that has not made them disappear forever. Globalization has ensured that small arms get sucked into conflicts far more easily than was the case in the Cold War, where at least the Superpowers acted as some sort of gatekeepers for violence and where states were propped up, not, as they now are, in a state of collapse. This is not to minimize the horrible effects of Superpower actions in many parts of the world, with the obvious after-effects of such actions still very visible in South-East Asia, the Horn of Africa and elsewhere – we must be careful not to fall into Cold War nostalgia.

Plate 15 'No weapons' sign at the entrance to a UN facility: DDR are often among the thorniest of post-war problems.

In the case of Northern Ireland, Colin McInnes (2000) has identified a number of issues that have to be addressed when decommissioning weapons. They can act as shorthand for other cases (McInnes 2000; Fitz-Gerald and Mason 2005):

> WHEN to decommission? Before or after the final political agreement? Should this be a prerequisite for negotiation?
> LINKAGE to other issues? In Northern Ireland, those issues included release of prisoners. Problems included 'inequality' of concessions.
> HOW to decommission – who provides data on weapons held; phasing of weapon surrender; to whom?
> WHOSE weapons – those of 'insurgent groups'; 'Government' or 'occupying forces'?
> WHAT weapons. Small arms, explosives...? N.B. 'the real "weapon" is the skill and experience of those who made them'.
>
> (McInnes 2000: 89)

In general, what we can say is that disarmament is a phase that overlaps demobilization and reintegration. This is because the signing of a peace agreement often leads to large numbers of young men using their weapons in a freelance way – usually through preying on their own or other communities. In parts of Africa, where there are few if any effective border controls (or indeed clear borders), armed gangs of former combatants regularly cross into areas that have low policing on missions of loot and rapine. This post-accord criminality is thus both a national and an international feature of efforts at DDR (Muggah 2005; Mac Ginty 2006).

How can ex-combatants be 'demobilized' and 'reintegrated'?

Proposed solutions to the problems outlined earlier have included the absorption of ex-combatants into existing police and military forces, even if DDR and SSR should clearly, and ideally, lead to far greater reintegration at other levels of society. It should therefore aim to 'help them develop alternative income-generating activities so they can provide for themselves and their families', though de Zeeuw comments that '[i]n many cases, the real socio-economic needs of the rank and file of former rebel organisations ... are not addressed' (de Zeeuw 2008: 12–13). The World Bank and major Western governmental agencies have in particular urged 'best practice' on IGOs trying to implement such schemes. Surveys of related practice (as with Fitz-Gerald and Mason 2005) are as yet inconclusive, with evidence clearly tending to vary according to a number of factors. These can be said to include the following:

- The commitment of donor countries to the process
- The conditions on the ground. So (we might say) in [country x] and [country y], there are reasons to be optimistic. In [country z], these conditions do not prevail, so [the following conclusions can be drawn]. We clearly have to be careful about over-generalization
- The commitment to and the ability to supervise 'SSR'

The Global Facilitation Network for Security Sector Reform' identified four particular areas that need to be addressed in SSR:

1. Core security actors (police, gendarmerie, civil defence, militias, etc.)
2. Management and oversight bodies
3. Justice and the rule of law
4. Non-statutory security forces ('liberation armies', private security organizations, etc.

Sad to relate GFR-SSN has been closed down since 2010, a victim of UK government spending cuts (https://www.ssrnetwork.net 2023). It has to be said that in some cases, the record of the international community in SSR has been very patchy.

Some of the worst examples in terms of levels of violence can be found in the linked conflicts of Angola and the DRC as well as in Sudan, though there are many other candidates (a good source is the Geneva-based Small Arms Survey organization, http://www.smallarmssurvey.org/), which publishes an annual report. Angola was struggling to emerge from a civil war which dated back to independence from Portugal in 1975 (for a good overview, see Cramer 2006: Chapter 4). The DRC has fallen into total and heavily armed chaos since the arrival of huge numbers of Rwandan *génocidaires* after the massacres of 1994 in eastern Zaire (now the Democratic Republic of Congo), some of whom have been supported by local and international actors with free supplies of cash and arms. The international community's efforts, through the United Nations Angola Verification Missions, UNAVEM I, II and III in the 1990s, to disarm the combatants in Angola was a humiliating failure, blamed by some for its inadequate budget and force levels (Gamba in Darby 2006: 61) but by others for its fundamental misreading of the nature of the civil war in which it was trying to intervene. Cramer (2006: 143) puts this down to the impossibility of rounding up the huge numbers of guns that have always flooded into Angola from the West since the seventeenth century and 'international linkages, political and economic' which are part of '500 years of violent conflict' (Cramer 2006: 147).

In effect, Cramer asserts that the spread of capitalism, through globalization, is the problem, and liberally minded IGOs can do little about this. We might also reiterate the point made earlier: war is a profitable enterprise, and combatants have to be given very clear reasons why they would want to substitute that earning potential for the economic uncertainties of peace (Nordstrom 2004).

But even if it could be argued that such generalizations ignore the very real benefits that capitalism has brought for many previously poverty-struck developing countries, we also have to admit that the record of capitalism as a force for non-violent beneficial change is hard to show in much of Africa (see Box 5.4 on the DRC), but that is demonstrably not the case in much of Asia, for example. In a globalizing world, the boundaries of where such spoiler violence can be found are unpredictable in the extreme. The same could be said of externally generated violence for more venal reasons in the DRC,

Box 5.4 DDR in the DRC

The civil war in the DRC erupted after the genocide in Rwanda in 1994 forced many hundreds of thousands of Hutu extremists and refugees to flee to the DRC from Rwanda. The 'génocidaires', or 'Interhamwe', took their arms with them and set up enclaves from which they embarked on raids into neighbouring countries and preyed on the local populations, the national army being in no state to stop them. Other local militias formed, and the area descended into anarchy, which was not helped by predatory capitalist entrepreneurs making the most of the anarchy to enrich themselves, as did the armies of Rwanda, Angola, Zimbabwe and Uganda, whether invited or not by the government in the capital, which had very little effective control over its eastern provinces. An 'All-Inclusive Accord' was signed between most of these warring parties in 2002 and agreed on terms of reference for DDR. UN peacekeepers were deployed to the eastern (Kivu and adjoining) areas, and a semblance of national control was re-established. The situation, however, remains very tentative, so although there have been serious attempts at DDR by the UN and other organizations, the situation in the east of the country is still chaotic, with local warlords regularly challenging the Blue Berets and DRC official forces, often successfully. Over four million people are considered to have been direct victims of the fighting and accompanying destruction. The record of DDR is mixed, with Kölln (2011: 18) reporting that of 2,400 child soldiers identified and processed by the Belgian Red Cross (one of many NGOs involved), only 553 had been demobilized and only 223 reunited with their families.

Sources: Boshoff; in Fitz-Gerald and Mason (2005); Pugh and Cooper (2004); Kölln (2011)

where freelance groups, multinational companies and different 'allied' or 'enemy' governments use the opportunity of the chaos to line their pockets with diamonds, coltan (a vital ingredient in mobile phones, see Pugh and Cooper 2004) and other raw materials. Without a viable state, it seems clear that no war ever 'ends'.

The challenge of public health

Developing countries generally have a very low ability to provide even basic levels of health care for their citizens, a problem exacerbated by conflict, which drives out both the central government and IGO and NGO provision. Therefore a clear potential beneficiary of the bringing of stable government to an area or region is that it will hopefully not only reduce levels of crime and violence but also help in the stabilization of the institutions of the state, which generally include educational, medical and other social welfare facilities. The re-establishment of these institutions is often overlooked in writings on reconstruction, but by far, the greater proportion of deaths and other suffering come from public health failures. Of the over four million victims so far of the civil war in the DRC, '[r]ather than battlefield deaths, most of these fatalities have been the result of disease and malnutrition as the state has failed to provide public health care or maintain sanitation and related infrastructure' (Mac Ginty and Williams 2005: 173). It is certain that in the calculation of 'war-related deaths', we have to take into account those caught in the crossfire of different groups, as well as the increase in morbidity due to people being displaced by the fighting, the spread of disease and the reduction of spending on health care by governments fighting for survival against insurgent groups (Muggah 2005: 240).

Plate 16 Ensuring global equitable access to COVID-19 vaccines.

Source: UNICEF, https://www.unicef.org/supply/covax-ensuring-global-equitable-access-covid-19-vaccines

In some cases, failures of the health-care system in developing countries can be the *cause* of men joining an insurgent group in the first place. One story we were told by a researcher in Sierra Leone was that a young man whose pregnant wife was refused treatment without a bribe at the (supposedly free) local hospital, as a result of which she died, used the first bullets from his new Kalashnikov to shoot up the maternity ward and its head doctor (Ahorsu 2007). It must also be understood that with the widespread use of rape as a weapon of war and the systematic abuse of women in war zones, huge mental and physical health problems are caused that can take decades to heal, if they ever can (Nordstrom 1999; Brown 2012).

The challenge of the presence of outside actors

The success or failure of the liberal peace has at its heart the dilemma of *who* or *what* is to do the reconstructing. The assumption is that this will be done by the 'international community' working together in the UN and other IGOs or subcontracting that to NGOs (Richmond and Carey 2005). This is what happened in the cases of Germany and Japan in the 1940s (Williams 2005, 2024), and in those cases, the local populations generally accepted the legitimacy of the actions taken. The same cannot necessarily be said of more contemporary examples.

The UN, IGOs and NGOs

The problem in many contemporary conflicts is that the international community acting through IGOs, like the UN, are rarely the only players. Their military muscle is generally weak or absent, and like NGOs, they are reliant on peacekeeping forces to be able to operate (for the experiences of NGOs in East Timor and Sierra Leone, see Jackson in Richmond and Carey 2005). In some cases, NGOs and the UN are reliant on tacit or overt militia (even insurgent) support and are therefore subject to often intolerable pressures to conform to what can be seen as 'alien' cultural practices by the local populations. Hence in Afghanistan during the period of rule by the Taliban, the UN and NGOs were in effect forced to comply with extreme *Sharia* law principles banning the unveiling, or even touching, of women. In Iraq, NGOs and the UN were forcibly put on notice that their presence would not be tolerated (Monshipouri in Richmond and Carey 2005). In this, albeit extreme, case, the dissuasion was by the expedient of a large bomb which destroyed most of the UN HQ in Baghdad and killed its representative, Sergio Vieira de Mello, in 2004.

In less violent cases, the presence of 'alternative' reconstruction groupings can directly challenge both the operations and the logic of liberal peace 'official' players. One example can be found in Lebanon, where Hezbollah has its own 'reconstruction' arm, Jihad Al-Bina and the Gulf states sponsor rival groupings, often to counter the ideological leanings of other groups. Such groups can rapidly marginalize the UN and conventional NGOs. The UN cannot, for example, distribute $12,000 in cash per household as Jihad Al-Bina did after the July 2006 war in Lebanon (Mac Ginty 2007: 458). Hamas' initial impulse came from its welfare role in Gaza after the Israeli occupation of 1967, later replacing the largely absent Palestinian Authority. Its military wing assumed greater importance to the point where it is now designated as a terrorist organization, especially after the attacks on Israel on 7 October 2023 (for an overview, see Robinson/CFR 2023). The geopolitical implications of the Hamas – Hezbollah – Iran alliance against Israel are clearly worrying for the whole area.

We would further suggest that there is an overall problem that has been better understood in the past few years. A growing trend towards the 'local' has become more marked in recent times, as there has been an increase in the role taken by local communities and a corresponding loss of faith by these communities in traditional leaders or 'outsiders'. The lack of agency felt by many people who have been the victims of war is often due to the perception that the IGOs, capitalist institutions and firms and foreign governments are too self-interested to really care about local people. As has been said, '[l]ocal peace-building agency is not a new phenomenom … [but] it has become more prominent, and there is a greater willingness in some quarters to take it more seriously' (Mac Ginty and Richmond 2013: 773–4, 777). To put it briefly, local people know the circumstances of their suffering more deeply than could ever be the case with NGOs or IGOs from outside the conflict, and they are increasingly demanding to be heard.

Conclusions

As the final manuscript of this book was being finalized, a new and even more appalling example of the problems of reconstruction was being demonstrated on the battlefield of Gaza. This was in addition to the horrors that have been nightly TV viewing in the comfortable West in Ukraine. Faced with such seemingly intractable conflicts and seemingly unsolvable problems of who can pay for such reconstruction it would be easy and understandable to shrug one's shoulders and confess to having no idea about how to proceed. How in particular can we hope to reconstruct a country in the middle of an ongoing war when any physical effort will probably be denied by the next missile to hit the building sites? How can reconstruction this time be better than previous attempts? How can past mistakes be learned from? We have no definitive answers to offer, only the opinions of some intelligent observers and the lessons of history.

We have been here before. At the end of his walk across Afghanistan in 2002, Rory Stewart found a piece of paper lying on the ground in Kabul. It was, he recounts, 'a draft proposal intended for the Afghan government, written in English [entitled 'Afghanistan: Rebuilding a Nation: 6 national priority sub-programs' Afghanistan's National Program for Reconstruction']

> There is a consensus in Afghan society: violence must end, respect for human rights will form the path to a lastin peace and stability across the country. The people's aspirations must be represented in an accountable, broad-based, gender-sensitive, multi-ethnic, representative government that delivers daily value.
>
> (Stewart 2014: 341)

Stewart recounts how the UN Assistance Mission in Afghanistan published similar laudable platitudes: it's objective was 'the creation of a centralized, broad-based, multi-ethnic government committed to democracy, human-rights and the rule of law. Its authors were in their late twenties or thirties, with at least two degrees.... They came from Western middle-class backgrounds, and in the evenings, they dined with each other and swapped anecdotes about corruption in the government and the incompetence on the UN. They rarely drove their 4WDs outside Kabul because their security advisors forbade it' (Stewart 2014: 290–1). Since August 2021, it is self-evident that none of this agenda has been fulfilled. The Taliban have no respect for any of the values expressed in this document. Was it all in vain, and what could be done to improve matters in the future?

In Stewart's view, in Afghanistan and Iraq '[m]any of [the international communities'] strategies and plans were simply lifted from other countries on the assumption that there were "universal lessons" for statebuilding, which did not depend on a particular culture and history' (Stewart 2014: 350). In other words, there was profound ignorance of the societies that were being reconstructed. A liberal peace interpretation might well be to say that there is a need for *better* institutional responses (for example, Gamba, in Darby 2006, or Paris 2004). Chabal and Dalloz, in an echo of Stewart, would point to an embedded enthnocentricity in such statements, an unwillingness to accept that we need to take cultural and historical context into account in condemning local practices or trying to impose 'a linear, when not a singular, form of modernization resulting in Westernization' (Chabal and Daloz 2006: 9). A less extreme critical analytical position might be to say that we need to listen to local populations more (Mac Ginty 2006). A more 'structural' approach would point to the underlying economic problems that assail many developing countries. As we saw with the earlier discussion of Sierra Leone, DDR and SSR will only work if employment can be found for the vast underclass that is now a feature of so many developing countries. War, to put it bluntly, is profitable; peace is not (Nordstrom 2004).

There have been some efforts to ascertain what 'lessons learned' there have been in the disastrous unfolding of military and reconstruction efforts in Afghanistan and Iraq. The United States has failed in both kinds of enterprise before. The practice was established after the war in Vietnam but only made public in the notorious '*Pentagon Papers*' leaked to the press in 1971. In that case, there were 47 volumes of testimony and analysis, 7,000 pages in all (Whitlock 2021: xvi). This process was renewed during the Afghan war after 2001. It has been particularly highlighted in the aftermath of US and NATO Coalition forces' withdrawal from Afghanistan in August 2021. There it has been revealed that for a large proportion of the war, the US government had set up an Office of the Special Inspector General for Afghanistan (SIGAR), which interviewed 428 individuals involved in the war, varying from Afghan officials to aid workers and military personnel at a variety of levels of command. A commentator on this process, who was able to get access to most of the interview data, summed up the general view: '[M]any senior U.S. officials privately viewed the war as an unmitigated disaster'. Individual commentary included such statements as '[t]here was no campaign plan. It just wasn't there' (Army General Dan McNeil, who twice served as President Bush's commander in Afghanistan). Among many other similar statements, perhaps the most damning was by 'Army Lt. Gen. Douglas Lute.... White House war czar under Bush and Obama'... '[w]e didn't have the foggiest notion of what we were undertaking' (Whitlock 2021: xv–xv).

More widely, it is clear that the misuse of the term 'reconstruction' in the 2000s may well, in the long run, serve as the most hubristic of all misuses of a term, simultaneously devaluing a major historical success and damaging hopes of any further use of the model in the future, a much greater potential problem as it is a basic ingredient of the policy toolbox of advocates of the 'liberal peace'. We assert that the distinction that dates back to the American Civil War between accusations of 'carpetbagging' and the more noble term 'nation-building' used about Germany and Japan in 1945 still has some force in the latest attempt at 'reconstruction' in Afghanistan and Iraq in the 2000s (Williams 2006: 139–45).

We therefore hope to have shown that the political arguments for reconstruction lie on unstable historical foundations and have been further damaged by historical and recent experiences, especially in Afghanistan and Iraq. In addition, economic arguments for intervention of all kinds have also been more fundamentally challenged from two key perspectives. Writers such as Cramer (2006) and Chabal and

Daloz (1999, 2006) argue that the current debate on development in general and reconstruction in particular tends to downplay the importance of culture as a variable. This debate is arguably part of a much wider one, which suggests that there is no longer a coherent field of 'development studies' with 'competing schools of theory or paradigms' but that it has become subsumed within other disciplines (as with 'area studies') or hacked to pieces by those ('post-structuralists', 'post-modernists', 'post-marxists') who have undermined all the 'positivist', for which read 'rational', bases for the study of development (Hoogvelt 1997: xi). One lasting impact of such epistemological uproar has been to make analysts (but not necessarily policymakers and politicians) focus more on what is specific to any reconstruction 'event' and less on what is generic, what might be termed a 'sociological turn'. We need to look at cases and their specificities, as well as to find 'one-size-fits-all' solutions.

It might also be remembered that some strategists such as Edward Luttwak (1999) and Stephen Van Evera (1999) think that the best and surest path to peace in most developing countries is to 'give war a chance', in other words to not interfere in any way or form, but to let the naturally dominant party emerge through war. This might be put differently by asking: *when does a conflict 'end'?* As we have stressed, the concept of bringing about a 'sustainable' peace is at the heart of thinking about the end of a conflict. But, as a growing number of writers have pointed out, the 'ending' of a conflict is a problematic issue, and therefore, the concept of 'post-conflict development' and reconstruction is also. Both these perspectives are joined by one key commonality – the belief that war is the midwife of stable nation states, and it is the lack of such stability in various parts of the world that provides humanity with most of its problems. So we must look in every case at the underlying rationale for development and what makes for a successful nation in itself as a prelude to seeing what might go wrong with the process of 'state-' or even 'nation-building'.

The botched 'reconstruction' of Iraq will stand as the key case study that will inform all attempts at nation-building for the foreseeable future. It has been pointed out (Stiglitz 2008) that the cost of the war to the American taxpayer (with the attendant lost opportunity cost for American education, health care and other policy areas) was, after five years of fighting, over $3 trillion. The cost to Iraqis, in terms of their development and levels of life, has been arguably even worse since the first American intervention in 1991. One calculation has Iraqi per capita income dropping from $2,279 in 1984 to $627 in 1991 and $450 by 1995, and levels like Madagascar and Rwanda immediately before the 2003 war (Ismael and Ismael 2005: 613). Toby Dodge has pointed to the 'neo-conservative' belief that the United States would inherit a functioning Iraqi state, while it in fact collapsed with the invasion (Dodge 2006: 188). Charles Tripp goes further and suggests that the Iraqi state had in effect collapsed far earlier than 2003, and the looting of ministries was just the final act. But he also postulates that the Iraqi 'shadow state' organized along local lines of patronage has since 2003 been able to organize a 'headless' insurgency that has tied Coalition troops down and further degraded the lives of ordinary Iraqis to catastrophic levels. There is, he postulates, no democracy to defend (Tripp 2008, quoted in a lecture at St Andrews, 22 February 2008).

So what might be done to address these dilemmas? One way is to look back in the 'toolbox' and ask what we have used in the past. Should the aim be to allow statebuilding in the developing world to progress independently of outside interference if Münkler's point about the need for 'clinical' conditions is to be met? One idea from the Cold War period is that of 'self-reliance', a decoupling from globalization and a deliberate development of indigenous industry at the expense of manufactured imports, as was suggested in the 'New International Economic Order' of the 1970s in the United Nations Conference on Trade and Development (UNCTAD). But how would self-reliance stop the determined 'new' warlord intent on extracting his coltan or diamonds from a decrepit African state (Pugh and Cooper 2004)? On the other hand, there is much to be said for Galtung's point that the terms of trade that this exchange implies (southern commodities for manufactured products from the West) mean there is 'an enduring acceptance of a long-term inferior position in which it will be difficult to satisfy the basic needs of local people' (Galtung, paraphrased by Jeong 1999: 36).

In the next chapter, we hope to take on these challenges and suggest how peace might conceivably be brought to developing countries by surveying approaches that attempt to examine the root, and especially the economic, causes and consequences of conflict to try and resolve deep-rooted processes both locally from below and from above through institutional action.

Summary

- The concept and practice of 'reconstruction' over many years has left it with a problematic image that cannot easily be repaired.
- Concepts and practices like 'state-' and 'nation-building' give rise to questions about the motivations and practices of those who try and implement them.
- One-size-fits-all approaches to reconstruction are unlikely to lead to optimal peacebuilding outcomes.
- DDR and SSR lie at the heart of most, if not all, attempts at post-conflict reconstruction and are in themselves problematic.

- The international community needs to take more notice of the specific cultural and other needs and constraints of any particular situation before embarking on reconstruction attempts.

Discussion questions

- Why has reconstruction had such an uneven reputation for effectiveness in the past two decades?
- *Should* the international community or powerful states attempt to 'reconstruct' economies and polities after wars?
- What is wrong, or right, with the proposition that democratizing a country will bring about a stable peace there?
- What are the main features of the practices known as 'DDR' and 'SSR'?
- How might the international community approach the question of reconstruction in Ukraine or Gaza?

Further reading

Some of the best discussions of the problems and potential of reconstruction can be found in: R. Caplan (2005) *International Governance of War-Torn Territories: Rule and reconstruction*, Oxford: Oxford University Press; and A. Williams (2006, 2024 *forthcoming*) *Liberalism and War: The victors and the vanquished*, London: Routledge, Chapters 4 and 5. A more critical view is that of O. Richmond (2014) *Failed Statebuilding: Intervention, the state, and the dynamics of peace formation*, New Haven, NJ: Yale University Press. The best short explanation of why reconstruction attempts in Afghanistan and Iraq were conducted as they were can be found in J. Dobbins, J. G. McGinn et al. (2003) *America's Role in Nation-Building: From Germany to Iraq*, Santa Monica, CA: Rand Corporation, as well as J. Dobbins, S. G. Jones, K. Crane and B. Cole DeGrasse (2007) *The Beginner's Guide to Nation-Building*, Santa Monica, CA: Rand Corporation. One of the best critiques is T. Dodge (2013) *Iraq – From War to a New Authoritarianism*, London: Adelphi/Routledge, as well as his (2006) 'Iraq: the contradictions of exogenous state-building in historical perspective', *Third World Quarterly*, 27(1), pp. 187–200. The classic insider's view of what went wrong is R. Chandrasekaran (2007) *Imperial Life in the Emerald City: Inside Iraq's Green Zone*, London: Bloomsbury. This book was a key inspiration for Paul Greengrass's film *Green Zone* of 2010, starring Matt Damon. There are few defenders of what happened in practice, though one recent example of a sympathetic view is E. Sky (2015) *The Unravelling: High hopes and missed opportunities in Iraq*, London: Atlantic Books.

The best introductions to DDR and SSR are A. Özerdem (2008) *Becoming Civilian: Disarmament, demobilization and reintegration*, London: I.B. Tauris; A. Özerdem and S. Podder (eds) (2011) *Child Soldiers: From recruitment to reintegration*, Basingstoke: Palgrave; and J. R. McMullin (2013) *Ex-combatants and the Post-conflict State: Challenges of reintegration*, Houndmills: Palgrave Macmillan. See also A. M. Fitz-Gerald and H. Mason (eds) (2005) *From Conflict to Community: A combatant's return to citizenship*, Shrivenham: Global Facilitation Network for Security Sector Reform.

Useful websites

A number of other websites are cited in the chapter, so please also refer to those. Data on current conflicts and wars can be found at the Peace and Conflict website, University of Uppsala, https://www.pcr.uu.se/research/ucdp/. Details of some important peace accords can be found at 'Taif Accord' in Lebanon (1989, https://peacemaker.un.org/lebanon-taifaccords89) and the Good Friday Agreement in Northern Ireland of 1998: (http://webarchive.nationalarchives.gov.uk/20130128134220/http://www.nio.gov.uk/the-agreement). Some key reconstruction agencies can be found at the following sites: EU: 'Ukraine Recovery and Reconstruction Needs Estimated $349 Billion' 9 September 2022: https://ec.europa.eu/commission/presscorner/detail/en/ip_22_5428: Foreign Commonwealth and Development Office (UK) https://www.gov.uk/government/organisations/foreign-commonwealth-development-office. On the UN's thinking on DDR, see the United Nations Institute for Disarmament Research (UNIDIR, http://www.unidir.org)

Bibliography

AFP (2023) Imray, G, Mogomotsi, M and Gambrell, J., 'Iran and Saudi Arabia are among 6 nations set to join China and Russia in the BRICS economic bloc', 24 August: https://apnews.com/article/brics russia china summit b59 00168d165cc78b36d5d5c068b7a50 (accessed 26 August 2023).

Ahorsu, K. (2007) The political economy of post Cold War conflicts in Sub Sahara Africa: The natural resources factor. The case of the 1991 Sierra Leone civil war. PhD thesis, University of Kent.

Allawi, A. (2007) *The Occupation of Iraq: Winning the War, Losing the Peace*. New Haven, CT: Yale University Press.

Angell, N. (1910) *The Great Illusion: A Study of the Relation of Military Power in Nations to Their Economic and Social Advantage*. London: William Heinemann.

Badie, B. (2014) *Le temps des Humiliés*. Paris: Odile Jacob; *Humiliation in International Relations: A Pathology of Contemporary International Systems*. London: Bloomsbury (2016).

Barr, J. (2018) *Lords of the Desert: Britain's Struggle with America to Dominate the Middle East*. London: Simon and Schuster.

Batchelor, P. and Kenkel, K.M. (2013) *Controlling Small Arms: Consolidation, Innovation and Relevance in Research and Policy*. London: Routledge.
Beah, I. (2008) *A Long Way Gone: Memoirs of a Boy Soldier*. New York: Sarah Crichton Books.
Bell, D. (2007) *The Idea of Greater Britain: Empire and the Future of World Order, 1860–1900*. Princeton, NJ: Princeton University Press.
Bellamy, A. (2014) *The Responsibility to Protect: A Defence*. Oxford: Oxford University Press.
Beretekeab R. (2000) *Eritrean Making of a Nation 1890–1991*. Uppsala: Uppsala University.
Beschloss, M. (2002) *The Conquerors: Roosevelt, Truman and the destruction of Hitler's Germany, 1941–1945*. New York: Simon & Schuster.
Betancourt, T.S. et al. (2008) High hopes, grim reality: Reintegration and the education of former child soldiers in Sierra Leone. *Comparative Education Review 52*(4): 565–87.
Betancourt, T.S. et al. (2010) Sierra Leone's former child soldiers: A follow-up study of psychosocial adjustment and community reintegration. *Child Development 81*(4): 1077–1095.
Bahtia, M.V. and Muggah, R. (2009) 'The Politics of demobilization in Afghanistan', pp. 126–164, *in* (Muggah, 2009).
Blainey, G. (1988) *The Causes of War*, 3rd edn. New York: Free Press.
Boutros-Ghali, B. (1992) *An Agenda for Peace, Preventive Diplomacy, Peacemaking and Peace-keeping*. New York: United Nations.
Brown, C. (2012) Rape as a weapon of war in the Democratic Republic of the Congo. *Torture 22*(1): 24–37.
Butler, E. (2012) *Friedrich Hayek: The Ideas and Influence of the Libertarian Economist*. London: Harriman House.
Caplan, R. (2005) *International Governance of War-Torn Territories: Rule and Reconstruction*. Oxford: Oxford University Press.
Caplan, R. (2007). From collapsing states to neo-trusteeship: The limits to solving the problem of 'Precarious Statehood' in the 21st century. *Third World Quarterly 28*(2): 231–244.
Carnegie Commission on Preventing Deadly Conflict (1997) *Preventing Deadly Conflict*. New York: Carnegie Corporation of New York.
Chabal, P. and Daloz, J.-P. (1999) *Africa Works: Disorder as a Political Instrument*. Oxford: James Currey.
Chabal, P. and Daloz, J.-P. (2006) *Culture Troubles: Politics and the Interpretation of Meaning*. London: Hurst.
Chandler, D. (2000) *Faking Democracy after Dayton*. 2nd edn. London: Pluto.
Chandrasekan, R. (2008) *Impertial Life in the Emerald City: Inside Baghdad's Green Zone*, London: Bloomsbury.
Chernobrov, D. (2018) Digital volunteer networks and humanitarian crisis reporting. *Digital Journalism* 6(7): 928–944.
Collier, P. (1999) *Doing Well Out of War*. Washington, DC: World Bank.
Cramer, C. (2006) *Civil War Is Not a Stupid Thing: Accounting for Violence in Developing Countries*. London: Hurst.
CSSF, (2022) https://www.gov.uk/government/organisations/conflict-stability-and-security-fund/about#:~:text=The%20Conflict%2C%20Stability%20and%20Security,promote%20international%20peace%20and%20stability (accessed 22 August 2023).
Crocker, C.A., Hamson, F.O., and Aall, P. (eds) (1996) *Managing Global Chaos: Sources of and Responses to International Conflict*. Washington, DC: USIP.

Crocker, C.A., Hampson, F.O., and Aall, P. (eds) (2007) *Leashing the Dogs of War: Conflict Management in a Divided World*. Washington, DC: United States Institute of Peace Press.
Darby J. (ed.) (2006) *Violence and Reconstruction*. Notre Dame, IN: University of Notre Dame Press.
De Goede, M. (2015) *Consuming Democracy: Local agencies and liberal peace in the Democratic Republic of Congo*. Leiden: Africa Studies Centre.
De Zeeuw J. (ed.) (2008) *From Soldiers to Politicians: Transforming rebel movements after civil war*. Boulder, CO and London: Lynne Rienner.
Dobbins, J. (2021) *Donald Rumsfeld: Anti-Nation-Builder*. July 6. https://www.rand.org/pubs/commentary/2021/07/donald-rumsfeld-anti-nation-builder.html (accessed 5 November 2023).
Dobbins, J., Jones, S.G., Crane, K. and Cole DeGrasse, B. (2007) *The Beginner's Guide to Nation-Building*. Santa Monica, CA: Rand Corporation.
Dobbins, J., McGinn J.G. et al. (2003) *America's Role in Nation-Building: From Germany to Iraq*. Santa Monica, CA: Rand Corporation.
Dodge, T. (2006) Iraq: The contradictions of exogenous state-building in historical perspective. *Third World Quarterly* 27(1): 187–200.
Dodge, T. (2013) *Iraq – From War to a New Authoritarianism*. London: Adelphi/Routledge.
Duffield, M. (2001) *Global Governance and the New Wars: The merging of development and security*. London: Zed.
Duffield, M. (2007) *Development, Security and Unending War: Governing the world of peoples*. Cambridge: Polity.
Durch, W., Holt, V., Earle C. and Shanahan, M. (2003) *The Brahimi Report and the Future of UN Peace Operations*. Washington, DC: Henry L. Stimson Center.
ECLAC (United Nations Economic Commission for Latin America and the Caribbean) (2002) *Situación y Perspectivas 2002. Estudio Económico de América Latina y el Caribe, 2001–2002*, 1 August. Santiago: ECLAC. [Quoted in Augé, C. (2002), *Le Monde Diplomatique*] [English edition], September.
Ehrenberg, J., McSherry, J.P., Sanchez, J.R. and Sayej, C.M. (eds) (2010) *The Iraq Papers*. Oxford: Oxford University Press.
Etzioni, A. (2015) COIN: A study of strategic illusion. *Small Wars & Insurgencies* 26(3): 345–376.
Fitz-Gerald, A.M. and Mason, H. (eds) (2005) *From Conflict to Community: A combatant's return to citizenship*. Shrivenham: Global Facilitation Network for Security Sector Reform.
Foner, E. (1989) *Reconstruction: America's unfinished revolution, 1863–1877*. New York: HarperCollins.
Fukuyama, F. (ed.) (2006) *Nation-Building: Beyond Afghanistan and Iraq*. Baltimore, MD: Johns Hopkins University Press.
Gawthorpe, A. (2018) *To Build as Well as Destroy: American Nation Building in South Vietnam*. Ithaca, NY and London: Cornell University Press.
GFN-SSR (2007) *A Beginner's Guide to Security Sector Reform*. Birmingham: GFN-SSR.
Gilmour, D. (2019) *The Long Recessional: The Imperial Life of Rudyard Kipling*. London: Penguin.
Ginifer, J. with input from Oliver, K. (2004) *Evaluation of the Conflict Prevention Pools: Country/Regional Case Study 3, Sierra Leone*. Bradford University, Channel Research Ltd, PARC & Associated Consultants.
Greene, O. and Taylor, N. (2012) *Small Arms, Crime and Conflict*. Abingdon: Routledge.

Hammond, P. (2007) *Framing Post-Cold War Conflicts: The media and international intervention*. Manchester: Manchester University Press.

Hayek, F. (1944, 2001) *The Road to Serfdom*. London: Routledge.

Hayek, F. (2001) *The Road to Serfdom*. London: Routledge.

Hazareesingh, S. (2012) *In the Shadow of the General: Modern France and the myth of de Gaulle*. Oxford: Oxford University Press.

Hehir, A. (2012) *The Responsibility to Protect: Rhetoric, reality and the future of humanitarian intervention*. London: Palgrave Macmillan.

Hoffman, D. (2011) *The War Machines: Young men and violence in Sierra Leone and Liberia (the cultures and practice of violence)*. Durham, NC: Duke University Press.

Hogan, M. (1987) *The Marshall Plan: America, Britain and the reconstruction of Western Europe, 1947–1952*. Cambridge: Cambridge University Press.

Hoogvelt, A. (1997) *Globalization and the Post-Colonial World*. London: Macmillan.

International Crisis Group Africa (2014) Curbing Violence in Nigeria (II): The Boko Haram Insurgency. Report 216, April.

International Monetary Fund/European Bank for Reconstruction and Development (2019) *Reassessing the Role of State-Owned Enterprises in Central, Eastern, and Southeastern Europe*. Washington DC: IMF.

Ismael, T.Y. and Ismael, J.S. (2005) Whither Iraq? Beyond Saddam, sanctions and occupation. *Third World Quarterly* 26(4–5): 609–29.

Jackson, M.G. (2005) A Necessary Collaboration: NGOs, Peacekeepers and Credible Military Force. The Case of Sierra Leone and East Timor. In Richmond O. C. H. (ed.) *Subcontracting Peace: The challenges of NGO peacebuilding*. Aldershot: Ashgate, pp. 109–18.

Jacoby, T. (2007) Hegemony, modernisation and post-war reconstruction. *Global Society: Journal of Interdisciplinary International Relations* 21(4): 521–37.

Japan, Ministry of Foreign Affairs (MOFA) (2007) Japan's efforts on peacebuilding: Towards consolidation of peace and nation-building. Available at: http://www.mofa.go.jp/policy/un/pko/effort0704.pdf (accessed 19 July 2015).

Jeong, H.-W. (1999) *Conflict Resolution: Dynamics, Process and Structure (Studies in Peace and Conflict Research)*. Farnham: Ashgate.

Jeong, H.-W. (2005) *Peacebuilding in Post-Conflict Societies: Strategy and process*. Boulder, CO and London: Lynne Rienner.

Kaldor, M. (2012) *New and Old Wars*. 3rd edn. Cambridge: Polity.

Keen, D. (2007) *Complex Emergencies*. Cambridge: Polity.

Kennan, G. (1962) Kennan to Rostow, 15 May 1962, Foreign Relations of the United States, 1961-63, Volume 8, *National Security Policy*, Doc. 85.

Keranen, O. (2014) Building states and identities in post-conflict states: Symbolic practices in post-Dayton Bosnia. *Civil Wars* 16(4): 127–46.

Ker-Lindsay, J. (2011) *Kosovo: The Path to Contested Statehood in the Balkans*. London: I.B. Tauris.

Ker-Lindsay, J. and Fabry, M. (2023) *Secession and State Creation: What Everyone Needs to Know*. Oxford: Oxford University Press.

Killick, J. (1997) *The United States and European Reconstruction, 1945–1960*. Edinburgh: Edinburgh University Press.

King, D. (2013) *Kurdistan on the Global Stage: Kinship, land, and community in Iraq*. New York: Rutgers University Press.

Kölln, A. (2011) *DDR in the Democratic Republic of Congo*. London: Peace Direct.

Lederach, J.P. (1997) *Building Peace: Sustainable reconciliation in divided societies*. Washington, DC: United States Institute of Peace Press.

Ledwidge, F. (2011) *Losing Small Wars: British Military Failure in Iraq and Afghanistan*. New Haven, CT: Yale University Press.
Lemarchand, R. (2009) *The Dynamics of Violence in Central Africa*. Philadelphia, PA: University of Pennsylvania Press.
Lewis, J. (2007) Nasty, brutish and in shorts? British colonial rule, violence and the historians of Mau Mau. *Round Table: The Commonwealth Journal of International Affairs* 96(389): 201–23.
Loewenberg, S. and Bonde, B. (eds) (2007) *Media in Conflict Prevention and Peacebuilding Strategies*. Bonn: DW Media Services.
Lowe, K. (2013) *Savage Continent: Europe in the aftermath of World War II*. London: Picador.
Luttwak, E. (1999) Give war a chance. *Foreign Affairs* 78(4): 36.
Mac Ginty, R. (2006) *No War, No Peace: The Rejuvenation of Stalled Peace Processes and Peace Accords*. London: Palgrave.
Mac Ginty, R. (2007) Reconstructing post-war Lebanon: A challenge to the liberal peace. *Conflict, Security and Development* 7(3): 457–82.
Mac Ginty, R. and Richmond, O. (2013) The local turn in peace building: A critical agenda for peace. *Third World Quarterly* 34(5): 763–83.
Mac Ginty, R. and Williams, A. (2005) Editorial: The Commonwealth and the inheritance of armed conflict. *Round Table: The Commonwealth Journal of International Affairs* 94(379): 173–5.
Manela, E. (2009) *The Wilsonian Moment: Self-Determination and the International Origins of Anticolonial Nationalism*. Oxford: Oxford University Press.
Mayers, D. (2018) *America and the Postwar World: Remaking International Society, 1945 – 1956*. London and New York: Routledge.
McInnes, C. (2000) A Farewell to Arms? Decommissioning and the Peace Process. In Cox, M., Guelke, A., and Stephens, F. (eds) *A Farewell to Arms? From 'long war' to Long Peace in Northern Ireland*. Manchester: Manchester University Press, pp. 78–92.
McMullin, J.R. (2011) Reintegrating young combatants: Do child-centred approaches leave children – and adults – behind? *Third World Quarterly* 32(4): 743–64.
McMullin, J.R. (2011a) Exclusion or Reintegration? Child Soldiers in Angola. In Özerdem, A. and Podder, S. (eds) *Child Soldiers: From Recruitment to Reintegration*. Basingstoke: Palgrave, pp. 359–87.
McMullin, J.R. (2013) *Ex-combatants and the Post-conflict State: Challenges of Reintegration*. Houndmills: Palgrave Macmillan.
Malo, S. (2022) The Green Dream to Rebuild a Sustainable Ukraine from the Rubble of War: 9 November. https://www.politico.com/news/magazine/2022/09/11/ukraine-russia-sustainability-00054910
Meredith, M. (2011) *The State of Africa: A History of the Continent Since Independence*. New York: Simon & Schuster.
Mkutu, K. (2008) Disarmament in Karamoja, Northern Uganda: Is this a solution for localised violent inter and intra-communal conflict? *Round Table: The Commonwealth Journal of International Affairs* 97(394): 99–120.
Muggah, R. (2005) No magic bullet: A critical perspective on disarmament, demobilization and reintegration. *Round Table: The Commonwealth Journal of International Affairs* 94(379).
Muggah, R. (2009) *Security and Post-Conflict Reconstruction: Dealing with Fighters in the Aftermath of War*. London: Routledge.
Münkler, H. (2004) *The New Wars*. Cambridge: Polity.

Nettlefield, L. (2010) *Courting Democracy in Bosnia and Herzegovina: The Hague Tribunal's Impact in a Postwar State*. Cambridge: Cambridge University Press.

Nordstrom, C. (1999) Girls and War Zones: Troubling Questions. In Indra, D. (ed) *Engendering Forced Migration: Theory and Practice*. New York: Berghahn, pp. 63–82.

Nordstrom, C. (2004) *Shadows of War: Violence, Power, and International Profiteering in the Twenty-first Century*. Berkeley, CA: University of California Press.

OECD (2007) *Handbook on 'Security Sector Reform'*. Paris: OECD.

OHR [Bosnian Office of the High Representative] (2023) HR Schmidt meets COM Chairwoman Krìsto, emphasises urgency for reforms and EU integration. 31 October. https://www.ohr.int/hr-schmidt-meets-com-chairwoman-kristo-emphasizes-urgency-for-reforms-and-eu-integration/ (accessed 6 November 2023).

Özerdem, A. (2008) *Becoming Civilian: Disarmament, Demobilization and Reintegration*. London: I.B. Tauris.

Özerdem, A. and Podder, S. (eds) (2011) *Child Soldiers: From Recruitment to Reintegration*. Basingstoke: Palgrave.

Paris, R. (2004) *At War's End: Building Peace After Civil Conflict*. Cambridge: Cambridge University Press.

Peters, K. (2007) Reintegration support for young ex-combatants: A right or a privilege? *International Migration*, *45*(5): 35–59.

Pitkin, H.F. (1967) *The Concept of Representation*. Berkeley, CA: University of California Press.

Plamenatz, J. (1960) *On Alien Rule and Self-Government*. London: Longmans.

Podder, S. (2012) From recruitment to reintegration: Communities and ex-combatants in post-conflict Liberia. *International Peacekeeping 19*(2): 186–202.

Pugh, M. and Cooper, N. with Goodhand, J. (2004) *War Economies in a Regional Context: Challenges of Transformation*. Boulder, CO: Lynne Rienner.

Richmond, O. and Carey, H. (eds) (2005) *Subcontracting Peace: The Challenges of NGO Peacebuilding*. Aldershot: Ashgate.

Robinson, K. /CFR (2023) What is Hamas? 31 October: https://www.cfr.org/backgrounder/what-hamas (accessed 6 November 2023).

Roeder, P.G. and Rothchild, D. (eds) (2005) *Sustainable Peace: Power and Democracy after Civil War*. Ithaca, NY and London: Cornell University Press.

Rogers, P. (2013) Lost cause: Consequences and implications of the war on terror. *Critical Studies on Terrorism* 6(1): 1–17.

Schneider, S. (2021) *The Apocalypse and the End of History: Modern Jihad and the Crisis of Liberalism*. London: Verso.

Seitz, T. (2018) *The Evolving Role of Nation-Building in US Foreign Policy: Lessons Learned, Lessons Lost*. Manchester: Manchester University Press.

Shepler, S. (2005) The rites of the child: Global discourses of youth and reintegrating child soldiers in Sierra Leone. *Journal of Human Rights* 4: 197–211.

Simms, B. (2014) *Europe: The Struggle for Supremacy, 1453 to the Present*. London: Penguin.

Sisk, T.D. (2013) Power-sharing in civil war: Puzzles of peacemaking and peacebuilding. *Civil Wars 15 Supplement 1*, 7–20.

Sky, E. (2015) *The Unravelling: High Hopes and Missed Opportunities in Iraq*. London: Atlantic Books.

Slomp, G. (2008) On Sovereignty. In Salmon, T., Imber, M. (eds) *Issues in International Relations*. London: Routledge.

Small Arms Survey (2015) *Global Burden of Armed Violence 2015: Every Body Counts*. Geneva: Small Arms Survey.
Snyder, S. (2003) *The Reconstruction of Nations: Poland, Ukraine, Lithuania, Belarus, 1569-1999*. New Haven, Conn: Yale University Press.
Stewart, R. (2014) *The Places in Between*. London: Picador.
Stewart, R. (2006) *Prince of the Marshes: And Other Occupational Hazards of a Year in Iraq*. London: Houghton Mifflin Harcourt.
Stiglitz, J. (2003) *Globalization and its Discontents*. New York: Norton.
Stiglitz, J. (2008) *The Three Trillion Dollar War: The True Cost of the Iraq Conflict*. New York: W.W. Norton.
Thomas, M. (2014) *Fight or Flight: Britain, France and their roads from empire*. Oxford: Oxford University Press.
Toal, G. and Dahlman, C.T. (2011) *Bosnia Remade: Ethnic Cleansing and Its Reversal*. Oxford: Oxford University Press.
Tripp, C. The local in Iraq. *Le Monde*. Available at: https://mondediplo.com/2008/01/02iraq (accessed on 9 December 2015).
United Nations Development Programme (2015) Post-2015 Sustainable Development Agenda. Available at: http://www.undp.org/content/undp/en/home/mdgoverview/post-2015-development-agenda/ (accessed on 19 July 2015).
United Nations Human Rights Commission (2015) *Report of the Commission of Inquiry on Human Rights in Eritrea*, A/HRC/29/42. New York: UNHRC.
Van Evera, S. (1999) *Causes of War: Power and the roots of conflict*. Cornell, NY: Cornell University Press.
Voller, Y. (2014) *The Kurdish Liberation Movement in Iraq: From insurgency to statehood*. London: Routledge.
Watson, A. (2015) 'Resilience is its own resistance: The place of children in post-conflict settlement', *Critical Studies on Security*, *13*(1): 47–61.
Wepundi, M., Nthiga, E., Kabuu E., Murray, R. and Alvazzi del Frate, A. (2012) *Availability of Small Arms and Perceptions of Security in Kenya: An assessment*. Geneva: Small Arms Survey.
Werbner, R. and Ranger, T. (eds) (1996) *Postcolonial Identities in Africa*. London: Zed.
Whitlock, C. (2021) *The Afghanistan Papers: A Secret History of the War*. New York: Simon and Schuster.
Wiafe-Amoako, F. and Mazrui, A. (2014) *Human Security and Sierra Leone's Post-Conflict Development*. Lanham, MD: Lexington.
Widner, J. (2005) Constitution writing and conflict resolution. *Round Table: The Commonwealth Journal of International Affairs* 94(381): 503–18.
Williams, A.J. (2005) 'Reconstruction' before the Marshall Plan. *Review of International Studies 31*: 541–58.
Williams, A.J. (2006, 2nd revised ed. 2024?) *Liralism and War: The victors and the vanquished*. London: Routledge.
Williams, A.J. (2007) Reconstruction: The bringing of peace and plenty or occult imperialism? *Global Society: Journal of Interdisciplinary International Relations 21*(4): 539–51.
Williams, A.J. (2020). *France, Britain and the United States in the 20th Century: A Reappraisal*. London: Palgrave Macmillan.
Winn, N. (ed.) (2004) *Neo-Medievalism and Civil Wars*. London: Frank Cass.

6 Development, aid and violent conflict

Introduction: how are aid and conflict prevention linked concepts?

The growth in public access to information, together with a rise in the number of people displaced by conflict and natural disasters (see for example, https://ucdp.uu.se/), have reified changes in policy making and political directions predicted in the second edition of this book. In 2015, we foresaw that following the end of the Cold War the link between aid, development and conflict would become increasingly significant. In May 2016, the UN Secretary-General convened the World Humanitarian Summit, attended by 55 heads of state, 'to deliver better for people caught in humanitarian crises' in response to the increasing inability of the international arena to protect populations in environments of contemporary conflict and complex emergencies.

First, this chapter aims to discuss the conundrum of just how difficult the delivery of aid can be. Contemporary wars are complex in the extreme in terms of the actors, the geography and the economic factor involved and many more factors. Counter-intuitively, many parties to conflict and war do not necessarily wish to 'end' it. As we have stressed in previous chapters, war can be extremely profitable for some actors, be they governments or 'rebels'. So humanitarian aid often has to be delivered whilst wars are ongoing, maybe with no end in sight. This is true of the wars in the Middle East (such as in Palestine, Iraq, Syria and Yemen) or elsewhere, such as Ukraine. The official governments may not wish aid to be delivered, as it is seen as helping 'rebels' to resist or because one or more sides in the conflict see this delivery as economically beneficial (Keen 2008). Whilst seeking to resolve dysfunctions in the humanitarian aid and development sectors that were increasingly seen to be excluding the voices of local actors, such considerations advanced the notion of a 'triple nexus' – an extension of the aid-development double nexus that added 'peacebuilding' to the continuum.

A second consideration of the chapter comes from changes in the 'official mind' of the international system as articulated through the

DOI: 10.4324/9781003353485-7

UN in the last few years. Since the start of the twenty-first century, reports from institutions such as the World Bank and the IMF have noted the economic progress of developing or 'low-income' countries to the status of developed or 'middle-income' countries. The continued prevalence of protracted conflicts has come despite this economic progress in line with the Sustainable Development Goals (SDGs) that the UN set itself in 2015 (UN/SDGs 2015). This in turn has led to the call for a donor focus that includes peace as part of the interlinked process for the provision of public goods, leading towards a decrease in violence and 'public bads' that accompany protracted conflict, such as the spread of disease, exploitation, and population movement (Carbonnier 2018).

Third: 'Peacebuilding' has long been core to the strategies of the UN and partners but was, in most cases, absent from the aid and development policies of the twentieth century, which linked relief, rehabilitation and development (LRRD). This was often because of the legacy of the post-war era were the UN emphasis was on 'rehabilitation'. Indeed, the first UN agency of any kind was the United Nations Relief and Rehabilitation Agency (UNRRA), set up under the aegis of the Roosevelt administration, even before the founding conference of the UN at San Francisco in 1945 (Williams 2006; Mayers 2018).

The 'humanitarians'

The inclusion of peacebuilding in the humanitarian lexicon is not without controversy. In the lead-up to the WHS, several of the major IGOs and INGOs operating on the frontlines of conflict, such as the International Committee of the Red Cross (ICRC) and Médecins Sans Frontières (MSF), presented objections to the potential blurring of lines between a humanitarian response to populations affected by conflict and the political dimensions of peacemaking (ICRC 2016; MSF 2016). Many academic and policy papers have indicated the importance of clarifying the divisions and similarities between the approaches (Carbonnier 2018; Tronc, Grace and Nahikian 2019; Dubois 2020; Südhoff, Hövelmann and Steinke 2020). A holistic approach that links aid, development and peace is the direction most donors in the Global North have chosen to follow, though the channels to direct assistance vary in policy and practice. Some prefer to follow the traditional patterns of Official Development Assistance (ODA) that operate from government to government. Others are choosing the 'New Way of Working' promoted at the WHS, which calls on aid and development actors to work together towards the Strategic Development Goals. Together, these different 'Humanitarians' represent heterogeneous ways of working in terms of their identity and approach even when they have broadly speaking the same ethos and goals.

A good example to illustrate this can be seen by looking at a very practical issue for many people who find themselves in a different country to the one where they originate, whether that be by 'voluntary' migration, 'illegal' migration or even 'forced' migration. In the case we have in mind, recent evaluations have highlighted the successes of more direct forms of aid delivery, notably cash transfers see (Box 6.1).

Box 6.1 Cash transfers

The increased availability of mobile technology, such as smartphones and other internet-linked devices, has led to innovative shifts in the way aid is delivered. Over past decades, the majority of emergency relief to populations affected by conflict or natural disaster was delivered as an 'in-kind' donation. For example, the US government would allocate a proportion of its cereal surplus to organizations such as World Food Programme (WFP) as a stockpile for the latter's response to humanitarian emergencies. Similar patterns of emergency stock-building occurred in non-food items, such as shelter materials (tents, blankets, etc.), public health and sanitation (medical equipment, pipes, water storage and purification units, etc.). Whilst the dynamics of war and displacement will never relieve the need for such in-kind support, when critical infrastructures have been lost or damaged or are absent, there is a recognition that critical services can be provided by local communities if affected communities have the financial resources to access these. Since the beginning of 2000, cash transfers have become increasingly used by UNHCR and WFP as a means to provide the means for access to goods and services, as well as developing self-reliance. Cash transfer projects were tested in response to refugee movements from conflict in Lebanon in 2006 and to displacement of populations following drought 2006–2010. Evaluations from such responses to Syrian refugees in Jordan and Lebanon indicated that the flexibility of cash-based assistance allowed families to meet self-identified needs and avoid negative coping strategies. As well as supporting local economies, people said they preferred cash, as it was more dignified and allowed them to prioritize and plan for the future, invest in economic opportunities and adapt to climate change (ALNAP, 2022).

This pattern of relief is increasingly the method of choice promoted by agencies and organizations such as Give Directly https://www.givedirectly.org/about/ the British Red Cross and members of IFRC https://cash-hub.org/

Sources: ALNAP (2022) The State of the Humanitarian System. ALNAP Study. London: ALNAP/ODI: 142–145; World Food Programme, 'WFP Evidence Summary Cash-Based Transfers Lessons from Evaluations' (Rome: WFP, 2021)

The characteristics of the wars that we have identified have now put humanitarian agencies on the front lines, both as distributors of conventional aid, such as foodstuffs, and also as potential participants, along with conventional government agencies and military forces, as well as paramilitary actors, in rebuilding 'failed states' (Ghani and Lockhart 2008). The actors that try to help alleviate the suffering of those increasingly caught up in current wars are collectively the 'humanitarians' in Hoffman and Weiss's term. 'External assistance' has become vitally important for governments and IGOs, and it has also become big business for firms who disburse or build infrastructure (Boyce and O'Donnell 2007). One important question that all who study the conflict-development-peace nexus are now asking is whether they can provide any kind of solution to the evident suffering of developing countries' populations or whether they are indeed part of the problem.

Consultants?

The same might be said about the reliance on consultancies by aid-donor countries. Can they really count as 'humanitarians'? After all, they explicitly 'do it for the money', do they not? There has been a growing belief that aid and capitalism have become too tied up with each other, a main theme of this chapter. However, the record is ambiguous when it comes to 'consultant' activity. A British organization called Carbon Brief found that over £2 billion of the British aid budget, representing 10 per cent of the total, was given to major consultancy companies such as KPMG and Adam Smith International as fees. The official aid figure was one already reduced by the Conservative government as part of its cost-saving plans in recent years. Many on the receiving end, such as the International Centre for Climate Change and Development in Bangladesh, have long harboured doubts about such use of resources 'which have not resulted in any real benefits after the international consultants leave the country' (Taylor 2023). In addition, many international organizations within the UN system charge a fee that is used to maintain their HQ or permanent local organizational structures, and in some cases, this can be a major part of the overall cost. Aid-donor governments also funnel a great deal of their aid budgets through the UN and other organizations, further reducing the actual amount of aid that is delivered. This is often in the name of accountability or 'value for money', laudable no doubt for the taxpayer. Charities have to run their organizations as well. So a much smaller part of the budget is spent on assistance to the country concerned than indicated by the headline figure. Does that necessarily have to be the case?

And of course, not all 'consultants' are the same, and some of their advice has been very welcome, even if not earthshattering in its effect. In addition, many small organizations and individuals deliver their services in the field at cost. One of the authors of this book went to Eritrea with a group of fellow academics over three proximate summers to advise on joining international organizations in the early 1990s at a cost to the country of $100 a day. UN *per diem* expense rates are several times higher. Whatever benefits to the country were rapidly overcome by the war with Ethiopia that ensued in 1998. Eritrea is a member state of most UN agencies. It belongs to the African Union and the Common Market for Eastern and Southern Africa. It has observer status in the Arab League. It suspended its membership in the regional IGO, the Intergovernmental Authority on Development (IGAD), in 2007. The UN reported both 'challenges' and a strong 'resilience' of the Eritrean state and economy. The country has very little Foreign Direct Investment ($73.8 million in 2021) and an appalling human rights record. It mainly trades with China, India and Iran and has been under US sanctions since 2021. This is in line both with Eritrea's long-standing obsession with self-reliance and its attitude towards what it sees as Western-dominated IGOs; the Eritrean government, led by Isaias Afeworki, has always expressed a dislike for IGOs and INGOs, possibly because the UN only recognized Eritrea's right to self-determination once it had comprehensively defeated Ethiopia in 1993 and have been seen as over-paid decadent foreigners ever since. (author observations; UN/Eritrea 2023).

More importantly for the theme of 'humanitarians' is that many doctors have given up their holidays with their Western health services to pass on their surgical or other skills to local medical practitioners on an entirely expenses-only basis. Neurosurgeon Henry Marsh spent many months in Ukraine treating locals for free (Marsh 2014). Many developed (and developing) country doctors have worked in the field for aid organizations like MSF and ICRC, often at great personal risk. David Nott is one such hero, often being forced to operate on patients whilst under fire (Nott 2019). These are good intentions by 'humanitarians' that never get mentioned in government statistics, as they come from civil society sources. The spirit of Henri Dunant is not dead yet.

The most obvious victims of wars in developing countries are civilians, but increasingly, so too are those who attempt to help them.[1] Hoffman and Weiss point out that not only are 'the victims of war … the actors' intended targets' but also '[h]umanitarian organizations have reacted to the new wars but have not adapted' (Hoffman and Weiss 2006: xvi–xvii). Other writers go further and see the aid agencies as part of a new pattern of global governance by the North whereby the populations of the South are made to feel their 'exclusion' in new ways and to enforce the ideological precepts of the liberal peace, in effect as agents of a new 'imperialism'. According to this view, underdevelopment has come to be seen as a security threat, not just a humanitarian disgrace, and consequently, all those who try to correct this insecurity are in effect complicit in such a logic (Duffield 2001, 2007; Easterly 2006). Equally problematic, these and other observers see aid as actually *fuelling* wars, keeping them going. Carolyn Nordstrom reports a conversation she had in Angola with a local youth during the civil war: 'Peace? Forget it, there's too much money being made here' (Nordstrom 2004: 191) as one example of such feelings. In short, aid since about 2001 has had a heavy security component; it has been 'securitized' (Duffield 2007) and is no longer seen as solely the charitable action it once was, and still is for many.

There is an increasing awareness that an unfocussed and indiscriminate distribution of aid can actually harm the state-building process, which is the stated prime objective of many IGOs and NGOs (OECD 2010). Since the withdrawal of interventionary forces in Afghanistan and Iraq and the upsurge of fighting in Syria, Yemen and Myanmar, all of which have happened in the period since 2014, there is also the problem that aid to these protracted conflicts has been drastically reduced, leaving societies made vulnerable by war even more exposed. Aid agencies face a crisis of access. They cannot access populations in need to assess the situation or deliver food or medicine.

We have therefore chosen to extract the idea of 'aid' from our wider discussion of conflict in developing countries because we believe that, like with issues of gender, health and other 'personal' issues, the problematic of aid's motivations and delivery gives us a powerful series of perspectives on what is right or wrong with current approaches to conflict and development. So aid provides another link to understanding the dynamics of conflict. It is undeniable that aid is a key part of the political economy of wars, and as the nature of wars changes, so does that of aid, as well as comments about it.

The purpose and history of aid

The *purpose* of aid has been defined, firstly, as an

> international social contract ... a broad understanding amongst developed countries that, in order for the world to be, or to be seen to be, a moderately equitable place, or at least to alleviate some of the worst suffering, there needs to be some form of international assistance.
>
> (Hunt, in Kingsbury et al. 2004)

Janet Hunt points out that, as well as this laudable aim, '[m]any donors provide aid not only for humanitarian reasons, but to enhance their own economic and political interests, through encouraging their own exports, or shaping the economic policies or political persuasion of recipient countries (Hunt, in Kingsbury et al. 2004: 67). It could be seen as yet another weapon in the arsenal of 'economic statecraft' to use the famous phrase coined by David Baldwin (Baldwin 1985).

It is usually seen as being part of a wider attempt to arm societies against negative influences – to develop their economies and in particular to help them pursue liberal democratic peace strategies. Since 1990, aid has been seen as using the 'anticipated 'peace dividend' to repair the ravages of the superpower competition in many war-torn and conflict-prone societies' (Forman and Patrick 2000: 2) and as an integral part of the liberal peace tool kit (Marriage 2008: 5–6).

Aid in historical perspective

Western scholarship is in general agreement that formalization of the modern humanitarian identity began with the foundation of the ICRC in 1863 and the promulgation of the first Geneva Convention the following year that confirmed international legal recognition of rights of protection and relief. The idea of mitigating human suffering occasioned by war and disaster evolved from ancient Classical expressions of communal responsibilities and later Renaissance, Reformation and Enlightenment theories of freedom, dignity and justice. These transformed into the humanitarian expressions we are familiar with today.

A Humanitarian Policy Group report into the history of the humanitarian system concurs that, in its broadest sense, 'humanitarian' action can be traced through hundreds of years of history across the world. It includes religion amongst the two most widely cited forces of influence, the other being around the articulation of the laws of war:

> Christian ideas of charity have been particularly important in Europe and North America, and scholarship has emphasised the importance of charitable gestures in other religions, including notably the tradition of *zakat* in Islam, one of several ways in which Islamic duty involves assisting others (Ghandour 2002; Benthall and Bellion-Jourdan 2003; Krafess 2005). Laws of war or limits on the acceptable conduct of war were adopted in ancient Greece and Rome; articulated in *The Art of War* ascribed to Sun Tzu in Warring States China; promoted by Saladin in the Middle East in the 1100s; taught to Swedish soldiers by Gustavus Adolphus in the 1600s; and recognised in the tenets of Hinduism, Islam and Judaism.
>
> (Davey, Borton and Foley 2013: 5)

The growth of the charity sector following the First and Second World Wars heralded the arrival of 'liberal humanitarianism'. In one of its first formulations as a result of the Marshall Plan (the full title of which was the 'European Recovery Program', initiated by the US Foreign Assistance Act of 1948), Marshall 'aid' was open to the same criticisms as those levelled against 'reconstruction' more recently. It was seen as politically motivated and had 'strings' attached which tied the recipient into an economic and political system that they did not necessarily want, and it was often only a sticking plaster on a big wound. More recently some scholars have suggested that aid became an extension of post-colonial power and even attitudes. Thomas MacCarthy has suggested that we can see the Western idea of development as intrinsically tied to liberal imperialism and more broadly about race. Hence the transition from a formal, land-based empire to a relationship with 'now formally independent, sovereign nations … [still left them] in fact unfree to control their own fates' (McCarthy 2009: x). Aid became part of the arsenal of weaponry in the Cold War, with both the West and the Soviet Union using aid to effectively buy allegiance in the developing world (Williams 2020; Engerman 2009).

Nonetheless, until the 1970s, 'aid' was seen by non-Marxists as a largely unproblematic extension of the idea of 'charity', helping out those less fortunate than oneself. As a child, I was told that any food I did not eat would be sent by parcel to help the 'poor children of Africa'. But in the 1970s, that critique got more intellectually problematic; it was now asserted that aid gave rise to 'dependency' and that was as bad a thing in LDCs as it was on the streets of London. Peter Bauer revived the idea that trade, not aid, was the key to development. He saw the positive examples of countries like Singapore and other 'newly industrialized countries' (NICs) forging ahead by opening up their economies and the progressive sclerosis affecting the socialist developing countries and those who still put their faith in handouts from the rich countries (Bauer 1991). This view was, and is, unpopular in some quarters – *The Guardian* described him as 'the shrillest Thatcherite spokesman against development aid for the third world' (*The Guardian* 6 May 2002)

Both Superpowers used the idea of 'aid' in their own ways to buy influence in the newly emerging developing world. The USSR set up 'radial' trade and development agreements (so called as they radiated from the hub of the Soviet Union), of which one of the most notorious was the provision of cultural, industrial and security assistance (including nuclear arms) to Communist Cuba, in return for Cuba providing much of the USSR's needs for sugar, tobacco and a security base in the Caribbean. The USA provided huge amounts of Marshall Aid to Western Europe and followed that up with technical assistance, soft loans, etc., as well as military 'advice' (Seitz 2012). This activity was predicated on the American success in rebuilding its own economy after the Great Depression. It also fell away after the debacle in

Vietnam in 1975 and the rise of the new economic orthodoxy of free markets epitomized by President Ronald Reagan and Prime Minister Margaret Thatcher (Ekbladh and Sutton in Fukuyama 2005).

With the growth of the charity sector came increasing professionalism of the humanitarian aid and development sectors. This gave rise to the emergence of minimum standards and codes of conduct to guide their action in the 1990s. Whilst these were mainly aimed at improving accountability, security and project quality for the benefit of communities affected by disaster, technical-level advice was frequently accompanied by messages around 'best practice' and 'good governance' that reflected the discourse of dominant stakeholders and donor interests of the Global North. The increase of actors and stakeholders in the sectors, together with a growth in the number and nature of international humanitarian responses, led to attempts by the UN and institutional donors to structure stable funding platforms, such as the Central Emergency Relief Fund (https://cerf.un.org) to which the UN agencies and its international NGO partners have access and the IFRC Disaster Relief Emergency Fund (https://www.ifrc.org/happening-now/emergencies/disaster-response-emergency-fund-dref), which channels emergency response funding to its member National Societies.

The economics and politics of aid: the case for intervention in developing countries

In the context of a book like this, a first set of questions about the case for intervention has to take into account both the economic and the political. These have tended to be put into separate disciplinary boxes, but such an approach can be seen as one-sided and prone to leading us to many misapprehensions and misconceptions.

The economic case for intervention: poverty leads to war

British Labour Party politician Jack Straw summed up why there is a widespread belief that poverty and conflict overlap:

> Look at where there are people living in poverty, on less than $3 a day and where there is conflict. The overlap is an exact fit. And look at where people live in prosperity and lack of conflict. Again the fit is exact.
>
> (Straw, BBC, 10 February 2007)

This has been progressively explored, especially since 2000, notably in the 'greed and grievance' debate developed by economist Paul Collier, who suggests that war is either fuelled by a sense of 'grievance' for real or perceived social, economic or political injustice or by a simple desire of one or more group within a conflict-torn society to benefit from the turmoil to amass the most resources possible. A classic example of this latter impulse was the accumulation of 'blood' diamonds by different groups in West Africa (Zoellner 2006; Bieri 2010).

Collier's contribution to the debate on aid includes, most notably, his book *The Bottom Billion*, which posited that aid had to be an important part of using 'carrots' and 'sticks' in encouraging developing countries towards better governance and development, the royal roads to better security (Collier 2007). As Collier put it in a talk at Monterey in May 2008, in his view, the mantra 'Aid – Trade – Security and Governance' [the 'ATSG Mantra'] must be followed in dealing with societies that have been stuck at the 'bottom', by which he means at the bottom of the economic and political pile, for 40 years (Collier 2008).

Collier's 'ATSG Mantra' was reinforced by major liberal economists like Jeffrey Sachs, whose equally uncompromising *The End of Poverty: Economic Possibilities for Our Time* (Sachs 2005) also picked up on the 'billion' souls to be saved from themselves. For both Collier and Sachs, the explicit model being emulated is that of the United States' actions in the 1940s. Europe was saved by aid, *ergo* Africa can be too. Sachs' work is best summarized in his own words: "War can … erupt as a result of the collapse of an impoverished society, one suffering the scourges of drought, hunger, lack of jobs, and lack of hope. Ending poverty is therefore a basic matter of our own security" (Sachs, Reith Lectures, BBC, 2007). Ending poverty is also the first of the UN SDGs' 15 goals. They are successors to the eight Millennium Development Goals that were formulated by the UN Secretary-General at the beginning of the twenty-first century, again with ending poverty as the top priority. The SDGs now frame progress towards the 2030 Agenda for Sustainable Development (UN 2022a). So there is no doubt that poverty eradication is the key idea.

But what is 'poverty'? The economist Amartya Sen developed arguments that saw access to a range of basic entitlements as a measurement of quality of life. Martha Nussbaum expanded on Sen's economic theories for development, formulating an approach 'focusing on what people are actually able to do and to be' (2006: 70). The goal for human welfare in her formulations of the approach is the right to a dignified life, and the goal for human welfare is the capability to achieve this. The right to dignity and respect (or, in Nussbaum's words, to 'a life worthy of living') form the core debates around contemporary expressions of human welfare that guide the development of the SDGs.

So does geography cause poverty and therefore war? In much of the writing on development, there is an implicit or explicit reference to the economic causes of conflict that has made it necessary, or to the potential healing power of economic action. Initially, all of these explanations ask what it is that makes for, in David Landes' (1998) term, the 'wealth and poverty of nations'. His essential point is incontestable – states and regions are differently endowed with economic assets and resources by virtue of nature's distribution of wealth. Landes points to the now somewhat discredited, but still valid, axiom that 'geography, especially climate, influence[es] human development' (Landes 1998: 3). It has been crudely suggested that hot climates can lead to less ardour in the workplace and the 'productivity of labor in tropical countries was reduced accordingly', a form of nineteenth-century environmental determinism that many would find distasteful. Disease, high morbidity, especially among infants, and poor water supplies have added to this obvious burden. (Landes 1998: Chapter 1 'Nature's Inequalities'. See also Chapter 1 for more discussion on the human side of underdevelopment.)

Classical economists (of whom Landes is one) give us to believe that states emerge from a natural process of a locally and internationally logical process of optimal allocation of resources, including land, capital and labour. Hence 'mature' economies have tended to go through a process of development. What is wrong with this? First, the 'mature' economies' all flourished in war and peace under a strong protective shield during their early development, whereas in contemporary times, most developing countries have been urged to compete in international markets without such a benefit. Second (and more importantly), societies do not just get to choose their economic policies in a rational way. Some societies, like Switzerland, have managed to survive and prosper in spite of having virtually no economic advantages – mostly low-grade and mountainous land areas endowed with poor natural resources, afflicted by linguistic and religious difficulties and surrounded by powerful and unfriendly neighbours, all factors that have on occasion engendered serious civil strife. Yet there are others, like Nigeria, the Democratic Republic of Congo and Russia, that are endowed with enormous natural resource bases and educated populations that seem to be in a state of perpetual strife. Some slightly 'populist' economists like Jared Diamond point to societies that implode through ecological hubris and even 'passivity' when faced with overwhelming challenges. Size is on occasion a benefit and in others a problem (Diamond 2005): a paradox of plenty (the 'resource curse') that can lead to lack of transparency, corruption and economic imbalance (the 'Dutch Disease': see also Further reading in this chapter and Chapter 1).

Of course, globalization might be said to have exacerbated both the differences and the potential benefits and disadvantages of such initial economic profiles. In the first wave of globalization in the nineteenth century, this became concretized in an imperial relationship often based on economic exploitation. In its earlier manifestations, in the nineteenth century, globalization has been identified by historians as being categorized as 'archaic, proto, modern and post-colonial' (Hopkins 2002). Another iteration of the same idea sees countries ranked as 'pre-modern', 'modern' and 'post- modern' based on how solid they are as nation states (Bradshaw 2016) Together, these categorizations link the economic and political. Russia, for example, is a 'modern' state with an arguably 'proto-modern' economic system based largely on the export of natural resources. The EU strives to be 'post-modern' in that it has, in some areas, pooled its sovereignty and has a vastly more diversified economy.

In the latest wave of globalization since the Second World War, the *international* economic relationship has arguably not changed much and the political has, and an ever more pronounced 'centre-periphery' relationship often based on the 'centre' being well endowed with capital and technology and the periphery with a wealth of primary resources and cheap labour. This is not a problem, or indeed a surprise to classical economists, but of course it does create tensions that often come out in apportioning blame for both underdevelopment and the ensuing conflict. Scholars have asked repeatedly if globalization therefore leads to 'convergence or divergence' of societies and economies (Hülsemeyer 2003).

Is it all about 'economic readjustment'?

The debate about whether the pursuit of market reform helps or hinders development in developing countries hinges on whether the market alleviates one of the main causes of conflict, widely identified as *poverty*, as noted earlier. The encouragement of trade plays a significant role in liberal thinking about comparative advantage (Adam Smith 1776), as well as that trade links encourage peace (Mill, Bright etc.). As Jacoby has succinctly put it, 'Living freely is thus trading freely' (Jacoby 2007: 524). So what is, potentially at least, wrong with the assumption that if we get the economics of a developing country emerging from war right, we will then get the politics right too?

To put it in a nutshell, if the recipients (or 'beneficiaries') of market reforms are of the clear opinion that they will not be better off as a result of them, they will tend to lump together their sense of grievance from whatever source that comes, with the message that their main persecution comes from a capitalist, globalized world system. If they are then told that they must comply with a policy of 'reconstruction', they will then reject the whole package. So, as with the *causes*, *escalation* and *maintenance* of conflict (Pugh and Cooper 2004) outlined in Chapter 1, so with the attempts to *end* it. We will explore the reasoning of why intervention might end up making conflict worse rather than better with a more in-depth examination of the following economic arguments.

In contrast to the 'Keynesian' interventionist, import-substitution, economic policies pursued by Western states and the IGOs alike until about 1980, a policy based on much more unfettered free-market principles prevailed after that date. This is often referred to as 'Reaganomics' (after President Ronald Reagan, 1980–8) and was also espoused by Prime Minister Margaret Thatcher (who served between 1979 and 1990). This way of thinking had profound effects on domestic economic policies across the West. It also had a huge impact on the thinking of international institutions like the World Bank and the IMF in the 1980s and led to the promotion of export-led growth by developing countries as opposed to the 'import-substitution' models previously fashionable. The result was that the 'self-reliance' models of the 1960s and 1970s popularized by economists like Raul Prebsich and institutions like the United Nations Conference on Trade and Development (UNCTAD) were jettisoned in the 1980s in favour of policies to integrate developing countries into the international trading system in a conscious imitation of the seeming successes of the NICs like Singapore and South Korea. This was the beginning of what we would now call 'neo-liberalism'. If the NICs could grow through exports and inward investment, why could not Ghana, India and Tanzania (for example) do the same (Toye 2014)? The answer was of course that the economic advantages of small entrepot states (Hong Kong and Singapore) and ones backed up by the might of the United States (such as Japan and South Korea) were quite different from those of poor, conflict-ridden states in Africa. But economic ideology and fashion did their work and the result was that aid (usually in the form of World Bank loans) was used to fund 'white elephant' investment projects across the Third World in the 1980s. Dams came to symbolize both the hope of this new vision of aid and also its failings (Usher 1997). Dams were often disastrously bad for the environment, drained foreign exchange reserves of the recipient states, encouraged corruption and often did not even produce much useful electricity, as with the British-sponsored Pergau Dam in Malaysia (Lankester 2012). The Aswan Dam in Egypt is one obvious example of both drawbacks, but there are many more. As we noted in Chapter 1, the Grand Ethiopian Renaissance Dam (GERD), which completed construction in 2022 and is estimated to take another year to fill, is a contemporary project causing concern among the neighbouring countries of Egypt and Sudan, which are highly reliant on water from the source of the Nile.

At much the same time, and for many of the same ideological reasons, forceful economic arguments against direct aid were made by writers like Bauer (Bauer 1991), who claimed it led to a form of welfare dependency similar to that experienced by benefit-dependent dwellers in Western inner-city areas, but the main drivers were political and came from Washington and London. The key effect was to limit the granting of direct aid and to make all loans (as for the dams) dependent on political and economic 'conditionality'. The result was a weakening of many developing countries' economies, so the end of the Cold War and the withdrawal of bilateral Soviet or American aid often led to their total collapse rather than the hoped-for transition to democracy. Rather than free governments and open economies, many developing countries have been left with crony capitalism, corrupt leaders, increased poverty and conflict.

At an institutional level, and to bring the story more up to date, the SDGs chart a path towards what is considered by the UN and its global partners to realize the links between welfare, the economy and security mentioned earlier and in the preceding chapters. In his foreword to the 2022 progress report, the UN Secretary-General closes by saying,

> To recover from the COVID-19 pandemic and deliver global sustainability, we need an urgent rescue effort for the SDGs. We must deliver on our commitments to support the world's most vulnerable people, communities and nations. Creating a global economy that works for all will require bold action. Greater investment in data infrastructure is needed to efficiently target investments now, anticipate future demands, avoid crises from descending into full-blown conflict and plan the urgent steps needed to achieve the 2030 agenda.
>
> (United Nations 2022b)

The call to tackle deprivation by external intervention is a reiteration of the classical view voiced by economists and political scientists in previous decades. There are many benign examples, such as the provision of educational programmes to all children in developing countries, with especially good impact on those targeting girls. Government-to-government support for sustainable development projects protecting health and the environment are necessary contributions. Direct international assistance in response to conflicts and natural disasters continues to play an important role. The shift in emphasis following the WHS in 2016 calling for 'a new way of working' has helped promote projects that engage greater local participation. Successes include the growth in cash assistance programmes which channel funds directly to vulnerable communities and have, to an increasing extent, replaced the in-kind assistance criticized by many as creating 'dependency' (pages 256-257). Also, on the negative side, increasing trends towards the engagement of expensive external consultancies to evaluate and advise on aid and development issues, with a focus on short-term projects, have been accused of devolving decision-making from the developing countries most affected by long-term threats such as climate change and migration (Taylor 2023).

Do no harm

Even before the WHS and the urge for a new way of working, writers like Mary B. Anderson believed that in certain circumstances, aid could indeed support peacebuilding activities but that carelessly applied aid would encourage damaging resource transfers and in effect disempower local people from taking control of their own destinies – summed up as 'do no harm' (Anderson 1999). Projects governed by the principle have received high-level backing from Western governmental institutions like Britain's Department for International Development (DFID, now incorporated into the FCDO). In one of these concentrating on 'conflict sensitivity', the logic is summed up as follows:

> Aid is not neutral in the midst of conflict. Aid and how it is administered can cause harm or can strengthen peace capacities in the midst of conflicted communities. All aid programmes involve the transfer of resources (food, shelter, water, health care, training, etc.) into a resource-scarce environment. Where people are in conflict, these resources represent power and wealth and they become an element of the conflict. Some people attempt to control and use aid resources to support their side of the conflict and to weaken the other side. If they are successful or if aid staff fail to recognise the impact of their programming decisions, aid can cause harm. However, the transfer of resources and the manner in which staff conduct the programmes can strengthen local capacities for peace, build on connectors that bring communities together, and reduce the divisions and sources of tensions that can lead to destructive conflict.
>
> (Conflict Sensitivity Forum, 2008–12)

The fact that this common-sense judgement has now been removed from the hybrid FCO/DFID website maybe says volumes about how the debate within the British government has moved in the period since our second edition in 2016.

Anderson's ideas have certainly struck a chord with those who believe that local solutions to local problems are more likely to be workable and effective than ones imposed by outsiders. It would also be true to say that the free marketeers (like Bauer) have largely won the day on one aspect of aid, in that it is now appreciated that throwing money into economies that are incapable of using it in productive ways is not very sensible. But it is also appreciated that aid that goes into countries that are deemed to have a track record of sound market-oriented policies can deliver effective, sustained policy reforms. Of course, this illustrates how we can see aid as another support for the overarching ideology of the liberal peace.

It also has been widely decided by Western policymakers and IGOs that, as we stated at the beginning of this chapter and earlier in the book, the experiences of the 1990s (especially the abortive intervention in Somalia) and the subsequent events after '9/11' show that development is now a security issue. There was 'a mounting perception and articulation that underdevelopment was dangerous and that – by implication – arising the level of development would increase security in the country and ultimately globally' (Schnabel and Carment 2004, Vol 2).

Aid and 'human security'

Much aid in the 1990s and since started to be directed towards DDR programmes with the aim of increasing 'human security', an ideal of seeing development and security as closely interlinked, which is an unexceptional statement but which needs much greater clarity to operationalize (Cooper, Pugh, and Turner 2010; Howe 2013). The concept is often explicitly linked to conflict prevention, with the idea that if the population's security needs can be provided for, the rest of their economic and social existence will be ensured. Aid and conflict prevention measures in general should therefore 'focus on long term, structural challenges to build safe, just, and stable societies' (Schnabel in Carment and Schnabel 2004: 109–131). One criticism that can be levelled against the concept of human security is that it also seems to have an inherent belief that basic 'human needs' can indeed be identified, a claim that we have seen in relation to conflict resolution techniques and proposals in Chapter 4. Other commentators on human security are not sanguine as to its usefulness. Marriage comments that it could be seen as so 'infinitely elastic' a concept as to be 'analytically unhelpful. None the less – or maybe because of this – its popularity in policy – making circles has been more enduring' (Cooper, Pugh, and Turner 2010; Marriage 2008: 4 and 6; Duffield 2001).

'Peacebuilding' has been added to the classical aid and development continuum, creating a 'triple nexus' to address the divide that sees armed conflict as an obstacle to development under the assumption that there can be no development without security (and vice versa). This addition is not without its critics. Many humanitarians working at the minimalist end of the spectrum, where the focus is protecting life and dignity, and who are already uncomfortable with an expanding interlinkage with the maximalists working on issues of structural violence, consider that the inclusion of peace to the spectrum increases the divide (Introduction and Further reading).

The organization of aid

The aid-development spectrum, with the addition of 'peace' since the WHS 2016, is usually framed as being part of a humanitarian system. Many academics and practitioners argue that a coherent system has never existed, with the organization of aid being organized through a loosely defined network of stakeholders, dominated in most of the modern period by clubs of humanitarian donors associated with the liberal peace. The most recent ALNAP report on the 'State of the Humanitarian System' indicates that, in 2021, 57 per cent of international humanitarian assistance came from just five public donors (the United States, EU institutions, Germany, the United Kingdom, Japan and 47 per cent of aid went directly to three UN agencies: WFP, UNHCR, UNICEF). Only 1.2 per cent of international aid funding went directly to national and local actors NGOs (ALNAP 2022).

ALNAP defines the humanitarian system as 'the network of interconnected institutional and operational entities that receive funds, directly or indirectly from public donors and private sources, to enhance, support or substitute for in-country responses in the provision of humanitarian assistance and protection to a population in crisis'. Whilst recognizing this functional diversity of stakeholder interests, a more encompassing perspective is of a shared 'spirit' to attend to the welfare and rescue of humans in distress, seen as a universal expression across history and cultures (Gordon-Gibson 2021).

In accordance with these changing paradigms, since the end of the Cold War, this 'humanitarian network' has evolved to deal with both the results of conflicts and aid and reconstruction efforts alike. Since the start of the 'War on Terror' in 2001, conflicts in Afghanistan, Iraq and elsewhere also show that the delivery of aid for purely humanitarian reasons has become in effect even more subordinate to geopolitics than it was during the Cold War. Marriage comments that 'there is abundant evidence that the motivations of donors are mixed, and that aid and other interventions are often tempered with blindness or misunderstanding' (Marriage 2008: 2). What, for example, has been the net result of the huge amount of aid that has been, and continues to be, pumped into the Palestinian territories for the local population? How can it be effective when the macro-political situation (the failed Oslo Accords, the Intifada and the brutal simmering war between Israel, Hamas and Hezbollah) means that the aid funds which were intended to bolster the peace process are often of limited effect? The same might be said about responses in Afghanistan and Iraq. Chris Wake paraphrases President Clinton on this case – 'no peace deal can be sustainable if it does not genuinely respond to the needs of everyday people' (Wake 2008: 109–130).

The actors in the delivery of aid are also those present in the full-blown reconstruction efforts discussed in Chapter 5. Donors and deliverers of aid can be bi- or multi-lateral, most usually charitable and sometimes denominational. Humanitarian aid can be delivered by international organizations, such as the UN, and international and local NGOs, as well as the Red Cross and Red Crescent Movement, which includes the ICRC, which provides assistance and protection of civilians affected by conflict, as well as prisoners of war, through the long-established Geneva Conventions, and the International Federation of Red Cross and Red Crescent Societies (IFRC), which responds to natural and man-made disasters through its membership of (currently) 191 National Societies across the world. The UN coordinates its activities through the United Nations Office for the Coordination of Humanitarian Affairs (OCHA; https://www.unocha.org).

Hoffman and Weiss (2006) refer to this 'humanitarian network' as a constellation of national, IGO, NGO and private contractors that surround a country in crisis, with its own government, local NGOs and victims/recipients of aid' (Hoffman and Weiss 2006: 122). The operation of this network obviously varies from case to case. In some areas, there will be a great deal of NGO activity; in others, there will be very little, as when there is a very poor security situation. To take a particular case, that of Sierra Leone, here all the above actors were present in a conflict which is both regional, in that it affects the whole of West Africa, but also with specific elements due to the nature of the country itself. The ending of the civil war was particularly engineered by the United Kingdom's armed forces and the post-war situation by its development agencies and aid disbursers, although the United States has funded and mainly ran the UN Special Court for Sierra Leone, seen as a major part of the post-war reconciliation process (see Chapter 4) NGOs in Sierra Leone and in Liberia have made great strides in bringing 'human security' to the populations of these two states. No one would dispute the need to get rid of the appalling regimes of Charles Taylor in Liberia and Foday Sankoh in Sierra Leone (Atkinson 2008). But surely, it would be prudent to say that the installation of democratic governments in these two places as a result of the expenditure of huge amounts of money by Western states has not really addressed the underlying problems of these countries, which are not so much about democracy as about economic and social hardship and inequality?

One sure conclusion that has to be drawn from the present stage of capitalist development is that where there is no viable state structure, the vacuum will tend to be filled by external and internal groups, usually armed to the teeth, and by the immense monetary capabilities of transnational corporations, who will do practically anything to ensure access to their raw material needs. The most obvious example of this is the trade in 'blood' or 'conflict diamonds', a subject that has spawned Hollywood films but has yet to produce any obviously effective codes of conduct for the extraction and end use of such products. The Kimberley Process, which was meant to certify the origin of diamonds and prevent conflict diamonds from entering the market, is widely seen as a failure. Diamonds, which often grace the necks and hands of the most beautiful women in the West, also lead to the hacking off of equally beautiful necks and hands of women in parts of Africa see (Box 6.2).

Box 6.2 Blood diamonds and Sierra Leone

The state of Sierra Leone, well endowed with natural resources, including diamonds, but with a succession of weak and corrupt governments, imploded with the end of the Cold War and the emergence of warlords like Foday Sankoh (and Charles Taylor in nearby Liberia) who used terror to extract and monopolize the exploitation of 'blood' diamonds. The international community recognized that markets in the West made such terror profitable and instigated the 'Kimberley Process' to identify and control such processes. This can be seen as being a useful and 'positive start ... but only a start'. The key variable has to be seen as the need for a regional, or even global, and not purely national, solution to the dissemination of such materials. The links between the illegal and violent export of such goods have even been explicitly linked to the funding of globally focussed groups like Al-Qaeda (Grant and Taylor 2004: 399). Maybe this will be what finally leads to their control, not the suffering of the Sierra Leonean and West African population in general. Sankoh was overthrown by Western-backed mercenaries and British troops, but an uneasy truce is all that can be said to be holding (source: Gberie 2005; Harris 2013; Keen 2005; Richards and Vincent in de Zeeuw 2008).

The problems of aid dispersal

From the 1960s, states and international organizations like the World Food Programme (http://www.wfp.org/about), the World Bank and the IMF became the key distributors of aid, increasingly backed up, and after 1990, they were increasingly supported by NGOs like OXFAM. ODA was intended to reach 1 per cent of Gross National Income (GNI), a UN target set in the 1980s. However, OECD countries' contributions had, by the end of the 1980s, only increased ODA to about 0.36 per cent of GNI, with the honourable exceptions of states, especially in Scandinavia and Britain, that managed to donate up to 0.7 per cent. These sums were often dependent on economic changes, such as the application of 'structural adjustment programmes', and political conditionalities, such as crackdowns on corruption, the implementation of election monitoring, emerging security concerns and re-balancing of budgets following unexpected occurrences, such as the COVID-19 pandemic. For example, in November 2020, the UK government cut the amount available to spend on international development from 0.7 per cent to 0.5 per cent of GNI as a "temporary measure" pending annual reviews against fiscal forecasts. A restoration of the aid budget is not expected for several years (UK Parliament 2022). In many cases, promises of aid were only partially delivered (Forman and Patrick 2000). Marc Williams' comment of 1994 that 'no clear standards exist by which to measure aid effectiveness and to determine under what conditions aid is likely to promote growth' (Williams 1994: 22–23) is still true today. In spite of this, it is often believed that aid will alleviate poverty and help to solve the conundrum outlined earlier by Jeffrey Sachs.

Why is this? First, aid cannot be delivered without security, usually in very difficult circumstances (see Plate 17) and is often accompanied by corruption. This applies as much to the providers of aid, like the United Nations, as to the receivers. It has been suggested that in effect

Plate 17 Relief convoy blocked by snow. Copyright ICRC. Photographer: Cédric Piralla. 01/02/1992.

an 'aid economy' grows up in parallel to a 'war economy', 'where the focus is not so much on benefiting from violence as it is on taking advantage of efforts to relieve suffering' (Hoffman and Weiss 2006: 107). In some conflicts, such as Bosnia and Afghanistan, maybe in most, aid can fall into the hands of warlords who exploit their position on the ground in an insecure environment to extort 'protection money' from aid agencies (Goodhand, in Barakat 2004). Similar dynamics are reported in the ongoing conflicts in Syria and Ukraine, where corruption, extortion and diversion of aid are a well-documented occurrence by both state and non-state authorities. Second, in purely economic terms, an obvious unintended result of providing (for example) many tonnes of free foodstuffs to alleviate a famine situation is to make local production of such foodstuffs uneconomic. Third, aid agencies inevitably create distortions in local employment patterns, as they can offer better salaries to drivers and translators as well as a more general workforce than the local economy can (Weiss 2013; Gordon-Gibson 2016). In some cases, like the Palestinian Authority (PA), virtually the whole economy has become dependent on outside assistance for public-sector salaries and, in that case, two-thirds of all government expenditures (Boyce and O'Donnell 2007: 200). Here aid is a positive incentive to corruption, which has arguably had unforeseen political results, including the rise of Hamas as an alternative to the PA in one section of Palestine (the Gaza Strip) and the consequent breaking up of the state.

Hoffman and Weiss argue that recent thinking among Western governments about the role of the private and public sectors (outlined earlier as a post-Keynesian consensus) has even led to aid being 'privatized' to some extent, to the point where for-profit organizations have been deployed in an effort to eradicate some of the 'inefficiencies' of the UN and public-sector organizations (see Box 6.4 PRTs). As disasters become progressively more frequent and complex, better collaboration through partnerships with private businesses becomes more important. Following the Indian Ocean tsunamis in 2004, engagement of the private sector in humanitarian operations grew substantially, notably in the provision of support services such as logistics, supply-chain management and information technology and communications (ICT) equipment (Cozzolino 2021). Critics fear that growing trends in private-sector engagement might result in a lack of accountability and weaken the application of humanitarian principles, as well as increase donor control, though competition for funding appears over-stated (Binder and Witte 2007). Commercial providers can respond more successfully and quickly, but they risk suffering from a lack of legitimacy as they are seen for what they are: profit-making enterprises (Hoffman and Weiss 2006: 152–3).

There is also a clear problem of 'turf wars' developing between IGOs, NGOs and governments and competition between stakeholders in the sectors. Aside from the differing approaches and operational policies across the aid-development-peace spectrum, described earlier, the quest for funding means actors jealously defending their place for visibility on the humanitarian landscape. Although this has been a constant presence, following the arrival of 24-hour news and instant messaging, this dynamic has become more acute. Field reports and policy evaluations have often referred to the 'media circus' that follows a high-visibility crisis (Gordon-Gibson 2016; Chernobrov 2018; Lough 2022). There are attempts to restructure the landscape, including the 'One United Nations Initiative', launched by the UN Secretary-General in 2006 to strengthen the management and coordination of the UN in the field of development, humanitarian assistance and the environment. This initiative has been largely taken over by a less integrated focus in pursuit of the SDGs. A shifting trend of stakeholder participation and donor agency from the Global North to the Global South may change the landscape, but similar patterns of competition are likely to remain.

In what David Keen aptly describes as 'complex emergencies' (Keen 2008), having so many players on the field is bound to lead to them duplicating each other, competing for scarce resources, in effect becoming part of the system that is a civil war. One solution that has been suggested that may help address this problem harks back to an older model. Rather than country-based policies, we arguably need regional policies, as with the Marshall Plan for Europe, or indeed the UNRRA, the first UN organization that dispersed aid and helped displaced people on a global scale in the period 1943–6 (see Williams 2006: 113–22 and above: 259). After all, the 'new' wars do not respect national boundaries any more than 'old' ones did, and the economic issues of the regions affected have much in common (as in Western Africa) (Pugh and Cooper 2004: 80 and *passim*).

Closer engagement with regional, national and sub-national structures was projected to be one of the main outcomes of the WHS in 2016, with declarations of the 'Grand Bargain' and a 'New Way of Working' that committed to getting more means into the hands of people in need and to improve the effectiveness and efficiency of humanitarian action. These included a target of 25 per cent of humanitarian funding to be channelled directly to local and national actors, and a series of 'collective outcomes' that marked progress towards a more coherent regional and country-based humanitarian strategy (see under Further reading). Realizing these goals has proven more problematic: to date, less than 2 per cent of funding is channelled through local actors, and equitable partnerships remain distant. However, regional platforms such as the Association of South-East Asian Nations (ASEAN) and the AU have become important negotiating partners with the UN and IOs to approve projects and gain humanitarian access see Box 6.3).

Box 6.3 Cyclone Nargis, 2008

Cyclone Nargis, a Category 4 storm that left over 130,000 dead, struck the south-east of Myanmar (Burma) on 2–3 May 2008. Myanmar was a country that had been economically and politically isolated by the Global North since the violent suppression by its military government of the pro-democracy uprising 20 years earlier. Concern for the humanitarian response at the time focussed on problems of access. Early reports from the UN agencies in the country indicated they had been able to reach only 30 per cent of those in desperate need of food and warned of an outbreak of infectious disease. The Inter-Agency Real-Time Evaluation observes that although there were certainly very real restrictions, local groups and some international organizations were able to access affected areas immediately after the cyclone.

A meeting of ASEAN foreign ministers shortly after the cyclone struck led to an agreement to form the Tripartite Core Group (TCG), comprising the government of Myanmar (chair), ASEAN and the UN, to act as an ASEAN-led mechanism to facilitate trust, confidence and cooperation between Myanmar and the international community in the urgent humanitarian relief and recovery work after Cyclone Nargis hit.

The involvement of ASEAN was a factor that played a critical role at the central level in opening channels of negotiation with the government of Myanmar. Nevertheless, strict controls over participation and access for international personnel and the definition of projects at the municipal level remained a serious constraint. Travel restrictions and bureaucratic regulations from the central government were reinforced to assure the status quo, which municipal authorities in the cyclone-affected townships were unable to circumvent. Through the Myanmar Red Cross Society (MRCS), the IFRC obtained a letter from the office of the Prime Minister in the capital Nay Pyi Taw that authorized the broad parameters of its proposed humanitarian response. Once this authorization was presented at the municipal level, a fear of noncompliance with central regulations was replaced by pro-active cooperation and an eagerness to define and approve humanitarian projects at an operational level in the affected townships according to the needs identified by the local communities, among which local government officials and the military authorities were part. This meant procedural restrictions to access were no longer an issue. Organization of the humanitarian response became a representation of the local community, managing projects that were relevant to them and to the authorities responsible for monitoring them.

Sources: Barber 2009; Gordon-Gibson 2021; Selth 2008; Turner, Baker, Myo Oo and Soe Aye (2008)

A final initial point for consideration might also be that it has to be asked if the weakest are not in effect the least protected. Tales of rape and child abuse have surfaced in a number of UN-run or sponsored camps in recent years, as far apart as Bosnia and Sierra Leone. The problem has been recognized as acute by the UN, with the report of an independent review on sexual exploitation and abuse by international peacekeeping forces in the Central African Republic published in June 2016 (UN – CAR 2016).

In addition, 'non-standard' refugees tend to get the least attention from aid disbursement agencies. Keen's research in an earlier Sudanese conflict (in 1984–5) led him to believe that pastoralists (who are a very common population group in much of Africa) and rural dwellers generally get the worst deal in terms of distribution and help, as do more generally disfavoured sections of any society. The telling parallel, he believes, is that of the relief so badly delivered to New Orleans in 2005 (where most people who were neglected were black) or to Thailand and Sri Lanka after the Tsunami Disaster of 2004, in which relief was much better delivered to town dwellers and those living in obviously popular tourist zones (Keen 2008: 121–5).

The problems of protecting aid workers

Another result of the 'new wars' and also of the 'war on terror' has been the breakdown of the idea that 'host' governments can or should protect aid workers to anywhere the same extent that used to be considered normal. A new phenomenon has emerged, that of the 'armed humanitarian', who is seen by some hostile locals as not only delivering aid but also a message of alien 'democracy' and Western mores. For a conservative Afghan elder, bringing education to the women of his area, as UNICEF has done, for example, is not a 'neutral' act; it is one that threatens his legitimacy as a lawmaker and the very culture of his society. More generally, we have seen a blurring of the distinction between civil and military actors, even if we must not forget that such distinctions have always to some extent existed. The most obvious example of this blurring is the use of private security companies to defend aid workers and in many cases to replace regular military forces altogether, as in Iraq (Kinsey 2009). In the case of Afghanistan, Provisional Reconstruction Teams (PRTs) often had mixed civilian and other non-military personnel with soldiers, as Michael McNerney puts it, not so much a 'model' as a 'muddle' (for more on this, see McNerney 2005–6).

Box 6.4 PRTs – Iraq

In Iraq, PRTs were civilian-military organizations designed to operate in semi-permissive environments, with an aim to achieve political objectives, counter-terrorism and promote social and economic development. US secretary of state Condoleezza Rice inaugurated the first PRT during a visit to Mosul on November 11, 2005.

In Iraq, PRTs were larger than the original design that originated in Afghanistan and almost entirely civilian, led by a senior State Department foreign service officer and composed of teams from USAID and the state, justice and agriculture departments, plus contractors and Iraqi experts. Military participation was limited to the deputy leader, liaison officers and civil affairs soldiers who performed various tasks. PRTs were located on US military bases and relied upon military escorts for security. At the programme's high point, there were 25 PRTs in Iraq, of which three were led by the United Kingdom, Italy and South Korea. The last PRT closed in late 2011 following the withdrawal of US military forces from Iraq.

In 2007, President Bush announced a 'surge' of US military forces to Iraq and the creation of ten new-style PRTs, known as "ePRTs," that would be embedded with the deploying combat brigade teams. These new PRTs were small, State Department–led units that included eight to 12 political advisers and development experts that supported counterinsurgency operations and helped build the capacity of local governments to foster development. Members of these ePRTs lived with their military counterparts, worked with mayors and grassroots representatives, and functioned as members of the brigade commander's staff. By 2008, the number of ePRTs had risen to 14, but the programme began a rapid drawdown with the departure of the 'surge' brigades. The last ePRT closed in August 2010.

Source: https://www.usip.org/publications/2013/03/provincial-reconstruction-teams-iraq

As a result, journalists, UN staff, and other 'neutrals' have found they are now, like it or not, seen as legitimate targets by warlords. This has been the case in Afghanistan and Iraq when the UN briefly withdrew its entire staff, humanitarian and otherwise, after the car bomb that killed over 40 UN staff in 2003. This meant they could only effectively operate from bases inside the security 'green zone' in Baghdad. Such 'bunkerization' may have the advantage of being more secure but also means the UN and partners are less comprehending of the local people and their circumstances (Chandrasekaran 2007). Violence against aid workers has also been acute in the conflict in Syria, with publicized displays of performative violence on journalists and aid workers performed most infamously by ISIS, as well as the bombing of protected facilities, such as hospitals and schools, by both state and non-state actors. The Aid Worker Security Report 2022, compiled from research including interviews with aid practitioners working in Myanmar, Ethiopia (Tigray), Ukraine, Syria and Yemen, finds that attacks on aid workers in 2021 claimed more lives than in any year since 2013. It notes that the risk of indiscriminate violence such as airstrikes, shelling, rocket attacks and remnants of war are likely to drive up casualty numbers in 2022, notably following Russia's invasion of Ukraine (Stoddard et al. 2021).

Many have been killed or kidnapped. Some have suggested that reducing the dissemination of small arms could help solve the problem (Howard 2008: 44), which seems unlikely given the difficulties that such an enterprise would pose in, say, Afghanistan, where possession of a weapon is seen as a sign of basic social status.

Does aid 'work'?

The economic debate since 2000 has started to move against aid as a good idea to help countries both develop and improve their security environments. The word 'trouble' crops up a lot – *The Trouble with Africa* (Calderisi 2006), *The Trouble with Aid* (Glennie 2008), as well as *Culture Troubles* (Chabal and Daloz 2006). Books with increasingly long and complicated titles have proliferated, like William Easterley's *The White Man's Burden: Why the West's Efforts to Aid the Rest Have Done So Much Ill And So Little Good* (Easterly 2007) and Giles Bolton's *Aid and Other Dirty Business: An Insider Reveals How Good Intentions Have Failed The World's Poor* (2007), and *Africa Doesn't Matter: How the West Has Failed the Poorest Continent and What We Can Do About It* (Bolton 2008).

These latter titles represent an impressive attack from authors who see aid as a form of 'postmodern imperialism', to use Easterly's phrase (Easterly 2006 Chapter 8). All of these writers see military intervention and aid as parts of a liberal peace package that has failed to solve the problems of development and conflict, although not always from the same political perspective.

The basic problem is the one articulated by Roger Riddell: *Does Foreign Aid Really Work?* He finds that development aid needs to be drastically 'recast' not eliminated (Riddell 2008 Chapter 15). And the basic solution is summed up in the much remarked upon book by Dambisa Moyo, that aid is somehow *Dead Aid*, that, in the approving words of a preface by Niall Ferguson, it 'encourages corruption and conflict, while at the same time discouraging free enterprise.' So, the logic continues, we need to reduce, or even eliminate it, will bring untold benefits to African (and other developing) economies and polities, cutting off the blood supply to the tumours that are poverty and war. Aid, says Moyo, kills off the potential for agricultural countries to develop viable export sectors by dumping huge quantities of 'free' produce on local markets, destroying the incentives for local market development. Worse still, the West's obsession with democracy and good governance, latterly reinforced by visits by rock stars of the 'glamour aid' variety has had the effect of 'atrophying' the debate on aid (Moyo 2009: 27).

Other critics of aid, like Linda Polman, go even further and assert that aid not only fuels wars but also prolongs them, an accusation that led to a furious polemic from Bob Geldof, the organizer of the 'Live Aid' concerts which helped deliver aid in the Ethiopian famine of the 1980s. Other songs released by Geldof since then have received unwelcome criticism along the same lines. One of Geldof's more printable comments was: 'It's a pop song, it's not a doctoral thesis. [Critics] can f*** off.' (Polman 2010, 2011; Geldof 2014).

Conclusion: does aid make conflict worse?

Many of the assumptions about peacebuilding take it as axiomatic that the development of each country takes a similar course, much in line with the classical view of development. Hence, the reconstruction that we looked at in Chapter 5 is to install, or re-install, proper systems of governance and economic development. Even those who believe that such peacebuilding methods are beneficial are aware of the difficulties involved. Roland Paris warns against believing that 'war-shattered states can be hurriedly rehabilitated' (Paris 2004: ix). We might even worry that aid can transform power structures and facilitate the emergence of warlord politics (see Do No Harm, earlier). This has been seen as the unhappy experience in Afghanistan in the 1980s and followed a similar sorry trajectory through the 1990s and into the 2000s (Goodhand, in Barakat 2004).

As we have seen earlier, others go much further and blame aid disbursement for the evolution of predatory capitalism itself and for the complicity of the international community in its worst excesses. Though these views (summed up in Does Aid Work) are certainly at the extreme end of the spectrum of criticism of current aid policies pursued by Western governments, IGOs and NGOs alike, the main critique has been of the underlying logic of a 'liberal peace' paradigm that only sees what it wants to see, supposedly governments fairly distributing aid to needy people without fear or ideological favour. This is seen as naïve and ultimately self-defeating. Mark Duffield, one major critic of aid, has also accused IGOs and development agencies more broadly of a crude 're-packaging of these aims over the past half-century' that has done nothing to improve the underlying logic or the delivery of aid on the ground. (Duffield 2007: 12). The underlying logic for him has been to provide a 'surplus population created through accumulation by dispossession [that] represents life belonging to capitalism'. In this logic, what is happening in Darfur, for example, is in the interests of capitalist development but not of the local population. The lens of inquiry is turning towards alternative patterns and partnerships, including those such as China's 'Belt and Road Initiative' and the emerging platforms of donors and practitioners in the Middle East, presented in Chapter 2. The question is whether or how far the evolving trajectories will differ from the pathways crafted by the liberal peace.

The critique bears a certain resemblance to that of other writers like Cramer (2006) and Keen (2008), who perceive a variety of errors that derive from the West seeing the problems of development as being those of the aftermath of wars rather than as due to the problems of a system of internal exploitation of one group by another in most countries that are examined. In effect, slavery has been replaced by a different form of rapacious global capitalism. No solution will, or can, therefore be found without realizing that we need to identify the issues within the political economies of developing countries before there can be any hope of addressing the long-term problems of both development and the conflicts that emerge from this, as the main underlying problems are ones of governance and dominance in developing countries. Neither, as both Cramer and Keen make very clear, can we get far in understanding these conflicts without an appreciation of the importance of violence in development. It is, in Keen's words, a difficult issue of walking the 'tight-rope between explaining and excusing' (Keen 2008: 5). Violence was a great part of the development of Western polities and economies; it is a part of those now undergoing a process of development. Maybe therefore aid can only ever be a sticking plaster on a necessary process of self-harm? Recent scholarship and policy debate have sought to identify the boundaries of humanitarian action and if there is a clearer case for defining 'red lines' than has been possible – or made to feel acceptable –in past constructions of the liberal peace narrative. The return of the Taliban to authority in Afghanistan and its edicts over the role of women has rekindled debates around the ethics of humanitarian support where abuse of globally recognized (UN) constructs of human rights are to the fore. These echo criticism of unclear red lines in humanitarian engagement with state authority during the conflict in Syria and remain an obstacle in response to recovery and reconstruction at the present time – including emergency response to communities in northern Syria affected by the recent earthquakes. Much of the analysis is around contemporary interpretations of the humanitarian identity in a social and political environment much changed from its formulation in the middle of the nineteenth and early twentieth centuries. The core principles of impartiality, neutrality and independence that have guided humanitarian action since then and which were adopted formally by the UN in 1991 have come increasingly under scrutiny in the swirl of political and populist forces of the twenty-first century. The idea of absolute neutrality has come under particular question when abuse of power and human rights from state and non-state authorities have been so publicly on display in many contemporary conflicts. The idea of 'solidarity' with affected communities is gaining ground in the modern humanitarian lexicon, with calls for a rethink of the humanitarian identity and its interface with politics and power. This searches for a place where the humanitarian can at once cooperate with power but challenge its authority when its principles are seen to be compromised (Gordon-Gibson 2021).

These writers all see the issues involved in development as part of a much wider nexus of liberal governance that perpetuates the North-South dependencies of the colonial period, except that now the main agents of dominance are not governments but Western multinational organizations and IGOs that in effect do their bidding, 'the suppliers, facilitators and cultural sustainers of inhumanity' (Duffield 2007, quoting Slim 1997). The emergence of the New Wars has both facilitated and been facilitated by the borderless nature of post–Cold War capitalism to create what Duffield calls 'network war' (Duffield 2001: 260) or Keen calls 'abusive war systems' (Keen 2008: 9). Aid is for them therefore an essential part of this 'network' or 'system' and must be reformed along with them if it is to help resolve or abate conflict.

The vital questions therefore have to be as to whether aid is only a palliative for much wider problems created by a post–Cold War capitalism that now finds itself unimpeded by Great Power interests and helped by the slow but sure disintegration of the local state structures into a system of local clientelisms that do not respect borders of national loyalties, but only the allure of money. The 'new' wars have created a new entrepreneurial class that serves powerful outside capitalist interests (for example, in the extraction of diamonds in West Africa) but is unimpeded by local state authorities. In such a situation, aid distribution and the organization of refugee assistance becomes just more grist to local warlord interests. It might help assuage the consciences of Western governments and their populations who want to 'do something' to help the starving peoples of the South, but in effect, it does the opposite. The botched intervention in Somalia in the early 1990s may be said to be a paradigmatic example of that problem.

May we therefore go back to the question that is asked from opposite perspectives by the conservative views of Luttwak and his urging to 'give war a chance' (Luttwak 1999), and the much more radical views of Cramer and Duffield, or the more moderate views of de Waal, Keen,

Hoffman and Weiss? Is aid making the conflicts in Africa and elsewhere worse? Would it not maybe be better not to distribute it at all? Is the problem not the one it has always been – outside interference?

Conversely, to follow such advice absolutely would be a counsel of despair, as countless more people would die as a result. In spite of their shortcomings, the international community has defined 'SDGs' to abolish poverty (or even number 1: 'End poverty in all its forms everywhere'). The Group of Seven (G7, formerly G8 until Russia was suspended in March 2014 over its involvement in the Ukraine crisis) has also made fulsome pledges to reduce poverty. Maybe most important of all, there is a growing awareness among ordinary people in the West that the problems of the developing countries are their problems and that they are willing to support political decisions to help their less fortunate sisters and brothers. More controversially, new donors are emerging, especially in the Middle East, China and South Korea (Mawdsley 2012: Chaturvedi 2012), who now hold many of the dollar surpluses as a result of globalized trade patterns and other factors. Just as there are new threats and new partnerships in the contemporary environments of aid and development, so too there must be a 'new way of thinking' to accompany the new way of working called for at the WHS in 2016 – one which disrupts the comfortable patterns of behaviour and 'rehumanizes the humanitarian narratives' (Gordon-Gibson 2021: 17). The consensus so far seems to be that these donors are making many of the same errors as the 'old' ones. We have to hope that the debate outlined in this chapter will throw up some better ideas for the aid debaters of the future.

Summary

- Aid – its delivery and logic – has become a key element in understanding contemporary conflicts, and their potential transformation, in the developing world.
- The discussion of aid elicits very different reactions from analysts of development, often depending on the cases they have studied but also as a result of their intellectual and policy approach.
- Aid is not 'neutral', either in its political logic or distribution.
- The period since the end of the Cold War has seen aid becoming 'securitized'.
- The delivery of aid often requires making difficult moral and political compromises.
- A 'new way of thinking' is beginning to emerge that is critical of the neo-liberal paradigms of the aid and development 'system', which calls for a closer engagement with non-traditional stakeholders and recognition of local capacities.

Discussion questions

- Can a consideration of the motives for and the delivery of aid help us in an understanding of the dynamics of contemporary conflicts and wars in the developing world?
- What moral and practical problems are there in the delivery of aid?
- Does aid help in peacebuilding?
- Why have aid workers now often become targets in developing country conflicts?
- Should we 'give war a chance'?

Note

1 According to the Aid Workers Security Database, 2020 marked the second year in a row of record attacks on aid workers (484), with 117 recorded deaths. By far the majority of victims are national staff and volunteers https://aidworkersecurity.org/ Projections for 2023-24 are higher.

Further reading

A comprehensive collection that covers important aspects of the discussion about aid is: Mac Ginty, R. and Petersen, J. (eds) (2015) *The Routledge Companion to Humanitarian Action*, London: Routledge. See also: Boyce, J.K. and O'Donnell, M. (2007) *Peace and the Public Purse: Economic Policies for Postwar Statebuilding*, Boulder and London: Lynne Reinner. For a general critique of aid see: Duffield, M. (2001) *Global Governance and the New Wars: The Merging of Development and Security*, London: Zed, Books; Duffield, M. (2007) *Development, Security and Unending War: Governing the World of Peoples*, Cambridge: Polity; Keen, D., (2008) *Complex Emergencies*, Cambridge, Polity, and Easterly, W. (2006) *The White Man's Burden: Why the West's Efforts to Aid the Rest Have Done So Much Ill and So Little Good*, London: Penguin, 2006. Works casting a critical gaze on humanitarian action include Terry, F (2002) Condemned to Repeat?: The Paradox of Humanitarian Action, Ithaca: Cornell University Press, Mukesh Kapila https://www.thenationalnews.com/opinion/comment/2022/12/28/why-is-modern-humanitarianism-failing-even-as-we-give-more-and-more-aid/ and studies presented by Médecins Sans Frontières, the Centre de Réflexion sur l'Action et les Savoirs Humanitaires (CRASH) https://msf-crash.org/en

On the 'Dutch Disease' and its relationship with aid, see: https://resourcegovernance.org/sites/default/files/nrgi_Resource-Curse.pdf

For background and information regarding the World Humanitarian Summit and 'The Grand Bargain' see https://interagencystandingcommittee.org/content/grand-bargain-hosted-iasc; https://www.un.org/jsc/ and https://interagencystandingcommittee.org/grand-bargain-official-website/letter-grand-bargain-ep-jan-egeland-january-2023

There are several platforms offering engaged discussions on the 'triple nexus, including the ICRC webpage 'Law and Policy' https://blogs.icrc.org/law-and-policy/2018/10/04/revisiting-nexus-numbers-principles-issue-social-change/ ALNAP https://www.alnap.org/the-nexus-current-status-and-discourse and the Centre for Humanitarian Action https://www.chaberlin.org/en/topics/triple-nexus-en/#:~:text=The%20declared%20goal%20of%20the,%2C%20development%20cooperation%2C%20and%20peacebuilding

Useful websites (all accessed as at 12 August 2023)

ICRC weblog 'Humanitarian Law and Policy' https://blogs.icrc.org/law-and-policy/features/ provides a broad selection of current debates around conflict and development, including links to related discussions on other websites. Articles on International Development Policy published by the Geneva Graduate Institute, such as: Gilles Carbonnier, "Official development assistance once more under fire from critics", *International Development Policy, Revue internationale de politique de développement*, 1 | 2010, 137–142 https://journals.openedition.org/poldev/141

On 'Do No Harm' and institutions that have signed up for the principle (this site includes many links to other organizations engaged in thinking about or delivering aid in conflict zones), see: http://www.conflictsensitivity.org. Also see (online and in print): OECD (2010), *Do No Harm: International Support for Statebuilding*, Paris, OECD. A good summary of the current disasters and responses to them can be found at: 'Relief Web': http://reliefweb.int. 'Human security' is best discussed at: http://www.humansecurityinitiative.org/definition-human-security. The United Nations Office for the Coordination of Humanitarian Affairs (OCHA) can be found at: https://www.unocha.org, and the World Food Programme at: (http://www.wfp.org/). Some useful websites on agencies that disburse aid: http://www.usaid.gov/locations/, [USAID] http://eur-lex.europa.eu/legal-content/EN/TXT/?uri=URISERV:dv0015 [EU] https://www.gov.uk/government/organisations/department-for-international-development [DFID, British Government agency] http://www.oxfam.org.uk [OXFAM]

Bibliography

ALNAP (2022) The State of the Humanitarian System, 2022 edition. https://sohs.alnap.org/sohs-2022-report (accessed 9 November 2023).

Anderson, M. (1999) *Do No Harm: How Aid Can Support Peace – or War*. Boulder, CO: Lynne Rienner.

Atkinson, P. (2008) Liberal interventionism in Liberia: Towards a tentatively just approach? *Conflict, Security and Development* 8(1): 15–45.

Baldwin, D. (1985) *Economic Statecraft*. Princeton, NJ: Princeton University Press.

Barakat S. (ed.) (2004) *Reconstructing War-Torn Societies: Afghanistan*. London: Palgrave Macmillan.
Barber, R. (2009) 'The responsibility to protect the survivors of natural disaster: Cyclone Nargis, a case study' *Journal of Conflict & Security Law* 14(1): 3–34.
Bauer, P. (1991) *The Development Frontier: Essays in Applied Economics*. Brighton: Harvester Wheatsheaf.
Benthall, J. and Bellion-Jourdan, J. (2003) *The Charitable Crescent: Politics of Aid in the Muslim World*. London: I. B. Tauris.
Bieri, F. (2010) *From Blood Diamonds to the Kimberley Process: How NGOs Cleaned up the Global Diamond Industry*. Farnham: Ashgate.
Binder, A. and Witte, J.M. (2007) Business engagement in humanitarian relief: Key trends and policy implications. A Humanitarian Policy Group Background Paper: ODI and GPPi.
Bolton, G. (2007) *Aid and Other Dirty Business: An Insider Reveals How Good Intentions Have Failed the World's Poor*. London: Ebury Press.
Bolton, G. (2008) *Africa Doesn't Matter: How the West Has Failed the Poorest Continent and What We Can Do About It*. London: Arcade Books.
Boyce, J.K. and O'Donnell, M. (2007) *Peace and the Public Purse: Economic Policies for Postwar Statebuilding*. Boulder, CO: Lynne Rienner.
Bradshaw, C. (2016) *Bloody Nations: Moral Dilemmas for Nations, States and International Relations*. London: Routledge.
Buchanan, C. and Muggah, R. (2005) *No Relief: Surveying the Effects of Gun Violence on Humanitarian and Development Personnel*. Centre for Humanitarian Dialogue and Small Arms Survey, UNDP. https://hdcentre.org/insights/no-relief-surveying-the-effects-of-gun-violence-on-humanitarian-and-development-personnel/ (accessed 8 November 2023).
Calderisi, R. (2006) *The Trouble with Africa: Why Foreign Aid Isn't Working*. London: Macmillan.
Carbonnier, G. (2018) Revisiting the nexus: Numbers, principles and the issue of social change. ICRC Humanitarian Law and Policy https://blogs.icrc.org/law-and-policy/2018/10/04/revisiting-nexus-numbers-principles-issue-social-change/
Chabal, P. and Daloz, J.-P. (2006) *Culture Troubles: Politics and the Interpretation of Meaning*. London: Hurst.
Chandler, D. and Hynek, N. (eds) *Critical Perspectives on Human Security: Rethinking Emancipation and Power in International Relations*. London: Routledge, pp. 83–96.
Chandrasekaran, R. (2007) *Imperial Life in the Emerald City: Inside Iraq's Green Zone*. London: Bloomsbury.
Chaturvedi, S. (2012) *Development Cooperation and Emerging Powers: New Partners or Old Patterns*. London: Zed.
Chernobrov, D. (2018) Digital volunteer networks and humanitarian crisis reporting. *Digital Journalism*, 6(7): 928–44.
Collier, P. (2007) *The Bottom Billion: Why the Poorest Countries Are Failing and What Can Be Done About It*. Oxford: Oxford University Press.
Collier, P. (2008) The bottom billion. https://www.ted.com/talks/paul_collier_the_bottom_billion?language=en (accessed 8 November 2015).
Cooper, N., Pugh, N. and Turner, M. (2010) Has human security had its day?
Cozzolino, A. (2021) Platforms enhancing the engagement of the private sector in humanitarian relief operations. *Sustainability*, 13, 3024.
Cramer, C. (2006) *Civil War Is Not a Stupid Thing: Accounting for Violence in Developing Countries*. London: Hurst.

Davey, E., Borton, J., and Foley, M. (2013) A history of the humanitarian system: Western origins and foundations. Humanitarian Policy Group, Working Paper. London: ODI.
De Zeeuw J. (ed.) (2008) *Transforming Rebel Movements After Civil War*. Boulder, CO: Lynne Rienner.
Diamond, J. (2005) *Collapse: How Societies Choose to Fail or Survive*. London: Penguin.
Dubois, M. (2020) The Triple Nexus – Threat Or Opportunity For The Humanitarian Principles? Centre for Humanitarian Action
Duffield, M. (2001) *Global Governance and the New Wars: The Merging of Development and Security*. London: Zed.
Duffield, M. (2007) *Development, Security and Unending War: Governing the World of Peoples*. Cambridge: Polity.
Easterly, W. (2006) *The White Man's Burden: Why the West's Efforts to Aid the Rest Have Done So Much Ill and So Little Good*. London: Penguin.
Engerman, D. (2009) *Know Your Enemy: The Rise and Fall of America's Soviet Experts*. Oxford; Oxford University Press.
Forman, S. and Patrick, S. (eds) (2000) *Good Intentions: Pledges of Aid for Postconflict Recovery*. Boulder, CO: Lynne Rienner.
Fukuyama, F. (2005) *State Building: Governance and World Order in the Twenty-First Century*. London: Profile.
Gberie, L. (2005) *A Dirty War in West Africa: The R.U.F. and the destruction of Sierra Leone*. London: Hurst.
Geldof, B. (2014) Bob Geldof tells band aid critics to 'F*** off!'. *Independent*, 9 December. [This page has been removed from the website, alas]
Ghandour, A.-R. (2002) *Jihad humanitaire: enquête sur les ONG islamiques*. Paris: Flammarion.
Ghani, A. and Lockhart, C. (2008) *Fixing Failed States: A Framework for Rebuilding a Fractured World*. Oxford: Oxford University Press.
Glennie, J. (2008) *The Trouble With Aid: Why Less Could Mean More for Africa*. London: Zed.
Gordon-Gibson, A. (2016) Goma 1994: Notes from the field. *Genocide Studies International*, *10*(2): 254–67.
Gordon-Gibson, A. (2021). *Humanitarians on the Frontier: Identity and Access along the Borders of Power*. Rowman & Littlefield.
Grant, J.A. and Taylor, I. (2004) Global governance and conflict diamonds: The Kimberley Process and the quest for clean gems. *Round Table: The Commonwealth Journal of International Affairs* *93*(375): 385–401.
Guardian (2002) *Lord Bauer*. http://www.theguardian.com/news/2002/may/06/guardianobituaries.obituaries (accessed 9 November 2023).
Harris, D. (2013) *Sierra Leone: A Political History*. London: Hurst.
Hoffman, P.J. and Weiss, T.G. (2006) *Sword and Salve: Confronting New Wars and Humanitarian Crises*. Lanham, MD: Rowman and Littlefield.
Hopkins, A.G. (2002) *Globalization in World History*. London: Pimlico.
Howard, A. (2008) Small arms vs humanitarian aid: Opportunities for action. www.aidandtrade.org/review-2007-2008 (accessed 18 July 2008).
Howe, B. (2013) *The Protection and Promotion of Human Security in East Asia*. Houndmills: Palgrave Macmillan.
Hülsemeyer, A. (ed) (2003) *Globalization in the Twenty-first Century: Convergence or Divergence*? London: Palgrave.
Hunt, J. (2004) Aid and Development. In Kingsbury, D., Remenyi, J., McKay, J., Hunt, J. (eds) *Key Issues in Development*. Basingstoke: Palgrave Macmillan, pp. 67–90.

ICRC (2016) ICRC's comment on the UN secretary-general's report, 22 April. https://www.icrc.org/en/document/world-humanitarian-summit-icrcs-comment-un-secretary-generals-report (accessed 9 November 2023).

Jacoby, T. (2007) Hegemony, modernisation and post-war reconstruction. *Global Society: Journal of Interdisciplinary International Relations 21*(4): 521–37.

Keen, D. (2005) *Conflict and Collusion in Sierra Leone*. London: James Currey.

Keen, D. (2008) *Complex Emergencies*. Cambridge: Polity.

Kinsey C. (2009) *Private Security and the Reconstruction of Iraq*. London: Routledge.

Krafess, J. (2005) L'influence de la religion musulmane dans l'aide humanitaire. Revue internationale de la Croix-Rouge, vol. 87, sélection française.

Landes, D. (1998) *The Wealth and Poverty of Nations: Why Some Are So Rich and Some Are So Poor*. London: Abacus.

Lankester, T. (2012) *The Politics and Economics of Britain's Foreign Aid: The Pergau Dam Affair*. London: Routledge.

Le More, A. (2008) *International Assistance to the Palestinians after Oslo: Political Guilt, Wasted Money*. London: Routledge.

Lough, O. (2022) *Social Media and Inclusion in Humanitarian Response*. HPG working paper. London: ODI https://odi.org/en/publications/social-media-and-inclusion-in-humanitarian-response-and-action/ (accessed 8 November 2023).

Luttwak, E. (1999) Give war a chance. *Foreign Affairs 78*(4).

Marriage, Z. (2008) Ambiguous agreements: Aid in negotiating processes. *Conflict Security and Development 8*(1): 1–13.

Marsh, H. (2014) *Do No Marm: Stories of Life, Death and Brain Surgery*. Weidenfield and Nicolson.

Mawdsley, E. (2012) *From Recipients to Donors: Emerging Powers and the Changing Development Landscape*. London: Zed.

Mayers, D. (2018) *America and the Postwar World: Remaking International Society, 1945–1956*. London: Routledge.

McCarthy, T. (2009) *Race, Empire and the Idea of Human Development*. Cambridge: Cambridge University Press.

McNerney, M. (2005–6) Stabilization and reconstruction in Afghanistan: Are PRTs a model or a muddle? *Parameters Winter*: 32–6.

Médecins Sans Frontières (2016) MSF to pull out of World humanitarian summit, 5 May. https://www.msf.org/msf-pull-out-world-humanitarian-summit

Moyo, D. (2009) *Dead Aid: Why Aid Is Not Working and How There Is Another Way for Africa*. London: Allen Lane.

Nordstrom, C. (2004) *Shadows of War: Violence, Power, and International Profiteering in the Twenty-First Century*. Berkeley, CA: University of California Press.

Nott, D. (2019) *War Doctor: Surgery on the Front Line*, London, Picador

OECD (2010) *Do No Harm: International Support for Statebuilding*. Paris: OECD.

Paris, R. (2004) *At War's End: Building Peace after Civil Conflict*. Cambridge: Cambridge University Press.

Polman, L. (2010) *The Crisis Caravan: What's Wrong with Humanitarian Aid?* New York: Metropolitan Books.

Polman, L. (2011) *War Games: The Story of Aid and War in Modern Times*. London: Viking.

Pugh, M. and Cooper, N. with Goodhand, J. (2004) *War Economies in a Regional Context: Challenges of Transformation*. Boulder, CO: Lynne Rienner.

Riddell, R.C. (2008) *Does Foreign Aid Really Work?* Oxford: Oxford University Press.

Sachs, J. (2005) *The End of Poverty: Economic Possibilities for Our Time*. London: Penguin.
Sachs, J. (2007) Economic solidarity for a crowded planet. Reith Lectures 2007. www.bbc.co.uk/radio4/reith2007/lecture4.shtml (accessed 27 July 2015).
Schnabel, A. and Carment, D. (eds) (2004) *Conflict Prevention for Rhetoric to Reality: Volume 1: Organizations and Institutions; Volume 2: Opportunities and Innovations; Volume 3: Pacific Settlement of International Disputes*. Lanham, MD: Lexington.
Seitz, T. (2012) *The Evolving Role of Nation-Building in US Foreign Policy: Lessons Learned, Lessons Lost*. Manchester: Manchester University Press.
Selth, A. (2008) Even paranoids have enemies: Cyclone Nargis and Myanmar's fears of invasion, *Contemporary Southeast Asia: A Journal of International and Strategic Affairs 30*(3): 379–402.
Slim, H. (1997) *Doing the Right Thing: Relief Agencies, Moral Dilemmas and Moral Responsibility in Political Emergencies and War*. Uppsala: Nordic Africa Institute.
Stoddard, A., Harvey, P., Czwarno, M. and Breckenridge, M.-J. (2021). Aid worker security report 2021: Crime risks and responses in humanitarian operations. *Humanitarian Outcomes*. https://www.humanitarianoutcomes.org/AWSR2021
Südhoff, R., Hövelmann, S., and Steinke, A. (2020). *The Triple Nexus In Practice: Challenges and Options for Multi-Mandated Organisations*. Centre for Humanitarian Action.
Taghdisi-Rad, S. (2011) *The Political Economy of Aid in Palestine: Relief from Conflict or Development Delayed?* London: Routledge.
Tartir, A. (2015) Book review: The political economy of aid in Palestine: Relief from conflict or development delayed? by Sahar Taghdisi-Rad. http://blogs.lse.ac.uk/lsereviewofbooks/2012/07/19/book-review-the-political-economy-of-aid-in-palestine-relief-from-conflict-or-development-delayed/ (accessed 8 November 2023).
Taylor, M. (2023) More than £2bn of UK foreign climate aid channelled through consultancies since 2010. *The Guardian*, 13 October. https://www.theguardian.com/environment/2023/oct/13/uk-foreign-aid-climate-through-private-consultancies
Toye, J. (2014) *UNCTAD at 50: A short history*. UNCTAD: Geneva.
Tronc, E., Grace, R. and Nahikian, A. (2019) Realities and myths of the "Triple Nexus". Local Perspectives on Peacebuilding, Development, and Humanitarian Action in Mali. Harvard Humanitarian Initiative/ATHA.
Turner, R., Baker, J., Oo Z.M. and Aye, N.S. (2008) *Inter Agency Real Time Evaluation of the Response to Cyclone Nargis*, UN OCHA.
UK Parliament (2022) 16 November 2022 Use of the aid budget to support refugees. https://committees.parliament.uk/committee/98/international-development-committee/news/174414/use-of-the-aid-budget-to-support-refugees-in-the-uk-new-inquiry-launched/
UN – CAR (2016) Report of an independent review on sexual exploitation and abuse by international peacekeeping forces in the Central African Republic: Note/by the Secretary-General: https://digitallibrary.un.org/record/840749?ln=en (accessed 9 November 2023).
UNdata (2023) (Palestine) http://data.un.org/countryData/Data/ShowDetail/PSE (accessed 23 October 2023).
United Nations (2015) The 17 goals. https://sdgs.un.org/goals (accessed 8 November 2023).
United Nations (2021) 2021 Annual Country Results Report: Eritrea: https://eritrea.un.org/sites/default/files/2022-04/2021%20UN%20Annual%20Country%20Results%20Report%20Eritrea.pdf (accessed 8 November 2023).

United Nations (2022a) Transforming our world: The 2030 agenda for sustainable development https://sdgs.un.org/2030agenda (accessed 8 November 2023).
United Nations (2022b) The sustainable development goals report, 2022. https://reliefweb.int/report/world/sustainable-development-goals-report-2022
US Department of State (2023) Eritrea: https://www.state.gov/countries-areas/eritrea/#:~:text=the%20United%20States.-,International%20Organizations,status%20in%20the%20Arab%20League (accessed 8 November 2023).
Usher A. (ed.) (1997) *Dams as Aid*. London: Routledge.
Wake, C. (2008) An unaided peace? The consequences of international aid in the Oslo peace process. *Conflict Security and Development 8*(1): 109–31.
Weiss, T. (2013). *Humanitarian Business*. Polity Press.
Williams, A.J. (2006) *Liberalism and War: The Victors and the Vanquished*. London: Routledge.
Williams, A.J. (2020) *France, Britain and the United States: A Reappraisal, Vol. 2, 1940–1961*. London: Palgrave Macmillan.
Williams, M. (1994) *International Economic Organisations and the Third World*. New York: Harvester Wheatsheaf.
World Humanitarian Summit (2016) https://ucdp.uu.se/ (accessed 7 November 2023).
Zoellner, T. (2006). *The Heartless Stone: A Journey Through the World of Diamonds, Deceit and Desire*. New York: St. Martin's Press.

Conclusion

The third edition of this book and its predecessors were intended to give those who are relatively new to the fields of both conflict and development a better idea about the debates that we consider to be the most important in understanding the relationship between them. We are aware that we cannot and will not have satisfied all those who read it, as the debates we have summarized are neither the only conceivable ones we could have considered nor by any means those that some would have chosen to emphasize. We are also aware that there are many different approaches that we could have taken from purely relating what the intergovernmental organizations (IGOs) have been doing in conflict zones in an uncritical manner through to a hard-line 'critical' approach which would take as its point of departure the premise that all Western governments and IGOs are primarily concerned with continuing their former imperialist ventures in a post-colonial environment. To be sure, we do rather veer toward the latter position than toward the former, a tendency that is shown by our privileging of certain discourses over others.

We are not alone in our change of emphasis. There is also a growing awareness among IGOs themselves of a need to rethink global and local development policies in the context of a changing geopolitical and moral landscape. The Ukraine and Gaza wars of 2022 and 2023 are one impetus behind this process, as are the destabilizing effects of new actors in development situations and macro problems like the climate crisis. Over the past decade, there has been a growing recognition of the shortcomings of the aid and development architecture constructed by the Global North, brought into a mainstream discussion following the World Humanitarian Summit in 2016, referred to extensively in this book. We also freely admit that without IGOs, there can be no delivery of better policies; without Western governments, there will not be access to the resources and organizational abilities made available to deliver those policies; without some form of self-interest being manifest, there will be no incentive to think about what these policies might be. Globalization has indeed produced many curious antipathies, synergies and disjunctures in thinking and action about

DOI: 10.4324/9781003353485-8

development. We all need to understand what those relationships are if we are to improve our game in both the analysis of conflict and the processes of development. In short, we are faced with a 'crisis of imagination' (Box 7.1) in the way we think about the issues we have discussed in this book. How the academic and policy community deal with this will determine the credibility of both for the near and more distant future.

Box 7.1 A crisis of imagination

When faced with a sudden crisis, the words 'do something' are invoked repeatedly, but the nature of the 'something' is rarely defined. When faced with threats to human life, the tendency is to invoke military intervention. Military intervention is lazy international relations and displays a lack of imagination. Its proponents ask superficial questions about the crises at hand and respond with what is often the only tool that has guaranteed funding and ideological support. This preference for militarism reflects the fact that politicians treat the military as superior to other state structures. The idea that the military is beyond civilian scrutiny is pervasive. The crisis of imagination begins in countries that always have money for war and guns but not for schools and hospitals, and it ends in a global politics where leaders are always ready to reach for soldiers. The writer and activist Nanjala Nyabola relates the aforementioned to the repeated failures in addressing the protracted humanitarian crisis in Haiti. In her opinion,

> [A] polycrisis like Haiti's requires a level of nuance that militaries are simply not equipped to deal with. The crisis is generational and therefore in-and-out interventions will not do, and the odds that anything less than a decade of engagement will be required are negligible. It is a crisis rooted in history that requires in-depth knowledge of the nation and its place in the world. The crisis requires environmental and social interventions – and not just the neoliberal belief that the market, if created, will provide material wealth that will magically redistribute itself throughout the society. It demands a level of humanism that cannot be found at the end of a gun. To address the issues facing Haiti would require a commitment to walk a long, demanding path with a traumatized people and fly in the face of the logics of statehood and white saviorism, let alone international relations, that we take for granted today. Military intervention cannot do that.

Source: Nanjala Nyabola (2023), The Crisis of Imagination in International Relations. *The Nation*. https://www.thenation.com/article/world/international-relations-haiti-militarism/

The approach we have taken in this book has been one premised on the idea that there are serious problems with the way that the twin processes of development and the conflict which is so often attendant on development interact with each other. We do not in any way deny that there has been a huge amount of goodwill and effort put in over many decades by development economists, workers in international organizations and even politicians at every level of the global system to try and ameliorate the conditions of under-development and war in which so many of the people on this planet exist. It is true that the 'Millennium Development Goals' and their successor, the 'Sustainable Development Goals', set by the UN and other agencies, can point to some successes. Absolute poverty levels are down worldwide and, in some places, like China, dramatically so. There is less violence if that is assessed across the developing world, but it is more concentrated than in 2009 (cf. Chapter 4).

But it is also clear that there is a growing chorus of criticism of many 'neo-liberal' development policies not just in the developing world. Aid policies are coming under increasing scrutiny for both their effect and for their intent, with accusations of a slow response to the emerging humanitarian impacts of climate change. This is also true of the continuing debate about the aftermath of the interventions and wars in Afghanistan, Syria, Yemen and Ethiopia, and the responses to forced displacement in Myanmar. Liberalism itself is more criticized than it was eight years ago, both globally and locally, as the rise of populist and nationalist movements in Europe also shows. But we also hope to have shown here that the reflexive nature of liberalism is its greatest strength. It is only because liberalism has been able to constantly reinvent itself that it still dominates global thinking about and the praxis of international relations (Williams 2006, 2024 *forthcoming*).

We certainly do not exempt our own profession of conflict analysis from our worries about some clear policy and intellectual failures. Those who study conflict have, until quite recently, tended to ignore the problems of development. They have also tended to be stuck in the study of very small-scale examples and to take a 'tourist' interest in what is often abject horror and suffering. Very few analysts of conflict are, for example, prepared to spend more than a few months studying a conflict, visiting the area and developing a thorough understanding of its dynamisms over many years. Understanding requires empathy, and empathy requires a lot of contact with the real world of conflict (something that risk-averse universities and insurers frown upon). So if this book has a primary purpose, it is to get a wider debate going between these different constituencies so that the delivery of aid, the arrangements of post-war conflict situations and thinking about how related issues might be improved upon in future.

A conclusion should not be the place to re-visit in detail the conclusions at the end of each chapter but rather to draw out some of the overarching concerns that could be useful in directing future research and action. What might be some of these overarching concerns?

The first is that we are very adamant in our feeling that we need to approach the pitfalls and huge advantages presented by a more rigorous use of historical examples in the study of conflict and development. Aficionados of the 2016 edition will notice that we have included more historical and contemporary literature, especially as it relates to the evolving crisis in the Middle East, and an appreciation of the interaction between the humanitarian sector and its political and institutional stakeholders, with the contested debates around the aid-development-peace 'triple nexus'. We have attempted to better establish clear genealogies of how and why certain terms are privileged in the theoretical and policy discourse, as we will never understand why these terms are seen in such different ways by their imposers and recipients. *Local* understandings of historical discourses play out in a conflict (Chabal and Daloz 2006), not just our *Western* readings of them. The obvious example that we have talked about at some length is that of 'reconstruction' (Chapter 4). The war in Iraq (2003–11) and its very troubled aftermath show us that the term can easily be linked to both the histories of the Iraqi people and that of those doing the 'reconstructing'. That can in turn evoke positive or negative historical connotations in both principal parties of the relationship. The alternative models that can be then generated can either exacerbate an existing conflict within a country or help it come to some kind of successful conclusion. There is already some suggestion that the people of the Middle East are now conceiving and practicing more indigenous forms of 'reconstruction' through grassroots organizations, such as Hamas or Hezbollah, which have local credibility but evoke hostile reactions from many governments and IGOs. But if the IGOs and governments' activities give the impression of 'colonialism' to the recipients, can they work? Equally, in trying to 'resolve' a conflict or 'mapping' it, there is no use in the outside world imposing its interpretation of how history should be interpreted. Again, in line with Chabal and Daloz, as well as de Goede, we believe the 'interpretation of meaning' has to be local.

On a negative note, we have also tried to show where an understanding of history can actually exacerbate a conflict. In many areas of the world, unscrupulous elites have stressed 'their' version of history to denigrate or even to incite their enemies to a hatred they did not previously feel. In the Middle East, for example, we can see that historical misunderstandings have become an embedded part of the discourse of conflict on all sides. But there is certainly a need for more investigation of how this might apply in many African countries, where the historical and anthropological literature needs to be far better understood by those who study conflicts in those areas. We are far from understanding how history 'plays' in many conflicts in developing countries, either as a force for peacebuilding or as one for peace destroying. One thing is clear however: the conflicts or humanitarian emergencies that we might read about in our newspapers did not begin with the last incident. Instead, we need to transcend the tendency toward ahistoricism and realize that current conflicts are often merely the latest instalment in a complex history of conflict, identity changes, population flows, border re-drawings and other long-term social, economic and cultural dynamics.

The next major concluding thought is that we are driven to believe that 'bottom-up' activities for peacebuilding are more likely to have positive long-term effects than 'top-down' approaches. The dangers of giving people the impression that they are having something passed down to them, rather than doing something for themselves, is bound to lead to a sense of disempowerment and alienation. If this is then reinforced by the feeling among a local population that the incoming ideas and power are not there 'for them', then the development or peace that is hoped for will not do much to help solve the conflict in question. The presence of peacekeepers is, for example, often seen as necessary but cannot be of much use to help a process of peace and development if those peacekeepers distort the local economy, suck out the best labour power, and generally lord it over the local population without contributing much to their security or well-being. The key point is that an appropriate balance between indigenous and external norms and capabilities needs to be struck. Of course, this is easier said than done, and the nature of the balance will differ from context to context. But a good starting point for potential interveners is to move beyond a position of assuming that they have all the answers. Instead, rather than viewing local populations as victims, recipients, dependents or troublemakers, it is important that they are viewed as change agents with substantial capabilities, with a recognized agency in the discourse around the construction and reconstruction of their society. We must be careful not to romanticize everything local and bottom-up. But, by and large, local approaches to peacebuilding and development can be sustainable and in tune with on-the-ground cultural norms.

Third, we would not be the only, but we would like to be the latest, thinkers to doubt the all-embracing truth of the 'liberal peace' thesis. It may well be true, even empirically, that 'democracies do not go to war with one another'. But the point is surely that in the case of all developing countries struggling with, or attempting to emerge from, conflict the basic principle does not pertain. Democracies in Europe took many hundreds of years to hone their conflict-resolution skills, ones that still desert them quite regularly. The countries of Africa and elsewhere have not, in many cases, ever achieved full 'state-ness', never mind democracy. Maybe they cannot without a long-drawn-out process of violence, as some claim from a variety of standpoints. But even if they can, they surely should be allowed to do it without undue outside interference. Whether it is democracy at gunpoint (as in the cases of Iraq and Afghanistan) or democracy by PowerPoint (as in the case of hectoring speeches by Western leaders when they visit developing countries), if democracy does not take root in villages, towns and cities across a country, then it will not thrive.

As a corollary to this point, we believe that many of the examples we have explored here show that the uncritical use of neo-liberal economic policies has done much to exacerbate the bad odour in which many 'good' Western principles are held in the developing world. Democracy is *per se* a system that most aspire to, but the relentless way in which the World Bank and other organizations have insisted on 'conditionality', 'big bang' changes and structural adjustment have done much to damage the view of those at the local level. This has been observed of course by World Bank officials themselves, with Stiglitz (2003) the most prominent among them. Others, like Amartya Sen (2001), have noted that poverty is the greatest enemy of democratic change and has lambasted the IGOs for their narrow emphasis on empowering the elite in developing countries. Importantly, poverty is relatively straightforward to deal with if there is generalized economic growth: everyone – rich and poor – can do well. Inequality, however, is another matter, and some (the rich or relatively rich) will have to give something up if those poorer than them are to benefit. That involves hard choices that rich countries and international organizations, by and large, have avoided.

Conflict theory may have something to say here in a general way to the development economists and governance experts. Conflict-resolution theorists (as was stressed in the Introduction), long derided as 'cranks' for advocating the empowerment of the grassroots of societies, are now seen as having seen the coming of elements of a 'post-state' world (through globalization) well before the realist/statists did. What once was considered equally crankish, such as the environment or even the study of unconventional warfare, are now seen as mainstream and form an increasingly important part of political debates and policy direction. Where only states were taken as the units of analysis in both conflict and strategic analysis as well as in economics sub-state actors, structures and individuals are now routinely evoked in suggesting both why problems occur and how to deal with them. IGOs will have to adopt this understanding to a far greater extent if they are to deal with today's and tomorrow's problems in the developing world. We may even have to break open that ultimate realist shibboleth of 'sovereignty' and contemplate re-parcelling the states of

Africa and elsewhere, or, alternatively, look to much more regional power sharing and identity-based policies. Kikuyu, Hutu, Tutsi or Xhosa people across Africa have to feel they can be given their recognition and dignity without only being referred to as 'Kenyan', 'Rwandan' or 'South African'. This is in much the same way as we have let the local populations of their respective areas sort out their identities in Ireland, the Former Yugoslavia or the Former Soviet Union without losing all of the benefits that those 'unitary' bodies once provided. How we keep what is best about old structures while allowing the emergence of new ones is not a problem in developing countries alone.

Fourth, we should be aware that the context in which development and conflict occur is changing rapidly. This change stems from many sources: an increasing global population and the consequent pressure on resources; the changing nature of power relations as China, India, Russia, Brazil and others stake their claim to be regional and world players; the declining importance of sovereignty in a globalizing world; the ever-present and growing dissatisfaction with the United Nations and other multilateral institutions; or the growing significance of China and some Gulf and Arab states as key players in development and humanitarian interventions. The key point is that many of the lenses that we have used to analyse conflict and development over the past decades will need to be radically re-assessed to take account of the changing nature of conflict and development processes. What we think we know now may not necessarily be fit for analytical or practical purposes in the future.

Crucially, travel to many non-Western contexts reveals an important truth: these societies are thriving, dynamic and bursting with inventiveness. It is not the case that the people of the developing world are somehow sitting around waiting for development to come to them. Many are fashioning their own preferred types of development and conflict avoidance through trade, cultural activities and family and community life. They are grasping opportunity and creating their own modernity that might be inflected with local and traditional ways of doing things. It may be a cliché, but travel to Somalia, and you will see camel traders with the latest smartphone. The 'remotest' trading centre in Uganda will have lots of cooking utensils for sale – made in China and transported through complex trading networks. If anyone is going to be 'left behind', it is those in the 'developed world' who do not update their thinking on the nature of the developing world.

Last, but not least, we hope we have demonstrated that the problems of conflict and development are ones that affect all of us, wherever we live and that there is a unity of the human spirit and of human suffering that makes us all morally responsible for each other's welfare. This is also true when we are attempting to impose democracy promotion and statebuilding in conflict or non-conflict situations. We have endeavoured throughout this book to see development as 'top-down' and the work of institutions (Chapter 2), as well as affecting ordinary people (Chapter 3 and elsewhere). That does not mean that we should be prescriptive about the moral universe others live in, but it does mean that we have to take responsibility for the negative effects of our actions in the West when we think about the impact they will have in developing countries. We need a new ethic of development as we need a new ethic of international relations. That will have to be for another book in this series to think about.

Bibliography

Chabal, P. and Daloz, J.-P. (2006) *Culture Troubles: Politics and the Interpretation of Meaning*. London: Hurst.
Sen, A. (2001) *Development as Freedom*. Oxford: Oxford University Press.
Stiglitz, J. (2003) *Globalisation and Its Discontents*. London: Penguin.
Williams, A.J. (2006; 2024 *forthcoming*) *Liberalism and War: The Victors and the Vanquished*. London: Routledge.

Index

For Product Safety Concerns and Information please contact our
EU representative GPSR@taylorandfrancis.com Taylor & Francis
Verlag GmbH, Kaufingerstraße 24, 80331 München, Germany